Thermoelastic Fracture Mechanics
Multiple Crack Interactions

Vera Petrova [1,2*] and Siegfried Schmauder [1]

[1] IMWF, University of Stuttgart, Pfaffenwaldring 32, D-70569 Stuttgart, Germany

[2] Voronezh State University, University Sq.1, Voronezh 394006, Russia

veraep@gmail.com *, Siegfried.Schmauder@imwf.uni-stuttgart.de

Published by **Materials Research Forum LLC**
Millersville, PA 17551, USA

Published as part of the book series
Materials Research Foundations
Volume 159 (2024)
ISSN 2471-8890 (Print)
ISSN 2471-8904 (Online)

Print ISBN 978-1-64490-294-3
ePDF ISBN 978-1-64490-295-0

Distributed worldwide by

Materials Research Forum LLC
105 Springdale Lane
Millersville, PA 17551
USA
https://www.mrforum.com

Printed in the United States of America
10 9 8 7 6 5 4 3 2 1

Table of Contents

Preface

The book deals with important problems of thermal and mechanical fracture of functionally graded materials on homogeneous substrate (FGM/H) structures with emphasizing on multiple crack interactions. FGMs are a special type of composites whose material properties vary in at least one spatial direction. This is achieved by changing the composition of the material and/or its structure. FGMs have wide engineering applications, in particular, in thermal barrier coatings, where ceramics are used on the coating top and then continuously vary to the metal in the substrate. Functionally graded coatings (FGCs) protect the inner metallic parts from overheating and melting. During manufacturing and service small defects appear in FGCs and then cracks can initiate from these defects; these defects and cracks change the fracture resistance of FGCs. In this context, a study of thermal fracture of FGC structures is of great demand.

The thermal and elastic properties of FGMs are modeled from the point of view of its practical application to the specific problem of cracks in FGMs, some functional forms, such as, exponential form and power law functions are used to simulate non-homogeneous properties of FGMs. The problem for FGM/H structures with pre-existing systems of cracks (among them are edge cracks, interface cracks between the coating and substrate, internal cracks in the FGM or FGC) are formulated in general by means of integral equations. The following methods for the solution of the integral equations are used: asymptotic analytical solutions for some special cases (including the method of small parameters), and numerical solutions based on Chebyshev polynomials. Different crack patterns (which are reported in experiments and available in the literature) are studied by carrying out numerical experiments with respect to the main characteristics of fracture, such as stress intensity factors, fracture angles (a deviation of cracks from the initial direction of propagation) and critical stresses when this propagation starts. This semi-analytical approach allows to correlate the structural material parameters (material gradation, crack parameters) and the thermo-mechanical loading parameters with the main fracture characteristics. The model in combination with a detailed parametric analysis can help to optimize the gradation of the FGCs and their structure in order to improve the fracture resistance of FGC/H systems operating under elevated temperatures. In this regard, potentially desirable thermal and mechanical properties of FGCs are analyzed as well as available real material combinations (e.g. (ceramic/ metal)/metal) for advanced thermal barrier coating applications.

The book consists of two main parts: Part I (Chapters 1 - 6) considers infinite functionally graded materials/homogeneous bimaterial structures (FGM/H) and Part II (Chapters 7 - 14) deals with semi-infinite, namely, functionally graded coatings on a homogeneous substrate (FGC/H). In Part I, solutions to the problems are obtained in an approximate analytical form (using the small parameter method) and in Part II - in a semi-analytical form.

Chapters 1 - 3 examines FGM/H bimaterial structures, which are subjected to a thermal flux (Chapter 1) and thermo-mechanical loads, namely, a thermal flux and a tensile load (Chapters 2 and 3). In Chapters 4 and 5, FGM/H bimaterials under thermal flux and shear load are studies with taking into account possible crack closure. Chapter 6 provides an approximate analytical solution for the antiplane shear problem for an interface crack (between two homogeneous materials) interacting with systems of internal cracks. The mathematical formulation for the antiplane shear stress (Mode III) problem is similar to the stationary thermal conductivity problem with respective boundary conditions and with respective material parameters. Thus, the solutions to both antiplane and thermal problems are also similar. As soon as a solution to the antiplane problem is obtained, it can be used in thermal crack problems.

Chapters 7 - 9 deal with FGC/H structures with edge cracks in the FGCs. Semi-analytical solutions are obtained for the thermo-mechanical loads (cooling by ΔT and tension). The effects of the interaction of edge cracks on further crack formation are studied with respect to main fracture characteristics, i.e., stress intensity factors (SIFs), the crack propagation direction (characterized by fracture angles) and critical loads, when this propagation is initiated. Chapters 10-14 consider edge and internal cracks in the FGCs. A series of computational experiments was performed to study the interaction of edge cracks with internal cracks. The influence of material and geometrical parameters of FGC/H structures on the fracture characteristics is investigated. Some special crack systems are studied, for example, edge cracks and multiple internal cracks imitating a curved interface (Chapter 13). Chapter 14 is devoted to the fracture of FGC/H structures, in which the FGM properties are modeled by formulas based on the rule of mixtures with a power-law coefficient λ as the gradation parameter. The shielding effect of a system of edge cracks is demonstrated, and the influence of an additional internal crack on this shielding effect is discussed. It is shown that the gradation parameter λ has a strong effect on the fracture characteristics, so it can serve as a key parameter for an optimization problem for increasing fracture toughness of FGCs. Thus, the model based on the rule of mixtures seems to be a good candidate for a theoretical evaluation of the desired material properties of FGCs for their further development in order to improve the thermal fracture resistance of coatings.

The book provides valuable knowledge and theoretical tools for studying the fracture of functionally graded materials and FGM coatings under various thermal and mechanical loading conditions. The book will be a useful resource for researchers and specialists in the field of fracture mechanics and materials science, as well as for other interdisciplinary specialists.

Vera Petrova and Siegfried Schmauder
Stuttgart, November 15, 2023

Acknowledgements

The investigations, described in this book, have been carried out in the framework of many research projects, funded by the German Research Society DFG (Deutsche Forschungsgemeinschaft), such as, Schm 746/80-1, Schm 746/92-1, Schm 746/106-1, Schm 746/113-1, Schm 746/131-1, Schm 746/139-1, Schm 746/139-2, and Schm 746/209-1.
The materials of the book are reproduced here with kind permissions from Elsevier and Wiley Online Library, and also open access with Springer and IOP Science.

References

This book includes the following publications of the authors and their colleagues.

1. Petrova, V., Schmauder, S. Thermal fracture of a functionally graded/homogeneous bimaterial with system of cracks, Theoretical and Applied Fracture Mechanics (2011) 55 (2), 148-157
2. Petrova, V., Schmauder, S. Interaction of a system of cracks with an interface crack in functionally graded/homogeneous bimaterials under thermo-mechanical loading, Computational Materials Science (2012) 64, 229-233
3. Petrova, V., Schmauder, S. FGM/homogeneous bimaterials with systems of cracks under thermo-mechanical loading: Analysis by fracture criteria, Engineering Fracture Mechanics (2014) 130, 12-20
4. Petrova, V., Schmauder, S. Mathematical modelling and thermal stress intensity factors evaluation for an interface crack in the presence of a system of cracks in functionally graded/homogeneous bimaterials, Computational Materials Science (2012) 52 (1), 171-177
5. Petrova, V., Schmauder, S. Crack closure effects in thermal fracture of functionally graded/homogeneous bimaterials with systems of cracks, ZAMM Zeitschrift fur Angewandte Mathematik und Mechanik (2015) 95 (10), 1027-1036
6. Petrova, V., Schmauder, S., Ordyan, M., Shashkin, A. Revisit of antiplane shear problems for an interface crack: Does the stress intensity factor for the interface Mode III crack depend on the bimaterial modulus? Engineering Fracture Mechanics (2019) 216, art. no. 106524
7. Petrova, V., Sadowski, T. Theoretical modeling and analysis of thermal fracture of semi-infinite functionally graded materials with edge cracks, Meccanica (2014) 49 (11), 2603-2615
8. Petrova, V., Schmauder, S., Shashkin, A. Modeling of edge cracks interaction, Frattura ed Integrita Strutturale (2016) 10 (36), 8-26
9. Petrova, V.E., Schmauder, S. Modeling of thermomechanical fracture of functionally graded materials with respect to multiple crack interaction, Physical Mesomechanics (2017) 20 (3), 241-249
10. Petrova, V.E., Schmauder, S. Fracture of functionally graded thermal barrier coating on a homogeneous substrate: Models, methods, analysis, Journal of Physics: Conference Series (2018) 973 (1), art. no. 012017
11. Petrova, V., Schmauder, S. Analysis of interacting cracks in functionally graded thermal barrier coatings, Procedia Structural Integrity (2020) 28, 608-618
12. Petrova, V., Schmauder, S. A theoretical model for the study of thermal fracture of functionally graded thermal barrier coatings with a system of edge and internal cracks, Theoretical and Applied Fracture Mechanics (2020) 108, art. no. 102605
13. Petrova, V., Schmauder, S. Thermal fracture of functionally graded thermal barrier coatings with pre-existing edge cracks and multiple internal cracks imitating a curved interface, Continuum Mechanics and Thermodynamics (2021) 33(4), 1487-1503
14. Petrova, V., Schmauder, S. Thermal fracture resistance of functionally graded thermal barrier coatings with systems of multiple cracks. Application of rule of mixtures, Procedia Structural Integrity (2022) 42, 1145-1152

CHAPTER 1

Thermal Fracture of a Functionally Graded/Homogeneous Bimaterial with System of Cracks

Vera Petrova [1,2]*, Siegfried Schmauder [1]

[1] IMWF, University of Stuttgart, Pfaffenwaldring 32, D-70569 Stuttgart, Germany

[2] Voronezh State University, University Sq.1, Voronezh 394006, Russia

veraep@gmail.com *, Siegfried.Schmauder@imwf.uni-stuttgart.de

Abstract

The thermal fracture of a bimaterial consisting of a homogeneous material and a functionally graded material (FGM) with a system of internal cracks and an interface crack is investigated. The bimaterial is subjected to a heat flux. The thermal properties of FGM are assumed to be continues functions of the thickness coordinate, while the elastic properties are constants. The method of the solution is based on the singular integral equations. For a special case where the interface crack is much larger than the internal cracks in the FGM the asymptotic analytical solution of the problem is obtained as series in a small parameter (the ratio between sizes of the internal and interface crack) and the thermal stress intensity factors (TSIFs) are derived as functions of geometry of the problem and material characteristics. A parametric analysis of the effects of the location and orientation of the cracks and of the inhomogeneity parameter of FGM's thermal conductivity on the TSIFs is performed. The results are applicable to such kinds FGMs as ceramic/ceramic FGMs, e.g., TiC/SiC, $MoSi_2/Al_2O_3$ and $MoSi_2/SiC$, and also some ceramic/metal FGMs.

Keywords

Functionally Graded Materials, Interface, Cracks, Thermal Fracture, Stress Intensity Factors

1. Introduction

Functionally graded materials (FGMs) are widely used in different engineering structures and are tailored so that to decrease bimaterial mismatch and residual stresses at the interface and to prevent delamination and debonding along the interface. Meanwhile, cracks and defects usually initiate and grow near interfaces and are the cause of additional residual stresses near the interface, so that crack interaction problems in FGMs and bimaterials are important for the investigation of fracture strength of materials.

Study of the thermal fracture of FGM/homogeneous bimaterials with defects includes different problems among them are the following. First - modelling of FGMs properties, thermal and elastic. For example, one of the popular ways is to assume that the properties have exponential form [1, 2] or linear law form [3], or other simple functional dependence; it makes the boundary value problem analytically tractable. The other way for FGM modelling is using a mixture rule with the volume fraction of the inclusion phase in the form of a power function [4]. Another often used method is to model FGMs as piece-wise multi-layered media, that is to divide the FGM into multiple layers with constant material properties in the gradient direction [5] with varying properties [6]. Self-consistent matricity models [7] are used in numerical simulations of FGMs.

The next main problem is the solution of classical crack problems for FGMs as well as the problem of interaction between cracks and interfaces and other non-homogeneities, such as, inclusions and dislocations. The crack interaction problems in homogeneous materials have been extensively investigated and many solutions have been obtained for different crack system configurations and different thermal and mechanical loadings [8, 9]. Crack interaction problems in FGMs are not so well examined because of the mathematical difficulty in the solution of constituent equations, but great progress is observed in this field.

The third problem is application of appropriate crack models taking into account the microstructure, i.e. cohesive crack models, atomistic models and others.

The main interest of the present work is the influence of material inhomogeneities, such as cracks, and material inhomogeneity parameters on the stress-strain state in the vicinity of the interface in FGM/homogeneous bimaterials subjected to a heat flux. Different aspects of deformation and fracture investigations of FGMs can be found in the literature, theoretical semi-analytical studies [1, 2, 10] and numerical simulations [7]. In a review by Birman and Byrd [11] the references devoted to the modeling and analysis of functionally graded materials and structures performed during the last ten years can be found. Review by Noda [10] is devoted to thermal crack problems in FGMs. The recent review by Ootao [12] is referred to basic thermoelastic and piezothermoelastic problems in FGMs. A multiple surface cracking analysis is performed and crack initiation behavior is discussed in a review by Wan and Mai [13] to study the thermal shock resistance behavior of ceramic/metal FGMs.

In view of the subject of the present work, some studies devoted to crack interaction problems in FGMs and thermal crack problems in FGMs should be mentioned. A general solution of a single and multiple radially oriented cracks embedded in a non-homogeneous infinite plate under mechanical loading was obtained in a paper by Shbeeb et al. [2]. It was assumed that the FGMs have a constant Poisson's ratio and the shear modulus is of an exponential form. Multiple collinear crack problems in FGMs with arbitrarily varying material properties were studied by Wang et al. [14]. The algorithm was applied to steady state or transient thermoelastic fracture problems. A laminated composite plate model was used to simulate the material non-homogeneity. In the paper by Guo and Noda [15] a new piecewise-exponential model is proposed to realize fracture mechanical investigations of FGMs with arbitrary properties. In this model the FGM is divided into some non-homogeneous layers along the gradient direction of the properties. By using this model the fracture problem of a functionally graded strip with arbitrarily distributed properties and a crack vertical to the free surfaces is studied. The review of crack problems in FGMs can also be found in this paper [15]. Crack spacing effect on the brittle fracture characteristics of semi-infinite FGMs with periodic edge cracks is studied in [4] using a mixture rule for FGM modelling. The FGM is cooled from sintering temperature and an incompatible eigenstrain is induced in the finite region due to mismatch in the coefficients of thermal expansion. It can be concluded that at present the investigations devoted to the crack

Materials Research Forum LLC
https://doi.org/10.21741/9781644902950

interactions in FGMs are restricted only to special cases of crack locations and in spite of existing numerous investigations in this field the interaction of arbitrary located cracks with an interface under the influence of a heat flux has not yet been investigated.

The presented model for the interaction of arbitrary located cracks in a FGM/homogeneous bimaterial under thermal loading is studied with the following assumptions: 1) The temperature and mechanical analysis are uncoupled, so that the solution consists of two steps: determination of the temperature distribution in the FGM/homogeneous bimaterial with cracks and the determination of the thermal stresses. 2) It is supposed that the FGM/homogeneous bimaterial is thermally non-homogeneous, but elastically homogeneous. It is applicable to some material combinations for which Young's modulus and Poisson's ratio may approximately assume as constant. Such models were used, for example, by Jin and Paulino [16]. 3) The thermal properties of the FGM are continuous functions of one coordinate and have exponential form.

2. General formulation of the problem

The geometry of the problem is shown in Fig. 1. A bimaterial is composed of a functionally graded material (denoted by number 1) located in the upper half plane and a homogeneous material (denoted by number 2) located in the lower half plane. The bimaterial is perfectly bonded with the exception of an interface crack of length $2a_0$. It is assumed that the FGM contains N cracks of length $2a_k$. Cartesian coordinates (x, y) are centered at the midpoint of the interface crack; the x-axis lies along the interface line. The local coordinate systems (x_k, y_k) are attached to each internal crack.

The crack position is determined by the defect midpoint coordinate (x_k^0, y_k^0), or in the complex variable $z_k^0 = x_k^0 + iy_k^0$ ($i^2 = -1$), and an inclination angle θ_k to the interface, i.e. to the x-axis (Fig. 1). The bimaterial is subjected only to thermal loading, by a remote heat flux of intensity q. The cracks are thermally isolated and traction free. The uncoupled, quasi-static thermoelastic theory is applicable to this problem so that the solution consists of the determination of the temperature distribution and the determination of the thermal stresses.

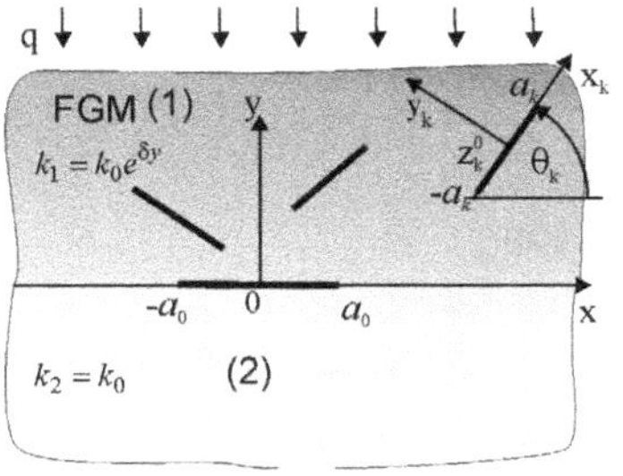

Figure 1. The geometry of the problem.

2.1 FGMs modeling

It is supposed that properties of functionally gradient materials depend only on the coordinate y. The thermal conductivity coefficient is

$$k_1(y) = k_0 e^{\delta y}, \tag{1}$$

where the constant k_0 is the thermal conductivity of the interface and of material (2), while δ is the so-called inhomogeneity parameter for FGM. With this assumption the thermal boundary value problem is simplified considerably.

The thermal expansion coefficient is also expressed in exponential form

$$\alpha_{t1} = \alpha_{t0} e^{\varepsilon y} \tag{2}$$

Here α_{t0} is the thermal expansion coefficient of a homogeneous material (2), and ε is the inhomogeneity parameter of the thermal expansion for the FGM.

The Young's modulus and Poisson's ratio are assumed to be constant, E_j=const, ν_j=const (j=1,2). Thus, the material is elastically homogeneous, but thermally nonhomogeneous. This kind of FGMs include some ceramic/ceramic FGMs such as TiC/SiC, $MoSi_2/Al_2O_3$ and $MoSi_2$/SiC, and also some ceramic/metal FGMs such as zirconia/nickel and zirconia/steel [16, 17].

The relation between global coordinates (x,y) and local coordinate systems (x_k,y_k) can be written using complex variables as follows

$$z = z_k^0 + z_k e^{i\theta_k},$$

where $z = x + iy, z_k = x_k + iy_k$. The parameter $z_k^0 = x_k^0 + iy_k^0$ is the origin coordinate of the system (x_k,y_k) in the global system, and at the same time it is a midpoint coordinate of a crack. In the local coordinate system connected with each arbitrary oriented crack the coefficient k_1 has the form

$$k_1(x_k, y_k) = k_0 e^{\delta y_k^0} e^{\delta_1 x_k + \delta_2 y_k}, \; \delta_1 = \delta \sin\theta_k, \; \delta_2 = \delta \cos\theta_k. \tag{3}$$

Similar expressions are written for the thermal expansion coefficient

$$\alpha_{t1}(x_k, y_k) = \alpha_{t0} e^{\varepsilon y_k^0} e^{\varepsilon_1 x_k + \varepsilon_2 y_k}, \; \varepsilon_1 = \varepsilon \sin\theta_k, \; \varepsilon_2 = \varepsilon \cos\theta_k. \tag{4}$$

2.2 Boundary conditions

The thermal boundary conditions for the full temperature $T_j^*(x,y)$ on the thermoisolated cracks in the bimaterial read as follows:

$$k_1 \frac{\partial T_1^*(x,+0)}{\partial y} = k_2 \frac{\partial T_2^*(x,-0)}{\partial y} = 0, \quad |x| \le a_0,$$

$$k_j \frac{\partial T_{jn}^*(x_n,\pm 0)}{\partial y_n} = 0, \quad |x_n| \le a_n. \tag{5}$$

and thermal fluxes and temperature are equal on the perfectly bonded interface. The mechanical boundary conditions for the traction-free cracks are

$$(\sigma_{1y} - i\tau_{1xy})^+ = (\sigma_{2y} - i\tau_{2xy})^- = 0, \ |x| \le a_0, y = 0,$$
$$(\sigma_{jy} - i\tau_{jxy})^+ = (\sigma_{jy} - i\tau_{jxy})^- = 0, \ |x_k| \le a_k, y_k = 0. \tag{6}$$

The continuity conditions at the interface are

$$(\sigma_{1y} - i\tau_{1xy})^+ = (\sigma_{2y} - i\tau_{2xy})^-, \ |x| > a_0, y = 0, \tag{7}$$

$$(u_1 - iv_1)^+ = (u_2 - iv_2)^-, \ |x| > a_0, y = 0,$$

and the condition at infinity is $\sigma_{ij} \to 0, \ \ x^2 + y^2 \to \infty$. Here σ_{jy} and τ_{jxy} are normal and shear stresses, v_j and u_j are normal and shear displacements, j = 1, 2 for materials 1 and 2. The signs '+' and '–' denote the limiting values of the functions on the upper and lower surfaces of the crack or the interface, respectively.

2.3 Methods used

For the solution of the problem a superposition principle is used. First, according to this principle, the problem for a crack in a bimaterial is equal to the superposition of the following two sub-problems: (a) The bimaterial without crack is subjected to remote heat flux, and the heat flux induced at the location x=0 is $q(y)$. (b) The bimaterial with crack is free of remote fluxes and only the crack faces are subjected by heat fluxes of intensity $-q(y)$. The analogous superposition scheme is applied to the thermoelastic problem. The problem (a) is a homogeneous problem of thermo-conductivity and this solution does not make a contribution to the singular fields at the crack tips. Problem (b) is called perturbation problem and it governs the singular crack-tip fields. In fracture mechanics the behavior of the singular crack-tip fields is important and considered firstly. In the following only the perturbation problem will be analyzed. The second step of the superposition scheme leads to decomposing the problem into a series of simple sub-problems, each of which contains either an internal crack or an interface crack.

Then the method of complex potentials and the theory of analytical functions are applied which lead to singular integral equations of the formulated problems. The singular integral equations are solved by the small parameter method for a special case where the interface crack is much larger than the internal cracks in the FGM.

3. Thermal problem

Due to the superposition principle the temperature field T_j^* (j=1,2) in the bimaterial with cracks is presented as

$$T_j^*(x,y) = T_j^0(x,y) + T_j(x,y) \quad (j=1,2), \tag{8}$$

where $T_j^0(x,y)$ – the temperature distribution in a bimaterial in the absence of cracks, $T_j(x,y)$ – the temperature perturbation caused by the cracks.

The thermal boundary conditions and continuity conditions for the temperature perturbation $T_j(x,y)$ read as follows:

$$k_1 \frac{\partial T_1(x,+0)}{\partial y} = k_2 \frac{\partial T_2(x,-0)}{\partial y} = q_0(x) \quad |x| \le a_0, \tag{9}$$

$$k_j \frac{\partial T_{jn}(x,\pm 0)}{\partial y_n} = q_n(x_n) \quad |x_n| \le a_n, \tag{10}$$

$$k_1 \frac{\partial T_1(x,+0)}{\partial y} = k_2 \frac{\partial T_2(x,-0)}{\partial y}, \;\; T_1(x,+0) = T_2(x,-0) \quad |x| \ge a_0, y=0, \tag{11}$$

$$T_1(\pm a_0,+0) = T_2(\pm a_0,-0), \;\; T_{jn}(\pm a_0,+0) = T_{jn}(\pm a_0,-0), \tag{12}$$

and the temperature perturbation vanishes at infinity. Here

$$q_0 = -(\partial T_j^0 / \partial y)\Big|_{y=0}, \quad q_n = -k_j (\partial T_j^0 / \partial y_n)\Big|_{y_n=0}. \tag{13}$$

The signs '+' and '–' denote the limiting values of the functions on the upper and lower surfaces of the crack or the interface, respectively.

The heat conduction equation for the steady state temperature in FGMs with thermal conductivity coefficient (1) is given by

$$\nabla^2 T_1 + \delta \frac{\partial T_1}{\partial y} = 0, \tag{14}$$

and for material 2 with $\delta = 0$ we have Laplace equation

$$\nabla^2 T_2 = 0, \tag{15}$$

where $\nabla^2 = (\partial^2 / \partial x^2) + (\partial^2 / \partial y^2)$.

Solving the temperature problem for an undamaged FGM/homogeneous bimaterial the thermal fluxes on the crack lines (13) are obtained as

$$q_0 = -k_0 q, \quad q_n = -k_0 q \exp(-\delta y_n^0)\exp(-x_n \delta \sin\theta_n)\cos\theta_n . \tag{16}$$

The perturbation problem is solved by the method presented in [9, 18, 19]. The system of N+1 singular integral equations for the unknown functions γ_k' is written as

$$\int_{-a_n}^{a_n} \frac{\gamma_n'(t)}{t-x}dt + \sum_{\substack{k=0\\k\neq n}}^{N} \int_{-a_k}^{a_k} \gamma_k'(t)P_{nk}(t,x)dt = \pi q_n(x) \quad (|x|<a_n), \tag{17}$$

$$n = 0,1,...,N.$$

The condition of the temperature continuity at the crack endpoints

$$\int_{-a_k}^{a_k} \gamma_k'(t)dt = 0 \tag{18}$$

should be taken into account. It follows from condition (12). Here the functions γ_k' are the derivative of temperature jumps on the crack lines

$$2\gamma_k = T_k^+ - T_k^- . \tag{19}$$

In Eqs. (17) q_n are the functions (16). The regular kernels P_{nk} (n,k =0,1,...,N), containing the geometry of the problem, are given by the following expressions:

$$P_{nk} = \mathrm{Re}[e^{i\theta_n} / (te^{i\theta_k} + z_k^0 - xe^{i\theta_n} - z_k^0)]. \tag{20}$$

"Re" denotes the real part of complex numbers.

The system of integral equations completely represents the problem of an arbitrary system of interacting cracks, but because of the complexity, an exact solution can be obtained only for a simple geometry and few cracks. Let as assume that all internal cracks have the same size $2a_k$=$2a$ (k=1,2,..,N), for example, they have the characteristic size of the grain size of the material. Suppose also that this size is much smaller than the size of an interface crack, i.e. $a << a_0$. In this case a solution of system (17) for an arbitrary system of internal cracks influencing the interface crack can be obtained by the small parameter method, where the small parameter is equal to the ratio of the internal crack size to the interface crack size, that is $\lambda = a / a_0$ and $\lambda << 1$.

Introducing nondimensional coordinates χ and τ:

$$x = a_n \chi, \; t = a_k \tau, \tag{21}$$

Eq. (17) can be rewritten in dimensionless variables. Applying the inversion formulas [20] to the Cauchy-type integrals in equations (17), we obtain the equation for the temperature jump on the interface crack line

$$\gamma_0'(\chi)=\frac{1}{\pi\sqrt{1-\chi^2}}\left\{-\int_{-1}^{1}\frac{\sqrt{1-\tau^2}}{\tau-\chi}\tilde{q}_0 d\tau+\sum_{k=1}^{N}\int_{-1}^{1}\gamma_k'(\tau)Q_{0k}(\tau,\chi)d\tau\right\} \tag{22}$$

and N equations for the temperature jump on the other cracks

$$\begin{aligned}\gamma_n'(\chi)=\frac{1}{\pi\sqrt{1-\chi^2}}\Bigg\{&-\lambda\int_{-1}^{1}\frac{\sqrt{1-\tau^2}}{\tau-\chi}\tilde{q}_n d\tau+\int_{-1}^{1}\gamma_n'(\tau)Q_{n0}(\tau,\lambda\chi)d\tau\\&+\sum_{k=1,k\neq n}^{N}\int_{-1}^{1}\gamma_k'(\tau)Q_{nk}(\lambda\tau,\lambda\chi)d\tau\Bigg\},\end{aligned} \tag{23}$$

where

$$\tilde{q}_0=a_0 q_0(\tau),\ \tilde{q}_n=a_n q_n(\tau) \tag{24}$$

and

$$Q_{nk}(\tau,\chi)=\frac{1}{\pi}\int_{-1}^{1}\frac{\sqrt{1-\xi^2}}{\xi-\chi}P_{nk}(\tau,\xi)d\xi$$

$$P_{nk}(\tau,\xi)=\operatorname{Re}\left[\frac{e^{i\theta_n}}{a_k\tau e^{i\theta_k}+a_k w_k-a_n\xi e^{i\theta_n}-a_n w_n}\right],\quad w_n=z_n^0/a_0. \tag{25}$$

w_n is nondimensional coordinate of the center of cracks.

The solution is sought as a power series with respect to the small parameter λ so that the derivative of a temperature jump on the interface crack γ_0' and the derivatives of temperature jumps on the microcracks γ_k' have the form:

$$\gamma_0'(\chi)=\sum_{p=0}^{\infty}\gamma_{0p}'(\chi)\lambda^p,\quad \gamma_n'(\chi)=\sum_{p=0}^{\infty}\gamma_{np}'(\chi)\lambda^p,\quad \lambda=a/a_0<<1. \tag{26}$$

The regular kernels Q_{nk} (25) are expanded in power series in λ

$$Q_{0k}=\sum_{p=0}^{\infty}Q_{0kp}(\tau,\chi)\lambda^p, Q_{n0}=\sum_{p=1}^{\infty}Q_{n0p}(\tau,\chi)\lambda^p, Q_{nk}=\sum_{p=1}^{\infty}Q_{nkp}(\tau,\chi)\lambda^p. \tag{27}$$

The expressions of coefficients in Eq. (27) are given in the Appendix A, Eq. (A.1).

Materials Research Forum LLC
https://doi.org/10.21741/9781644902950

For convergence of the series (27) the following inequality

$$\left|\lambda/(\tau - z_k^0/a_0)\right| < 1, |\tau| < 1$$

must be satisfied. It is fulfilled if the cracks are not intersected.

Inserting series expansions for the unknowns (26) and for the kernels (27) into Eqs (17) and equating the coefficients of corresponding powers of λ, the recurrent system of equations for the coefficients in series (26) is formed (see Appendix Eqs. (A.2), (A.3)).

The solution is obtained up to λ^2 so that only expressions needed for this calculation are presented in (A.2) and (A.3). It is worth to emphasize that the second approximation includes only terms representing the interaction between the interface crack and each of the microcracks and does not include any terms representing the mutual interaction of microcracks. This approximation is valid for distances between microcracks at least larger than half of their length [8, 9]. The interaction of microcracks is taken into account in fourth approximation.

It should be noted that in Eq. (A.2) due to the condition (18) $\gamma'_{01} = 0$ and the second integral in γ'_{02} also equals to zero. It can be shown that all coefficients with the odd second index are equal to zero. The second approximation for the derivative of the temperature jump γ'_0 on the interface crack line is written

$$\gamma'_0(\chi) = \gamma'_{00}(\chi) + \lambda^2 \gamma'_{02}(\chi), \tag{29}$$

where

$$\gamma'_{00}(\chi) = \frac{\tilde{q}_0 \chi}{\sqrt{1-\chi^2}}, \qquad \gamma'_{02}(\chi) = \frac{1}{\pi\sqrt{1-\chi^2}} \sum_{k=1}^{N} \int_{-1}^{1} \gamma'_{k1}(\tau) Q_{0k1}(\tau,\chi) d\tau. \tag{30}$$

The zeroth approximation of γ'_0 corresponds to an isolated interface crack.

The initial and the second approximations for each of the internal cracks have the following form:

$$\gamma'_{n0}(\chi) = 0, \qquad \gamma'_{n1}(\chi) = \frac{\tilde{q}_0}{\sqrt{1-\chi^2}} \left\{ f(\chi, \theta_n, w_n, \delta a_0) - \chi \operatorname{Re}\left[e^{i\theta_n} \left(1 - \frac{w_n}{\sqrt{w_n^2 - 1}} \right) \right] \right\}, \tag{31}$$

where

$$f(\chi,\theta_n,w_n,\delta a_0) = -\frac{1}{\pi} e^{-\delta a_0 \operatorname{Im} w_n} \cos\theta_n \int_{-1}^{1} \frac{\sqrt{1-\tau^2}}{\tau-\chi} e^{-\delta_1 a_0 \lambda \tau} d\tau . \tag{32}$$

Formulae (16) and (24) have been taking into account in the expressions (30) and (31). In Eq. (31) the calculation of the integral, denoted by the function $f(\chi,\theta_n,w_n,\delta a_0)$, is cited in Eqs. (A.5), (A.6). "Im" denotes the imaginary part of complex numbers.

4. Thermoelastic problem

We suppose that the material is elastically homogeneous so that the method presented in [9] can be used. Assuming that the traction free condition applies to the cracks (6), a system of singular integral equations for the unknown displacement jumps on the crack lines can be written as [9, 21]:

$$\int_{-a_n}^{a_n} \frac{G_n(t)}{t-x} dt + \sum_{\substack{k=0 \\ k\neq n}}^{N} \int_{-a_k}^{a_k} [G_k(t)K_{nk}(t,x) + \overline{G_k(t)}L_{nk}(t,x)]dt = 0, \, |x| < a_n , \tag{33}$$

$$\int_{-a_n}^{a_n} G_n(t)dt = -2i \int_{-a_n}^{a_n} \beta_t t\gamma'(t)dt_n = iA_n, \, n = 0,1,...,N . \tag{34}$$

$\overline{(...)}$ is the complex conjugate. The function G_k consists of two parts as

$$G_k(x) = g_k'(x) + 2i\beta_t\gamma_k(x), \tag{35}$$

where functions γ_k' were found from the thermoconductivity problem, while the unknown functions $g_n'(x)$ are the derivatives of displacement jumps on cracks lines

$$g_n'(x) = \frac{2\mu}{i(\kappa+1)} \frac{\partial}{\partial x}[(u_n^+ - u_n^-) + i(v_n^+ - v_n^-)] . \tag{36}$$

In Eqs. (34) and (35) β_t is βt = α_t E^*. The following applies to the plane strain case $E^* = E/(1+\kappa)$ and $\kappa = 3-4\nu$, while the plane stress condition corresponds to $E^* = E/(1+\kappa)/(1+\nu)$ and $\kappa = (3-\nu)/(1+\nu)$. $\mu = E/(2(1+\nu))$ is the shear modulus.

The condition (34) provides that the displacements are single-valued at the end points of cracks. The condition of temperature continuity at the crack tips $\gamma_n(\pm a_n) = 0$ is also taken into account in Eq. (34).

The regular kernels K_{nk} and L_{nk} are

$$K_{nk}=\frac{e^{i\theta_k}}{2}\left(\frac{1}{T_k-X_n}+\frac{e^{-2i\theta_k}}{\overline{T_k}-\overline{X}_n}\right),\ L_{nk}=\frac{e^{-i\theta_k}}{2}\left(\frac{1}{\overline{T_k}-\overline{X}_n}+\frac{T_k-X_n}{\overline{T_k}-\overline{X}_n}e^{-2i\theta_k}\right), \tag{37}$$

$$T_k=te^{i\theta_k}+z_k^0,\qquad X_n=xe^{i\theta_n}+z_n^0.$$

They describe the geometry of the problem.

The system of singular integral equations (33) is rewritten in dimensionless form using coordinates (21) and is regularized with taking into account the equation (34):

$$G_0(\chi)=\frac{1}{\pi\sqrt{1-\chi^2}}\left\{i\frac{A_0}{a_0}+\sum_{k=1}^{N}\lambda\int_{-1}^{1}[G_k(\tau)M_{0k}(\lambda\tau,\chi)+\overline{G_k(\tau)}N_{0k}(\lambda\tau,\chi)]d\tau\right\} \tag{38}$$

$$\begin{aligned}G_n(\chi)=\frac{1}{\pi\sqrt{1-\chi^2}}\Bigg\{&i\frac{A_n}{a_n}+\frac{a_0}{a_n}\int_{-1}^{1}[G_0(\tau)M_{n0}(\tau,\lambda\chi)+\overline{G_0(\tau)}N_{n0}(\tau,\lambda\chi)]d\tau\\&+\sum_{k=1,k\neq n}^{N}\int_{-1}^{1}[G_k(\tau)M_{nk}(\lambda\tau,\lambda\chi)+\overline{G_k(\tau)}N_{nk}(\lambda\tau,\lambda\chi)]d\tau\Bigg\}.\end{aligned} \tag{39}$$

The equation for an interface crack has been separated here.

Then, as in the thermal problem, the solution of equations (38) and (39) is sought in the series form in the small parameter $\lambda=a/a_0$

$$G_0=\sum_{p=0}^{\infty}G_{0p}(\chi)\lambda^p,\ G_n=\sum_{p=0}^{\infty}G_{np}(\chi)\lambda^p. \tag{40}$$

The kernels

$$\begin{aligned}M_{nk}(\tau,\chi)&=\frac{1}{\pi}\int_{-1}^{1}\frac{\sqrt{1-\xi^2}}{\xi-\chi}K_{nk}(\tau,\xi)d\xi,\\N_{nk}(\tau,\chi)&=\frac{1}{\pi}\int_{-1}^{1}\frac{\sqrt{1-\xi^2}}{\xi-\chi}L_{nk}(\tau,\xi)d\xi\end{aligned} \tag{41}$$

are expanded in the series with respect to λ

$$M_{nk}=\sum_{p=0}^{\infty}m_{nkp}(\tau,\chi)\lambda^p,\quad N_{nk}=\sum_{p=1}^{\infty}n_{nkp}(\tau,\chi)\lambda^p. \tag{42}$$

The constants A_k (34) are also presented in series of λ:

$$A_n = \sum_{p=0}^{\infty} A_{np}\lambda^p. \tag{43}$$

Using (43) and (29)-(31) for (34), we obtain

$$\begin{aligned} A_0 &= -2\int_{-a_0}^{a_0} \beta_t \gamma_0'(t)dt = a_0(A_{00} + \lambda^2 A_{02}), \\ A_n &= -2\int_{-a_0}^{a_0} \beta_t \gamma_n'(t)dt = a_n \lambda A_{n1}. \end{aligned} \tag{44}$$

The coefficients of the series (43), (44) are cited in the Appendix B, Eqs. (B.1), (B.2). The coefficients in the series (42) are also presented in Appendix B by Eqs. (B.3), (B.4) and only those which are needed for further calculations. The full expressions can be found in [9].

Substituting the series (40), (42) and (43) into equations (38) and (39) and equating the expressions at like powers of λ, the system breaks down into a recurrent sequence of integral equations (see Eqs. (B.5), (B.6)). After solving the recurrent equations, the solution is represented as a Maclaurin series over even powers of the small parameter λ. The second approximation for the function G_0 for the interface macrocrack is written

$$\begin{aligned} G_0(\chi) &= G_{00}(\chi) + \lambda^2 G_{02}(\chi) = \\ &-\frac{i}{\pi\sqrt{1-\chi^2}}\{A_{00} + \lambda^2\{A_{02} + \frac{1}{2}A_{00}\sum_{k=1}^{N}\{I_{k0}^T m_{0k1}'(\chi) - \overline{I}_{k0}^T n_{0k1}'(\chi) \\ &\quad + 2A_{k1}[m_{0k0}(\chi) - n_{0k0}(\chi)]\}\}. \end{aligned} \tag{45}$$

Here A_{00}, A_{02} and A_{k1} are given in Eqs. (B.1), (B.2), $m_{0k0}, n_{0k0}, m_{0k1}'$ and n_{0k1}' in Eqs. (B.3), (B.4) and I_{k0}^T is

$$I_{k0}^T = \frac{1}{2}\left[\frac{e^{-2i\theta_k} - 1}{\sqrt{\overline{w}_k^2 - 1}} + \frac{1}{\sqrt{w_k^2 - 1}} - \frac{e^{-2i\theta_k}(1 - w_k\overline{w}_k)}{(\overline{w}_k^2 - 1)^{3/2}}\right]. \tag{46}$$

5. Thermal stress intensity factors

The singular field at a crack tip in FGMs has the same form as in homogeneous media [10], and the concept of stress intensity factors can be applied directly to the cracks in FGMs. Besides, the interface crack between the FGM and the homogeneous material with a smooth transition between these materials is also a classical crack with square-root singularities at the crack tips. Therefore, the thermal stress intensity factors (TSIFs) are found by

$$k_I^{\pm} - ik_{II}^{\pm} = \mp \lim_{\chi \to \pm 1} \sqrt{a_0(1-\chi^2)}G_0(\chi), \tag{47}$$

where the upper part of the "$\pm$" or "$\mp$" sign refers to the right tip and the lower part to the left tip of the interface macrocrack.

Inserting the expression (45) into (47), the TSIFs are obtained up to λ^2 as

$$k_{I0}^{\pm} = \lambda^2 \beta_t q k_0 a_0 \sqrt{a_0} \frac{1}{2} \sum_{k=1}^{N} \{\mathrm{Re}(I_{k0}^T)\mathrm{Im}(m_{k1} - n_{k1}) + \mathrm{Im}(I_{k0}^T)\mathrm{Re}(m_{k1} + n_{k1}) - 2J_k^T(\delta)\mathrm{Im}(m_{k0} - n_{k0})\}, \tag{48}$$

$$k_{II0}^{\pm} = \beta_t q k_0 a_0 \sqrt{a_0} \{\pm 1 + \frac{\lambda^2}{2} \sum_{k=1}^{N} \{2J_k^T(\delta)\mathrm{Re}[e^{i\theta_k}(w_k / \sqrt{w_k^2 - 1} - 1)] + \mathrm{Re}[I_{k0}^T]\mathrm{Re}(m_{k1} - n_{k1}) + \mathrm{Im}(I_{k0}^T)\mathrm{Im}(m_{k1} - n_{k1}) - 2J_k^T(\delta)\mathrm{Re}(m_{k1} - n_{k1})\}\}. \tag{49}$$

where I_{k0}^T and J_k^T are Eqs. (46) and (B.2) respectively; m_{k1}, n_{k1}, m_{k0} and n_{k0} are obtained from Eqs. (B.3), (B.4) by setting $\chi = \pm 1$ and are cited by Eqs. (B.7) and (B.8).

The TSIF for a single crack (without microcracks) subjected by a heat flux normal to the crack line, first derived by Sih (1962) [22], is written as

$$k_{II}^{\pm} = \pm \beta_t q k_0 a_0 \sqrt{a_0}, \quad k_I^{\pm} = 0. \tag{50}$$

The interaction of cracks leads to mixed mode conditions in the interface crack surfaces, i.e. $k_I \neq 0$, this will be discussed in the next section.

6. Results and discussion

It can be seen from formula (48) and from the following Figs. 3a – 6a that the thermal stress intensity factors k_I have small values. Meanwhile, k_I is non-zero for most crack configurations, the non-zero k_I means that the stress-strain state in the vicinity of the interface crack is mixed – mode state in contrast to the pure shear mode in the stress- strain state for a single crack, Eq. (50). The inhomogeneous parameter δ affects the TSIF k_I and this influence depends on the microcrack locations and orientations (see Figs. 3a – 6a). Besides, the TSIF k_I at the interface crack tips can be negative. For example, the influence of a microcrack with midpoint coordinate $w_k = -0.25i$ (in the lower half-plane) on the interface crack leads to negative k_I as we can see in Fig. 3a. The negative value of k_I means that the interface crack surfaces could close. Therefore, it might be necessary to account for the closure of the crack surfaces for these crack configurations as it was done for a system of macro-microcracks in a homogeneous material [8, 9] or to consider one of the special interface crack models, for example, the cohesive crack model [23]. However, the results, considered in the paper, are presented for the influence of cracks located in the upper half-plane, that is, in the FGM. For these crack locations k_I is mostly non-negative. At the same time the crack closure means the possibility of interface crack healing. So it looks as the cracks in the lower half plane are less dangerous for the interface and the interface crack than the cracks in the FGM.

The asymptotic analytical formulae obtained for TSIFs, Eqs. (48) and (49), allow one to investigate the influence of different arrays of microcracks (arbitrarily located and orientated) on the interface crack. Assume that all microcracks have the same angle of inclination θ with the x-axis. Figures 3-6 show the influence of the inhomogeneity parameter δ of thermoconductivity and of the microcrack inclination angle θ on the thermal stress intensity factors k_I and k_{II} at the interface crack tips for different arrangements of multiple microcracks (Figs. 2a, b, c and d). The calculations were performed with $\lambda = 0.1$. The microcrack centers are presented by $x_n = a_0 n / r, y_n = a_0 m / s$ $(n, m = \pm 1, \pm 2, ...)$, with $r = s = 5$, and in the complex form $w_n = z_n^0 / a_0 = (x_n + i y_n) / a_0$. The TSIFs $k_{I,II}$ are normalized by $| k_{II} |$ (50) and denoted by K_1 and K_2 in the figures. The non-dimensional inhomogeneity parameter of thermal conductivity is δa_0, but in the figures we leave the designation δ.

Normalized TSIFs K_1 and K_2 as functions of inclination angle θ of a microcrack with midpoint coordinate $w_k = 0.25i$ (Fig. 2a) are shown in Fig. 3 for different values of the inhomogeneity parameter δ: $\delta = 0$ – large dashed line, 0.5 – small dashed line, 1 – solid line. The curves of TSIFs at the right interface crack tip are denoted by $K_{1,2}^+$ and at the left tip by $K_{1,2}^-$. In the Fig. 3a the curve below the x-axis is the influence of one crack below the interface crack line (with the midpoint coordinate $w_k = -0.25i$) on the TSIF K_1^+. We have negative K_1^+ at this case as it was mentioned above. In Fig. 3a the results are presented for K_1^+ at the right interface crack tip, in Fig. 3b – for K_1^- at the left tip and in Fig. 3c – for K_2^+. The variation of $| K_2^- |$ is similar to K_2^+, but K_2^- has negative values.

Figure 4 shows the influence of a system of cracks ahead of the interface crack (Fig. 2b) on the TSIFs $K_1^\pm$ (Fig. 4a, b) and $K_2^\pm$ (Fig. 4c). For this case the inhomogeneity parameter δ affects on $K_1^\pm$ stronger than on $K_2^\pm$. The TSIF K_2 at the right crack tip is greater than at the left crack tip, and also is greater than for a single crack because of normalized TSIF K_2>1 (K_2=1 corresponds to the value for a single crack).

The influence of a system of cracks above the interface crack on the TSIFs is studied in Fig. 5 for the system of cracks presented in Fig. 2c and in Fig. 6 for the system presented in Fig. 2d. Figures 5a and 6a show that K_1 at the right interface crack tip is greater than at the left tip for all inclination angles θ of microcracks and for all δ values. For the uniform distributed microcracks above the interface crack (Fig. 6a) we can see that the difference between K_1^+ and K_1^- is less than in Fig. 5a. Increasing the inhomogeneous parameter δ from 0 (corresponds to a homogeneous material) to 1 increases the K_1^+ at the right crack tip. The influence of δ on the TSIF K_2 is the following. At the right crack tip the tendency is the same as for TSIF Mode I K_1^+. At the left crack tip the increasing of δ leads the increase of the TSIF $| K_2^- |$ for microcrack inclination angles from 0 to $\pi / 6$, and δ does not much influence the TSIF $| K_2^- |$ for

Materials Research Forum LLC
https://doi.org/10.21741/9781644902950

$\pi/6<\theta<\pi/2$. The absolute value of $|K_2^{\pm}|$ is less than 1, i.e. $|K_2^{\pm}|$ is less than the TSIF Mode II for a single crack under the same loading conditions (Figs. 5b, c and 6b, c).

As it was mentioned above the solution can be applied for some material combinations such as some ceramic/ceramic and some ceramic/metal materials. The presented results for $\delta>0$ correspond to the case where the thermal conductivity k is increased with increasing the y coordinate. It could be, for example, for an FGM/homogeneous bimaterial ($MoSi_2/Al_2O_3$)/Al_2O_3, where the lower material is alumina Al_2O_3 with $k^{Al_2O_3}=25\,Wm^{-1}K^{-1}$ and the upper material is FGM $MoSi_2/Al_2O_3$ varying gradually from molibdenum disilicide $MoSi_2$ with $k^{MoSi_2}=52\,Wm^{-1}K^{-1}$ to Al_2O_3 [24, 25]. At the same time $MoSi_2$ and Al_2O_3 have similar thermal expansion coefficients, it corresponds to $\varepsilon=0$.

This solution contains also other important parameter, coefficient of thermal expansion, the influence of which on the TSIFs will be the subject of further investigation.

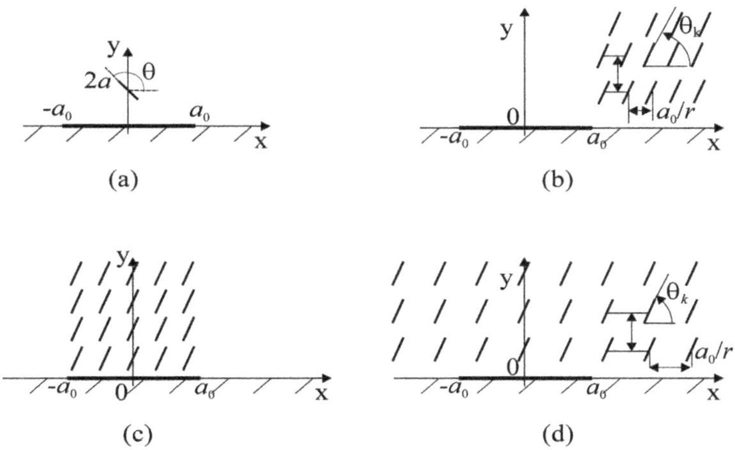

Figure 2. Schemes of locations of the interface crack and microcracks: (a) one crack above the interface crack $w_k=0.25i$; (b) non-symmetrically disposed microcrack system ahead of the interface crack; (c) and (d) microcracks above the interface crack.

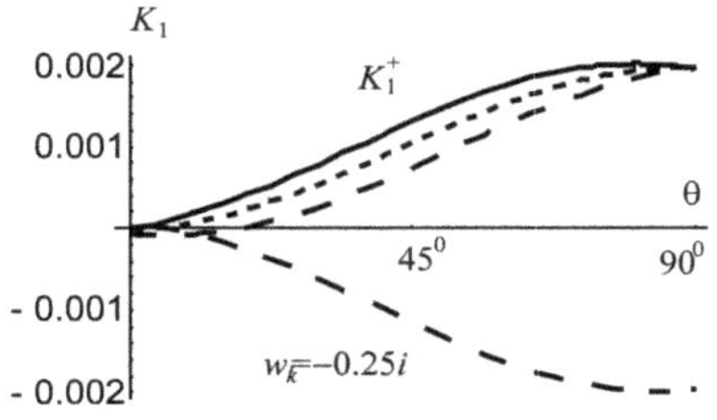

(a)

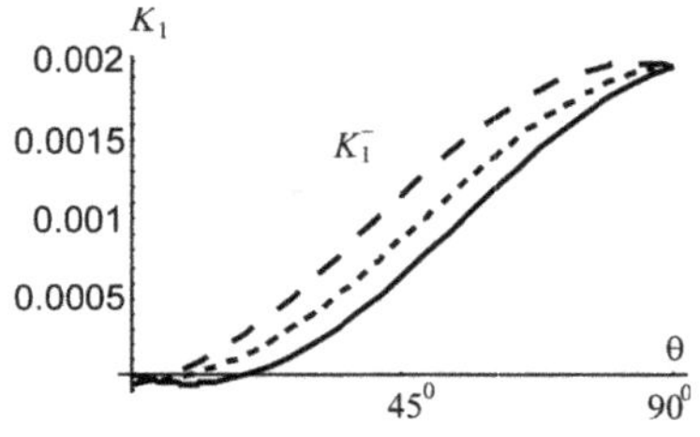

(b)

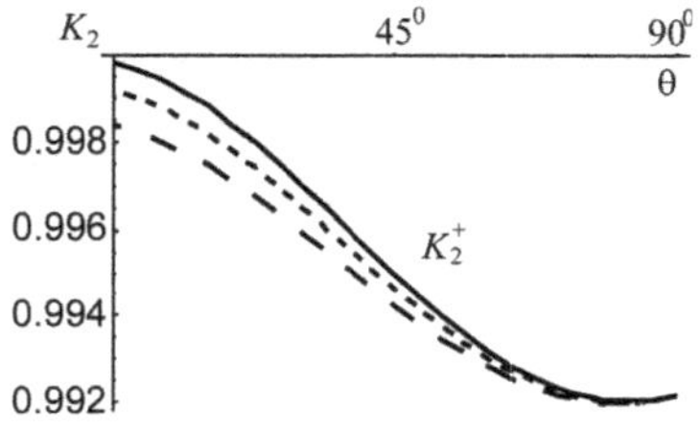

(c)

Figure 3. Normalized stress intensity factors K_1 (a, b) and K_2 (c) at the interface crack tips vs inclination angle θ, one crack above the interface crack ($w_k = 0.25i$, Fig. 2a) and below the interface crack, $w_k = -0.25i$ (a), for different values of inhomogeneity parameter δ: large dashed line – δ = 0, small dashed line – 0.5, solid line – 1.

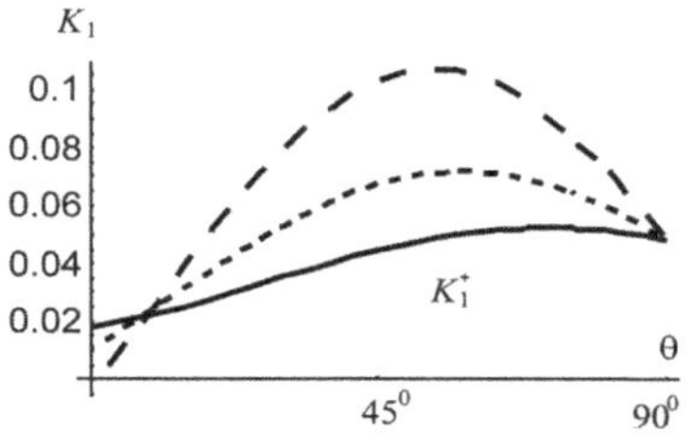

(a)

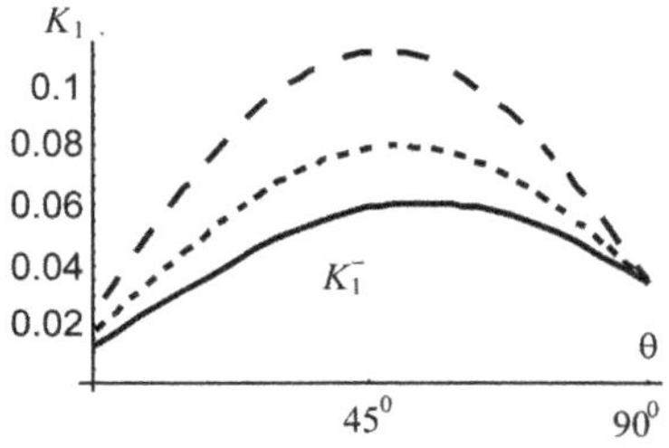

(b)

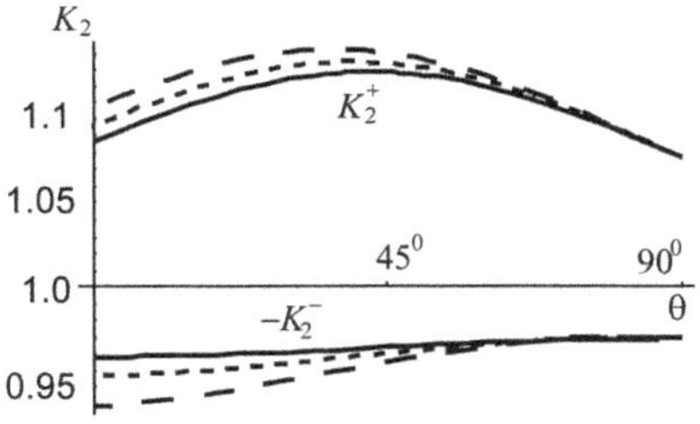

(c)

Figure 4. Influence of systems of microcracks (Fig. 2b) on TSIFs: (a) on K_1 at the right interface crack tip (b) on TSIF K_1 at the left interface crack tip; (c) on K_2 at the interface crack tips; for different values of inhomogeneity parameter δ. The designations are the same as in the previous Fig 3.

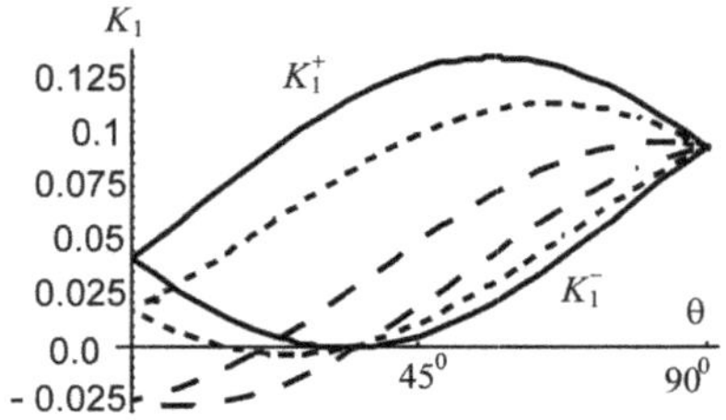

(a)

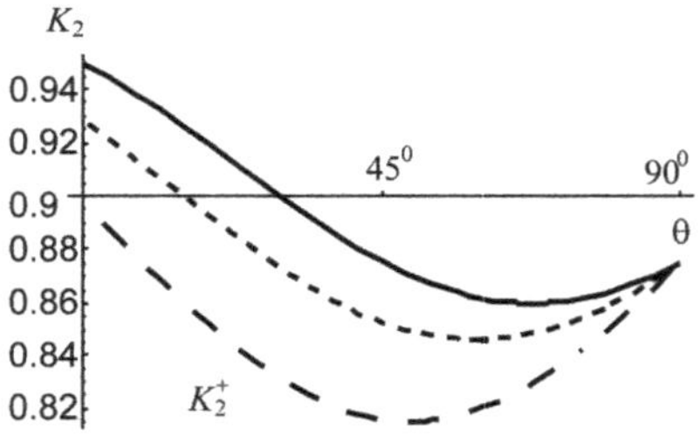

(b)

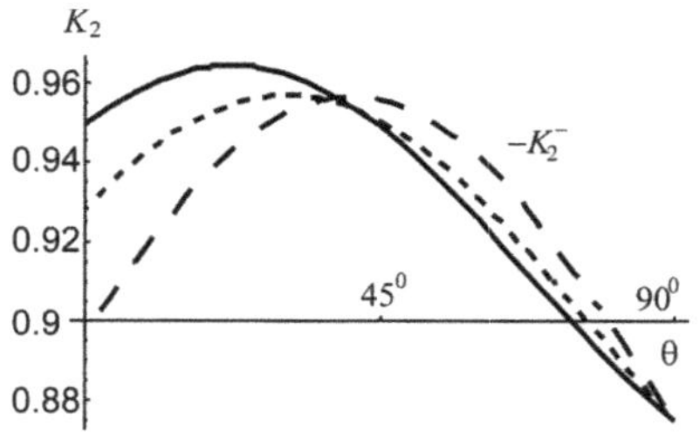

(c)

Figure 5. Influence of system of microcracks above the interface crack (Fig. 2c) on TSIFs K_1 (a) and K_2 (b, c) at the interface crack tips, for different values of inhomogeneity parameter δ. The designations are the same as in previous figures.

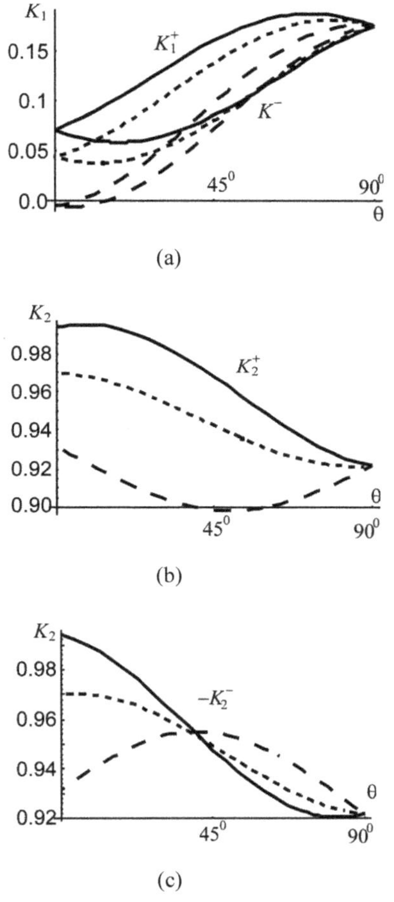

Figure 6. Influence of system of microcracks above the interface crack (Fig. 2d) on TSIFs K_1 (a) and K_2 (b) at the interface crack tips, for different values of inhomogeneity parameter δ. The designations are the same as in previous figures.

Conclusion

Theoretical analyses of the thermal interface crack problem in a FGM/homogeneous bimaterial with internal defects has been performed: formulation of the boundary conditions and analytical treatment by means of the integral equation method for the thermal and thermoelastic problems. The methods of superposition and small parameter method were applied as well as the theory of analytical functions. For a special case where an interface crack is significantly larger in size than internal cracks in the FGM the asymptotic analytical solution of the problem was obtained. Thermal stress intensity factors were derived as functions of geometry of the problem and material characteristics.

The effects of the location and orientation of the cracks and of the material inhomogeneity parameters, in particular, the inhomogeneity parameter of thermo-conductivity, on the thermal stress intensity factors at the interface crack in FGM/homogeneous bimaterials was demonstrated. It was shown that the TSIFs can be amplified or shielded by the system of microcracks, besides, the inhomogeneity parameter of thermo-conductivity notably affects the TSIFs of the interface crack.

The problem contains other important parameters, such as the inhomogeneity parameter of thermal expansion coefficient, whose influence on the TSIFs will be the subject of a further investigation. The other significant problem is prediction of extension of the interface crack and of the crack growth direction in FGM/homogeneous bimaterials. It is supposed to consider this problem later with using special theories of crack propagation, e.g. minimum strain energy density function [26].

The results can be used for modeling the residual stress-strain distribution near interfaces and interface cracks in graded bimaterial joints (FGM/homogeneous), for the purpose to improve material characteristics to resist thermal failure. The model is applicable for such kind of FGMs as: some ceramic/ceramic FGMs, e.g., TiC/SiC, $MoSi_2/Al_2O_3$ and $MoSi_2/SiC$, and also some ceramic/metal FGMs, e.g., zirconia/nichel and zirconia/steel.

Acknowledgement

V. Petrova acknowledges the support of the German Research Foundation under grants Schm 746/80-1 and Schm 746/92-1.

References

[1] F. Delale, F. Erdogan, The crack problem for a nonhomogeneous plane. ASME J. Appl. Mech. 50 (1983) 609-614. https://doi.org/10.1115/1.3167098

[2] N.I. Shbeeb, W.K. Binieda, K.L. Kreider, Analysis of the driving forces for multiple cracks in an infinite nonhomogeneous plate. Part I: Theoretical analysis. Part II: Numerical solutions, ASME J. Appl. Mech. 66 (1999) 492-506. https://doi.org/10.1115/1.2791075

[3] Y.-D. Li, K.Y. Lee, An antiplane crack perpendicular to the weak/micro-discontinuous interface in a bi-FGM structure with exponential and linear non-homogeneities, Int. J. Fracture 146 (2007) 203-211. https://doi.org/10.1007/s10704-007-9161-7

[4] A.M. Afsar, H. Sekine, Crack spacing effect on the brittle fracture characteristics of semi-infinite functionally graded materials with periodic edge cracks, Int. J. Fracture 102 (2000) L61-L66.

[5] B.L. Wang, J.C. Han, S.Y. Du, Thermoelastic fracture mechanics for nonhomogeneous material subjected to unsteady thermal load, ASME J. Appl. Mech. 67 (2000) 87-95. https://doi.org/10.1115/1.321153

[6] Y.-S. Wang, G.-Y. Huang, D. Gross, on the mechanical modeling of functionally graded interfacial zone with a Griffith crack: plane deformation, Int. J. Fracture 125 (2004) 189-

205. https://doi.org/10.1023/B:FRAC.0000021042.28804.f1

[7] S. Schmauder, U. Weber, Modelling of functionally graded materials by numerical homogenization. Arch. Appl. Mech. 71 (2001) 182-192. https://doi.org/10.1007/s004190000124

[8] V. Petrova, V. Tamuzs, N. Romalis, A survey of macro- microcrack interaction problems, ASME Appl. Mech. Rev. 53 (2000)117-146. https://doi.org/10.1115/1.3097344

[9] V. Tamuzs, N. Romalis, V. Petrova, Fracture of Solids with Microdefects, NOVA Science Publishers Inc., New York, 2000.

[10] N. Noda, Thermal stresses in functionally graded materials. J. Therm. Stresses 22 (1999) 477-505. https://doi.org/10.1080/014957399280841

[11] V. Birman, L.W. Byrd, Modeling and analysis of functionally graded materials and structures, ASME Appl. Mech. Rev. 60 (2007) 195-216. https://doi.org/10.1115/1.2777164

[12] Y. Ootao, Transient Thermoelastic and Piezothermoelastic Problems of Functionally Graded Materials, J. Therm. Stresses 32 (2009) 656-697. https://doi.org/10.1080/01495730902850918

[13] B.-L. Wang, Y.-W. Mai, On Thermal Shock Behavior of Functionally Graded Materials, J. Therm. Stresses 30 (2007) 523-558. https://doi.org/10.1080/01495730701273981

[14] B.-L. Wang, Y.-W. Mai, N. Noda, Fracture mechanics analysis model for functionally graded materials with arbitrarily distributed properties, Int. J. Fracture 116 (2002) 161-177. https://doi.org/10.1023/A:1020137923576

[15] L.-C. Guo, N. Noda, Modeling method for a crack problem of functionally graded materials with arbitrary properties - piecewise-exponential model, Int. J. Solids Struct. 44 (2007) 6768-6790. https://doi.org/10.1016/j.ijsolstr.2007.03.012

[16] Z.-H. Jin, G. Paulino, Thansient thermal stress analysis of an edge crack in a functionally graded material, Int. J. Fracture 107 (2001) 73-98. https://doi.org/10.1023/A:1026583903046

[17] J.F. Shackelford, W. Alexander, CRC Materials Science and Engineering Handbook, CRC Press, Boca Raton, 2001. https://doi.org/10.1201/9781420038408

[18] V. Tamuzs, V. Petrova, N. Romalis, Thermal fracture of a macrocrack with closure as influenced by microcracks, Theor. Appl. Fracture Mech. 21 (1994) 207-218. https://doi.org/10.1016/0167-8442(94)90034-5

[19] V. Petrova, K. Herrmann, Thermal crack problems for a bimaterial with an interface crack and internal defects subjected to a heat source, Int. J. Fracture 128 (2004) 49-63. https://doi.org/10.1023/B:FRAC.0000040967.13962.ec

[20] N.I. Muskhelishvili, Singular Integral Equations, Noordhoff, Groningen, the Netherlands, 1953.

[21] V.V. Panasyuk, M. P. Savruk, A. P. Datsyshin, Stress Distribution near Cracks in Plates and Shells (Russian), Naukova Dumka, Kiev, 1976.

[22] G. C. Sih, on the singular character of thermal stresses near a crack tip, Trans. ASME J.

Appl. Mech. 29 (1962) 587-589. https://doi.org/10.1115/1.3640612

[23] D.-J. Shim, G.H. Paulino, R.H. Dodds Jr., J resistance behaviour in functionally graded materials using cohesive zone and modified boundary layer models, Int. J. Fracture 139 (2006) 91-117. https://doi.org/10.1007/s10704-006-0024-4

[24] J.F. Shackelford, W. Alexander, CRC Materials Science and Engineering Handbook, CRC Press, Boca Raton, 2001. https://doi.org/10.1201/9781420038408

[25] O.P. Chakrabarti, P.K. Das, S. Mondal, Study of indentation induced cracks in MoSi2-reaction bonded SiC ceramics, Bull. Mater. Sci. 24 (2001) 181-184. https://doi.org/10.1007/BF02710098

[26] G. C. Sih, Mechanics of Fracture Initiation and Propagation: Surface and volume energy density applied as failure criterion, Kluwer Academic Publishers, Dordrecht, The Netherlands, 1991.

Appendix A

The coefficients in the series (27) are

$$Q_{0k} = -\frac{1}{\pi}\mathrm{Re}\left\{\sum_{p=0}^{\infty} (\lambda\tau)^{p-1}\int_{-1}^{1} \frac{\sqrt{1-\xi^2}}{\xi-\chi}\frac{e^{i(p-1)\theta_k}}{(\xi-w_k)^p}d\xi\right\} = \sum_{p=0}^{\infty} Q_{0kp}\lambda^p,$$

$$Q_{n0} = -\mathrm{Re}\left\{\sum_{p=1}^{\infty} \lambda^p \frac{e^{ip\theta_n}}{(\tau-w_n)^p}\left[\chi^p - \sum_{2r=2,\ldots}^{p} \frac{|2r-3|!!}{(2r)!!}\chi^{p-2r}\right]\right\} = \sum_{p=1}^{\infty} Q_{n0p}\lambda^p, \tag{A.1}$$

$$Q_{nk} = \mathrm{Re}\left\{\sum_{p=0}^{\infty} \lambda^p \frac{e^{ip\theta_n}}{w_{kn}^p}\frac{1}{\pi}\int_{-1}^{1} \frac{\sqrt{1-\xi^2}}{\xi-\chi}(\xi-\tau e^{i\theta_{kn}})^{p-1}d\xi\right\} = \sum_{p=1}^{\infty} Q_{nkp}\lambda^p.$$

The recurrent system of equations for coefficients in series (26) is following

$$\gamma'_{00}(\chi) = -\frac{1}{\pi\sqrt{1-\chi^2}}\int_{-1}^{1} \frac{\sqrt{1-\tau^2}}{\tau-\chi}\tilde{q}_0 d\tau,$$

$$\gamma'_{01}(\chi) = \frac{1}{\pi\sqrt{1-\chi^2}}\sum_{k=1}^{N}\int_{-1}^{1} \gamma'_{k1}(\tau)Q_{0k0}(\chi)d\tau, \tag{A.2}$$

$$\gamma'_{02}(\chi) = \frac{1}{\pi\sqrt{1-\chi^2}}\sum_{k=1}^{N}\int_{-1}^{1} [\gamma'_{k1}(\tau)Q_{0k1}(\tau,\chi) + \gamma'_{k2}(\tau)Q_{0k0}(\chi)]d\tau$$

and

$$\gamma'_{n0}(\chi)=0,$$
$$\gamma'_{n1}(\chi)=\frac{1}{\pi\sqrt{1-\chi^2}}\left[-\int_{-1}^{1}\frac{\sqrt{1-\tau^2}}{\tau-\chi}\tilde{q}_n d\tau+\int_{-1}^{1}\gamma'_{00}(\tau)Q_{n01}(\tau,\chi)d\tau\right],$$
$$\gamma'_{n2}(\chi)=\frac{1}{\pi\sqrt{1-\chi^2}}\left\{\int_{-1}^{1}[\gamma'_{00}(\tau)Q_{n02}(\tau,\chi)+\gamma'_{01}(\tau)Q_{n01}(\tau,\chi)]d\tau\right.$$
$$\left.+\sum_{k=1,k\neq n}^{N}\int_{-1}^{1}\gamma'_{k1}(\tau)Q_{nk1}(\chi)d\tau\right\}. \tag{A.3}$$

The expression (31) contains the following integral

$$f(\chi,\theta_n,w_n,\delta a_0)=-\frac{1}{\pi}e^{-\delta a_0\operatorname{Im}w_n}\cos\theta_n\int_{-1}^{1}\frac{\sqrt{1-\tau^2}}{\tau-\chi}e^{-\delta_1 a_0\lambda\tau}d\tau, \tag{A.4}$$

which is calculated as

$$\frac{1}{\pi}\int_{-1}^{1}\frac{\sqrt{1-\tau^2}}{\tau-\chi}e^{-b\tau}d\tau$$
$$=\frac{1}{\pi}\int_{-1}^{1}\frac{\sqrt{1-\tau^2}}{\tau-\chi}\left(1-b\tau+\frac{1}{2}b^2\tau^2-\frac{1}{6}b^3\tau^3+\frac{1}{24}b^4\tau^4+...\right)d\tau \tag{A.5}$$
$$=I_0-bI_1+\frac{1}{2}b^2I_2-\frac{1}{6}b^3I_3+...$$

with using the following formula [21, 23]

$$I_p(\tau,\chi)=\frac{1}{\pi}\int_{-1}^{1}\frac{\sqrt{1-\tau^2}}{\tau-\chi}\tau^p d\tau$$
$$=\sum_{q=0}^{p}(-1)^{q+p}C_p^q\left(-\chi^{q+1}+\sum_{2r=2,4,...}^{q+1}\frac{|2r-3|!!}{(2r)!!}\chi^{q-2r+1}\right),\ C_p^q=\frac{p!}{q!(p-q)!}. \tag{A.6}$$

Appendix B

The coefficients of the series (43), (44) are

$$A_{00}=\pi\beta_t a_0 q,$$
$$A_{02}=\pi\beta_t a_0 q\sum_{k=1}^{N}J_k^T\operatorname{Re}[e^{i\theta_k}(\frac{w_k}{\sqrt{w_k^2-1}}-1)], \tag{B.1}$$

$$A_{n1} = -\pi\beta_t a_0 q J_n^T,$$

where

$$J_n^T = \exp(-\delta a_0 \operatorname{Im} w_k)\cos\theta_n - \operatorname{Re}[e^{i\theta_n}(1 - w_n / \sqrt{w_n^2 - 1})]). \tag{B.2}$$

The coefficients in the expression (45) are the following:

$$m_{0k1}(\tau,\chi) = \tau\, m'_{0k1}(\chi),\ \ n_{0k1}(\tau,\chi) = \tau\, n'_{0k1}(\chi),$$

$$m'_{0k1}(\chi) = e^{i\theta_k} \operatorname{Re}\left[e^{i\theta_k} \frac{1 - w_k\chi}{(\chi - w_k)^2\sqrt{w_k^2 - 1}} \right], \tag{B.3}$$

$$n'_{0k1}(\chi) = \frac{e^{-2i\theta_k}}{2(\chi - \overline{w}_k)^2\sqrt{\overline{w}_k^2 - 1}} \left[(w_k - \overline{w}_k)\frac{\chi^2 - 2\overline{w}_k^3\chi + 3\overline{w}_k^2 - 2}{(\chi - \overline{w}_k)(\overline{w}_k^2 - 1)} + (e^{2i\theta_k} - 1)(1 - \overline{w}_k\chi) \right],$$

$$m_{0k0}(\chi) = \frac{e^{i\theta_k}}{2}\left[2 + \frac{\sqrt{\overline{w}_k^2 - 1}}{\chi - \overline{w}_k} + \frac{\sqrt{w_k^2 - 1}}{\chi - w_k} \right], \tag{B.4}$$

$$n_{0k0}(\chi) = \frac{e^{-i\theta_k}}{2}\frac{(w_k - \overline{w}_k)(1 - \overline{w}_k\chi)}{(\chi - \overline{w}_k)^2\sqrt{\overline{w}_k^2 - 1}}.$$

The recurrent system of equations for coefficients in series (40) is the following

$$G_{00}(\chi) = \frac{i}{\pi\sqrt{1 - \chi^2}}\frac{A_{00}}{a_0},$$

$$G_{01}(\chi) = \frac{1}{\pi\sqrt{1 - \chi^2}}\sum_{k=1}^{N} \int_{-1}^{1} [G_{k0}(\tau)m_{0k0}(\chi) + \overline{G_{k0}(\tau)}n_{0k0}(\chi)]d\tau, \tag{B.5}$$

$$G_{02}(\chi) = \frac{1}{\pi\sqrt{1 - \chi^2}}\left\{ i\frac{A_{02}}{a_0} + \sum_{k=1}^{N}\int_{-1}^{1}[G_{k0}(\tau)m_{0k1} + \overline{G_{k0}(\tau)}n_{0k1} + G_{k1}(\tau)m_{0k0} \right.$$

$$\left. + \overline{G_{k1}(\tau)}n_{0k0}\,]d\tau \right\}$$

and

$$G_{n0}(\chi) = \frac{1}{\pi\sqrt{1-\chi^2}} \int_{-1}^{1} [G_{00}(\tau)m_{n01} + \overline{G_{00}(\tau)}n_{0k1}(\chi)]d\tau\,, \tag{B.6}$$

$$G_{n1}(\chi) = \frac{1}{\pi\sqrt{1-\chi^2}} \left\{ iA_{n1} + \int_{-1}^{1}[G_{00}(\tau)m_{n02} + \overline{G_{k0}(\tau)}n_{n02}]d\tau \right.$$
$$\left. + \sum_{k=1,k\neq n}^{N} \int_{-1}^{1}[G_{k0}(\tau)m_{nk1} + \overline{G_{k0}(\tau)}n_{nk1}]d\tau \right\}.$$

In Eq. (48) m_{k1}, n_{k1}, m_{k0} and n_{k0} are obtained from Eqs. (B.3), (B.4) by setting $\chi = \pm 1$ and multiplying the functions by ± 1:

$$m_{k1} = e^{i\theta_k}\,\mathrm{Re}\left[\frac{e^{i\theta_k}}{(\overline{w}_k \mp 1)\sqrt{w_k^2 - 1}}\right],$$

$$n_{k1} = \frac{e^{-2i\theta_k}}{2(\overline{w}_k \mp 1)\sqrt{\overline{w}_k^2 - 1}}\left[(w_k - \overline{w}_k)\frac{2\overline{w}_k \pm 1}{\overline{w}_k^2 - 1} + e^{2i\theta_k} - 1\right], \tag{B.7}$$

$$m_{k0} = e^{i\theta_k}\left[1 - \mathrm{Re}\left(\frac{\sqrt{w_k^2 - 1}}{\overline{w}_k \mp 1}\right)\right], n_{k0} = \frac{e^{-i\theta_k}}{2}\frac{w_k - \overline{w}_k}{(\overline{w}_k \mp 1)\sqrt{\overline{w}_k^2 - 1}} \tag{B.8}$$

CHAPTER 2

Interaction of a System of Cracks with an Interface Crack in Functionally Graded/Homogeneous Bimaterials under Thermo-Mechanical Loading

Vera Petrova [1,2]*, Siegfried Schmauder [1]

[1] IMWF, University of Stuttgart, Pfaffenwaldring 32, D-70569 Stuttgart, Germany

[2] Voronezh State University, University Sq.1, Voronezh 394006, Russia

veraep@gmail.com *, Siegfried.Schmauder@imwf.uni-stuttgart.de

Abstract

The work is devoted to the investigation of fracture processes in the vicinity of an interface crack in functionally graded/homogeneous bimaterials with internal defects subjected to tensile loading and a heat flux. A previously obtained solution (Petrova, Schmauder, 2009, 2010) for the case of thermal loading on the same geometry is used as a part of the present solution. The solution is based on the integral equation method and it is assumed that thermal properties of the functionally graded material (FGM) possess exponential form. An asymptotic analytical solution is derived for a special case where an interface crack is larger than internal cracks in the FGM. The stress intensity factors are presented as asymptotic analytical functions of geometry of the problem and material properties. Analyses of the effects of the location and orientation of the cracks and the material non-homogeneity parameters on the stress intensity factors in FGM/homogeneous bimaterials is performed in the presence of thermal and mechanical loading. Examples of some FGM/homogeneous bimaterial are discussed.

Keywords

Functionally Graded Materials, Interface, Cracks, Crack Closure, Thermal Fracture, Thermal Stress Intensity Factor

1. Introduction

Advanced composites, such as functionally graded materials (FGMs), have attracted growing interest because of their wide application in numerous branches of modern technology. FGMs are designed in order to decrease bimaterial mismatch and residual stresses at interfaces and to prevent delamination along the interfaces. The material properties of FGMs are typically varied along one direction and the transition from one material constituent to the other is smooth. In order to improve the fracture resistance of FGMs, the study of fracture strengths of these materials is required.

Crack interaction problems in homogeneous materials have been extensively investigated and a large number of solutions have been obtained for different crack system configurations and different thermal and mechanical loadings [1, 2]. Many papers were also devoted to different models and semi-analytical solutions for cracks in FGMs [3, 4], but crack interaction problems were obtained only for special crack arrangements so far. For example, in Ref. [5] radially distributed cracks were considered under mechanical loading. In Ref. [6] a transient thermoelastic response of FGMs containing collinear cracks was investigated. A periodic array of cracks in FGMs subjected to thermo-mechanical loading is studied in [7] with taking into account a possible crack closure. Thermoelastic fracture analysis of FGMs is important for improving fracture characteristics of FGMs. The problem of thermal shock induced crack propagation was analyzed in [8] and a strong influence of residual stresses and the type of composition gradient on the crack length was shown.

The present work is a continuation of previous investigations [9-11] in which the problem of crack interactions with an interface crack in FGM/homogeneous bimaterials under the influence of heat flux was studied. Now a similar problem is considered for an FGM/homogeneous bimaterial subjected to both thermal and mechanical loading.

2. Formulation of the problem

2.1 Geometry of the problem and assumptions

Fig. 1 shows the geometry of the problem. A bimaterial is composed of a functionally graded material (denoted by number 1) and a homogeneous material (denoted by number 2). The bimaterial is perfectly bonded with the exception of an interface crack of length $2a_0$. It is assumed that the FGM contains N cracks of length $2a_k$. Cartesian coordinates (x, y) are centered at the midpoint of the interface crack; the x-axis lies along the interface line. Local coordinate systems (x_k, y_k) are attached to each internal crack. The crack position is determined by the defect midpoint coordinates (x_k^0, y_k^0) and an inclination angle θ_k to the interface, i.e. to the x-axis (Fig. 1).

The bimaterial is subjected to a heat flux of intensity q and a tensile stress P applied at infinity. The cracks are thermally isolated and traction free. It is assumed that the properties of the functionally graded material depend only on the coordinate y. The thermal conductivity coefficient and the thermal expansion coefficient are given as

$$k_1(y) = k_0 e^{\delta y}, \quad \alpha_1 = \alpha_0 e^{\omega y} \tag{1}$$

where, the constant k_0 is the thermal conductivity and α_0 is the thermal expansion coefficient of the interface and of material 2, while δ and ω are non-homogeneity parameters for the FGM. Young's modulus and Poisson's Ratio are assumed to be constant, Ej = const, ν_j = const (j = 1, 2). Thus, the material is elastically homogeneous, but thermally non-homogeneous.

The uncoupled, quasi-static thermoelastic theory is applicable to this problem; that is the temperature distribution is independent of the mechanical field, and the solution consists of the determination of the temperature field and the determination of the thermal stresses in the system.

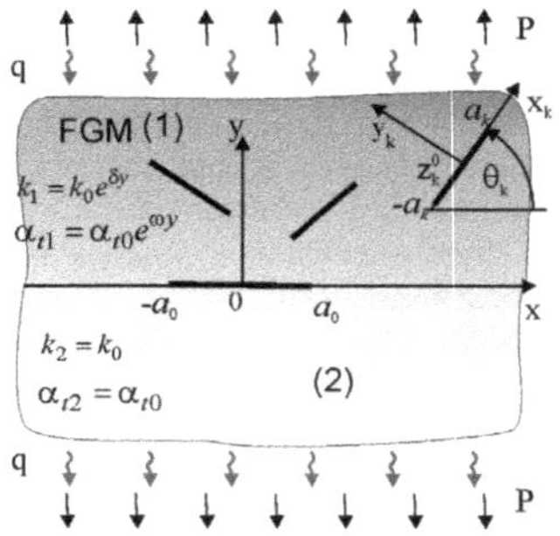

Figure 1. The geometry of the problem.

2.2 Thermal problem

The formulation and solution of the thermal problem is the same as in previous works [9-11] and is repeated (cited) here briefly. Due to the superposition principle the temperature field in the bimaterial with cracks is presented as a sum of two terms: the temperature distribution in a bimaterial in the absence of cracks and the temperature perturbation caused by the cracks. We are interested first of all in the determination of the temperature perturbation. It is supposed that the cracks are thermally isolated and continuity conditions for thermal fluxes and temperatures are fulfilled at the interface. Solving the temperature problem for an undamaged FGM/homogeneous bimaterial the thermal fluxes on the crack lines are obtained as

$$q_0 = -k_0 q, \quad q_n = -k_0 q \exp(-\delta y_n^0) \exp(-x_n \delta \sin \theta_n) \cos \theta_n \tag{2}$$

Supposing that the non-homogeneity of the FGM is revealed only in non-homogeneous thermal fluxes (2) on crack surfaces, the perturbation problem can be solved by the method presented in [1, 2] using similar integral equations. The system of *N+1* singular integral equations for the unknown functions γ_k' were used. The functions γ_k' are the derivatives of temperature jumps on the crack lines

$$2\gamma_k(t) = T_k^+ - T_k^- \tag{3}$$

2.3 Thermoelastic problem

The mechanical boundary conditions for the traction-free cracks are given as

$$(\sigma_{1y} - i\tau_{1xy})^+ = (\sigma_{2y} - i\tau_{2xy})^- = 0, \ |x| \le a_0, y = 0,$$

$$(\sigma_{jy} - i\tau_{jxy})^+ = (\sigma_{jy} - i\tau_{jxy})^- = 0, \ |x_k| \le a_k, y_k = 0,$$

and the continuity conditions at the interface are stratified, i.e. the stresses are equal and displacements are equal at the interface. Here σ_{jy} and τ_{jxy} are normal and shear stresses, j = 1, 2 for materials 1 and 2. The remote tensile stress P is applied perpendicular to the interface crack.

Because we suppose that the material is elastically homogeneous we can use directly the method presented in [1, 2]. Using the superposition principle the problem is transformed to the problem with boundary conditions on the crack lines:

$$\sigma^{\pm}_{ny} - i\tau^{\pm}_{nxy} = p_n(x), \quad p_n = -\frac{P}{2}(1+e^{-2i\theta_n}), |x| \le a_n, n = 0,1,...,N, \tag{4}$$

and with zero stresses at infinity $\sigma_{ij} \to 0, \quad x^2 + y^2 \to \infty$. The thermoelastic part of the solution (for stresses caused by the heat flux) is obtained from the system of singular integral equations for the traction-free cracks:

$$\int_{-a_n}^{a_n} \frac{G_n(t)}{t-x} dt + \sum_{\substack{k=0 \\ k\neq n}}^{N} \int_{-a_k}^{a_k} [G_k(t) K_{nk}(t,x) + \overline{G_k(t)} L_{nk}(t,x)] dt = 0, \quad |x| < a_n, \tag{5}$$

$$\int_{-a_n}^{a_n} G_n(t) dt = iA_n = -2i \int_{-a_n}^{a_n} \delta_t t \gamma'(t) dt_n, \; n = 0,1,...,N \tag{6}$$

$\overline{(...)}$ is the complex conjugate. The function G_k consists of two parts as

$$G_k(x) = g'_k(x) + 2i\delta_t \gamma_k(x), \tag{7}$$

where functions γ'_k (3) were found from the thermo-conductivity problem, while unknown functions $g'_n(x)$ are the derivatives of the displacement jumps on the crack lines

$$g'_n(x) = (2\mu / i(\kappa+1)) \partial[(u_n^+ - u_n^-) + i(v_n^+ - v_n^-)] / \partial x$$

In Eqs. (5) – (7) δ_t is $\delta_t = \beta_t/(\kappa+1)$. The following applies to the plane strain case $\kappa = 3\text{-}4\nu$ and $\beta_t = \alpha_t E$, while the plane stress condition corresponds to $\kappa = (3 - \nu)/(1 + \nu)$, and $\beta_t = \alpha_t E/(1 + \nu)$ while $\mu = E/(2(1 + \nu))$ is the shear modulus. v_n and u_n are normal and shear displacements. The condition of Eq. (6) provides that displacements are single-valued at the end points of the cracks. The condition of temperature continuity at the crack tips $\gamma_n(\pm a_n) = 0$ is also taken into account in Eq. (6). The regular kernels K_{nk} and L_{nk} contain the geometry of the problem.

The part of the solution due to mechanical load is determined by the integral equations

$$\int_{-a_n}^{a_n} \frac{g'_n(t)}{t-x} dt + \sum_{\substack{k=0 \\ k\neq n}}^{N} \int_{-a_k}^{a_k} [g'_k(t) K_{nk}(t,x) + \overline{g'_k(t)} L_{nk}(t,x)] dt = \pi p_n, \quad |x| < a_n$$

$$\int_{-a_n}^{a_n} g_n'(t)dt = 0,\ n = 0,1,...,N \tag{8}$$

where p_n is given in Eq. (4).

3. Solution of the problem

3.1 Solution by small parameter method

The solution is derived for a special case when an interface crack is significantly larger in size than internal cracks in the FGM. Let us assume that all internal cracks in the FGM are of the same size $2a_k = 2a$ ($k = 1, 2, ... , N$), for example, they have the characteristic size of the grain size of the material. Suppose also that $2a << 2a_0$. In this case the small parameter is $\lambda = a/a_0$ and $\lambda << 1$. Introducing non-dimensional coordinates χ and τ by $x = a_n\chi$ and $t = a_k\tau$, the equations are rewritten in dimensionless variables. The unknown functions are sought as a power series with respect to λ

$$f_0(\chi) = \sum_{p=0}^{\infty} f_{0p}(\chi)\lambda^p, \quad f_n(\chi) = \sum_{p=0}^{\infty} f_{np}(\chi)\lambda^p, \quad \lambda = a / a_0 << 1.$$

where functions f_n are γ_n', G_n and g_n' in Eqs. (5) and (8).

The regular kernels are expanded in series with respect to λ, too. For convergence of these series the following condition have to be satisfied

$$\left|\lambda / (\chi - z_k^0 / a_0)\right| < 1, \quad |\chi| < 1,$$

It is satisfied if cracks are not intersected, see Ref. [1]. The solution is obtained in closed form up to λ^2 (see [1] and [9] for details). The second approximation for the function G_0 is given as [9-11]

$$G_0(\chi) = G_{00}(\chi) + G_{02}(\chi)\lambda^2 \tag{9}$$

and for g_0' is

$$g_0'(\chi) = g_{00}'(\chi) + g_{02}'(\chi)\lambda^2 \tag{10}$$

The zero-th approximation G_{00} and g_{00}' in Eqs. (9) and (10) corresponds to an isolated interface crack and the second G_{02} and g_{02}' is taking into account the influence of each microcrack on the interface crack. Using the functions (9) and (10) the stress intensity factors (SIFs) are obtained.

3.2 Stress intensity factors

In this work we consider elastically homogeneous materials so that we can use the classical definition of the stress intensity factors. It should be noted, that the crack tip singular field in

FGMs has the same form as in homogeneous media and the concept of the stress intensity factors can be also applied directly to cracks in FGMs. Besides, the interface crack between the FGM and the homogeneous material with smooth transition between these materials is also classical crack with square-root singularities at the crack tips. Therefore, in all these cases the stress intensity factors (SIFs) are found by

$$k_{In}^{\pm} - ik_{IIn}^{\pm} = \mp \lim_{\chi\to\pm 1}\sqrt{a_n(1-\chi^2)}\left(G_n(\chi)+g_n'(\chi)\right),\quad n=0,1,2,\ldots,N, \tag{11}$$

where the upper part of the "$\pm$" or "$\mp$" signs refers to the right tip and the lower part to the left tip of the crack.

For a uniform heat flux and a tension load P acting at infinity the stress intensity factors at the interface crack tips are obtained up to λ^2 as

$$k_{I0}^{\pm} - ik_{II0}^{\pm} = \delta_t^h q k_0 a_0\sqrt{a_0}\left[\mp i + \lambda^2\sum_{k=1}^{N} F_1(w_k,\theta_k,\delta,\omega)\right] + P\sqrt{a_0}\left[1+\lambda^2\sum_{k=1}^{N} F_2(w_k,\theta_k)\right] \tag{12}$$

or writing full expressions we have the thermal part of SIFs

$$\begin{aligned} k_{I0}^{q\pm} &= \pm\lambda^2\delta_t^h q k_0 a_0\sqrt{a_0}\,\frac{1}{2}\sum_{k=1}^{N}\{\mathrm{Re}(I_{k0}^T)\,\mathrm{Im}[m_{k1}-n_{k1}] \\ &\quad + \mathrm{Im}(I_{k0}^T)\,\mathrm{Re}[m_{k1}+n_{k1}] + 2\exp(\omega a_0\,\mathrm{Im}(w_k))J_k^T(\delta)\,\mathrm{Im}[m_{k0}-n_{k0}]\}, \end{aligned} \tag{13}$$

$$\begin{aligned} k_{II0}^{q\pm} &= \pm\delta_t^h q k_0 a_0\sqrt{a_0}\{1+\frac{\lambda^2}{2}\sum_{k=1}^{N}\{2J_k^T(\delta)\,\mathrm{Re}[e^{i\theta_k}(w_k/(w_k^2-1)^{1/2}-1)] + \mathrm{Re}[m_{k1}-n_{k1}] \\ &\quad \mathrm{Re}[I_{k0}^T] + \mathrm{Im}(I_{k0}^T)\,\mathrm{Im}[m_{k1}-n_{k1}] - 2\exp(\omega a_0\,\mathrm{Im}(w_k))J_k^T(\delta)\,\mathrm{Re}[m_{k0}-n_{k0}]\}\}, \end{aligned} \tag{14}$$

and the mechanical part of SIFs

$$\begin{aligned} k_{I0}^{P\pm} &= P\sqrt{a_0}\left(1+\lambda^2\frac{1}{2}\sum_{k=1}^{N}\mathrm{Re}(J_k m_{k1}+\bar{J}_k n_{k1})\right), \\ k_{II0}^{P\pm} &= P\sqrt{a_0}\,\lambda^2\frac{1}{2}\sum_{k=1}^{N}\mathrm{Im}(J_k m_{k1}+\bar{J}_k n_{k1}), \end{aligned} \tag{15}$$

where I_{k0}^T, J_k^T and J_k are

$$I_{k0}^T = \frac{1}{2}\left[\frac{e^{-2i\theta_k}-1}{\sqrt{\bar{w}_k^2-1}} + \frac{1}{\sqrt{w_k^2-1}} - \frac{e^{-2i\theta_k}(1-w_k\bar{w}_k)}{(\bar{w}_k^2-1)^{3/2}}\right], \tag{16}$$

$$J_k^T(\delta) = \exp(-\delta a_0 \operatorname{Im} w_k)\cos\theta_k - \operatorname{Re}\left[e^{i\theta_k}\left(1 - \frac{w_k}{\sqrt{w_k^2 - 1}}\right)\right], \tag{17}$$

$$J_k = \frac{1}{2}\left[e^{-2i\theta_k} - 1 + e^{-2i\theta_k}\frac{\overline{w}_k - w_k}{(\overline{w}_k^2 - 1)^{3/2}} + \frac{w_k}{\sqrt{w_k^2 - 1}} + \frac{\overline{w}_k}{\sqrt{\overline{w}_k^2 - 1}}\right] \tag{18}$$

and m_{k1}, n_{k1}, m_{k0} and n_{k0} are

$$m_{k1} = e^{i\theta_k}\operatorname{Re}\left[\frac{e^{i\theta_k}}{(\overline{w}_k \mp 1)\sqrt{w_k^2 - 1}}\right],$$

$$n_{k1} = \frac{e^{-2i\theta_k}}{2(\overline{w}_k \mp 1)\sqrt{\overline{w}_k^2 - 1}}\left[(w_k - \overline{w}_k)\frac{2\overline{w}_k \pm 1}{\overline{w}_k^2 - 1} + e^{2i\theta_k} - 1\right],$$

$$m_{k0} = e^{i\theta_k}\left[1 - \operatorname{Re}\left(\frac{\sqrt{w_k^2 - 1}}{\overline{w}_k \mp 1}\right)\right], \quad n_{k0} = \frac{e^{-i\theta_k}}{2}\frac{w_k - \overline{w}_k}{(\overline{w}_k \mp 1)\sqrt{\overline{w}_k^2 - 1}}. \tag{19}$$

Besides

$$w_k = z_k^0 / a_0$$

is the non-dimensional complex coordinate of the midpoint of the cracks. In Eqs. (13) – (19) Re and Im denote the real and imaginary parts of complex numbers correspondingly. The formulae (12) – (15) were obtained by inserting Eqs. (9) and (10) into (11).

For a single crack (without microcracks) subjected to a heat flux normal to the crack surfaces the thermal stress intensity factors are given as [9, 12]

$$k_I^{\pm} = 0, \quad k_{II}^{\pm} = \pm\delta_t^h q k_0 a_0 \sqrt{a_0} \tag{20}$$

and for the crack under a tensile load P remotely applied normal to the crack the stress intensity factors are [12].

$$k_I^{\pm} = P\sqrt{a_0}, \quad k_{II}^{\pm} = 0. \tag{21}$$

The interaction of cracks leads to mixed mode conditions in the interface crack surfaces, i.e. $k_I \neq 0$ in the first case and $k_{II} \neq 0$ in the second one. The influence of both thermal and mechanical loading results in mixed-mode conditions near the interface crack.

4. Parameters of materials

The formulae for SIFs (12) – (19) contain geometrical parameters of the problem, such as, length of cracks, coordinates of the centers of cracks w_k and inclination angles θ_k of small cracks to the interface, and parameters of materials, the main of them are inhomogenety parameters of thermal conductivity and of the thermal expansion coefficient. The influence of these parameters on the SIFs at the interface crack tips can be investigated.

The values of the inhomogeneity parameters are estimated on the following reasons. From exponential form of the thermal conductivity Eq. (1) the inhomogeneity parameter δ is $\delta = (1/y)\ln(k_1/k_2)$. That means, the value depends on the ratio of material properties and the value of y. We consider an infinite domain and it is supposed that the value of δ changes slowly, we take $-1.0 \le \delta \le 1.0$. The same concerns the inhomogeneity parameter of the thermal expansion coefficient ω ($\omega = (1/y)\ln(\alpha_{t1}/\alpha_{t2})$.

Table 1 gives the thermal properties [13, 14] of some FGM/homogeneous material combinations and corresponding values of the inhomogeneity parameters δ and ω. The elastic moduli, Young's modulus and Poisson's ratio, of these materials are similar.

5. Results and discussion

The influence of different arrays of microcracks on the thermal SIFs at the interface crack was investigated in the case of a heat flux in Refs. [9-11]. In the present work, using formulae (12) – (19), two cases of the problem are considered: the case of a heat flux and the case of both a heat flux and a tensile load for the geometry shown in Fig. 2.

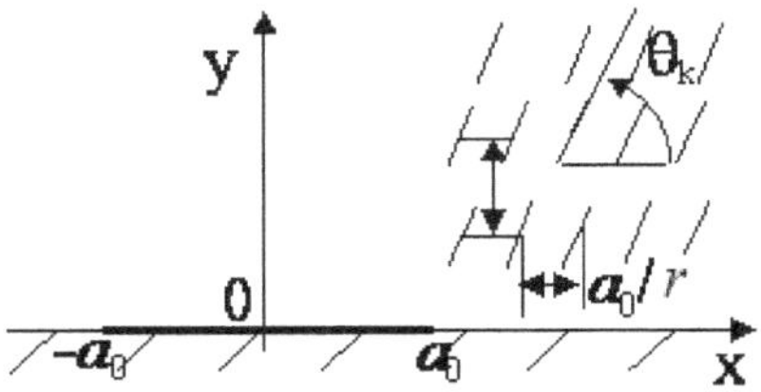

Figure 2. The scheme of locations of the interface crack and microcracks: non-symmetrically disposed microcrack system ahead of the interface crack.

Assume that all microcracks have the same angle of inclination θ to the x-axis. The microcrack centers are presented by $x_n = a_0 n/r=$, $y_n = a_0 m/s$ (n, m = ±1, ±2,...), with r = s = 5, $w_n = (x_n + i\ y_n)/\ a_0$ (Fig. 2). The calculation scheme of the system of microcracks in the FGM is shown in Fig. 2. SIFs $k_{I,II}$ are normalized by k_0 and denoted by k_1 and k_2 in the figures. k_0 is | k_{II} | Eq. (20) in the case of only thermal loading and k_I Eq. (21) in the case of tensile loading. It will be supposed that $k_0 = | k_{II} | = k_I$ in the case when both thermal and mechanical loads are applied. The calculations were performed with $\lambda = 0.1$. The non-dimensional inhomogeneity parameters of thermal conductivity and of thermal expansion are δa_0 and ωa_0 but in the figures we will leave the designation δ and w.

SIFs k_1 and k_2 at the right interface crack tip as functions of the inhomogeneity parameter w of thermal expansion coefficient and for different δ is presented in the Fig. 3 and 4 for the non-symmetrically disposed microcrack system ahead of the interface crack (Fig. 2) for $\theta = 0$ and the results for $\theta = \pi/4$ are cited in Ref. [11]. The same system of cracks but with different inclination angles θ produces different influences on the SIFs. The microcracks with $\theta = 0$ cause the interface crack closure $- k_1$ is negative for most parameters of δ and w. At the same time for $\theta = \pi/4$ k_1 is positive for all inhomogeneity parameters [11]. The examples of the material combinations are the following: parameters $\delta > 0$ and $w < 0$ correspond to the FGM/homogeneous material combinations (SiC/TiC)/TiC and (SiC/$MoSi_2$)/$MoSi_2$; and parameters $\delta < 0$ and $w > 0$ correspond to (TiC/SiC)/SiC and ($MoSi_2$/SiC)/SiC/ (see Table 1). These regions are indicated in Figs. 3 – 6.

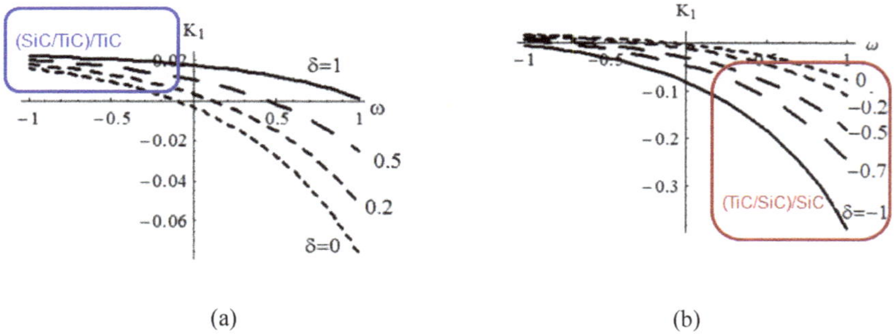

Figure 3. The case of a heat flux. Influence of a non-symmetrical system of microcracks (with θ=0) above the interface crack (Fig. 2) on thermal SIFs k_1 at the right interface crack for (a) δ >0 and (b) δ<0.

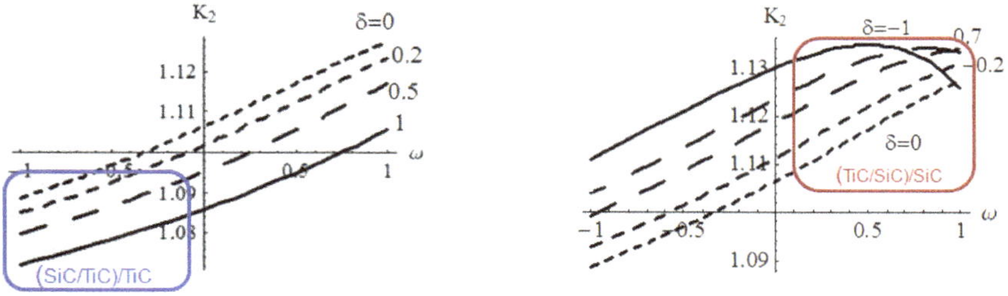

(a) (b)

Figure 4. The case of a heat flux. Influence of a non-symmetrical system of microcracks (with θ=0) above the interface crack (Fig. 2) on thermal SIFs k_2 at the right interface crack for (a) δ >0 and (b) δ <0.

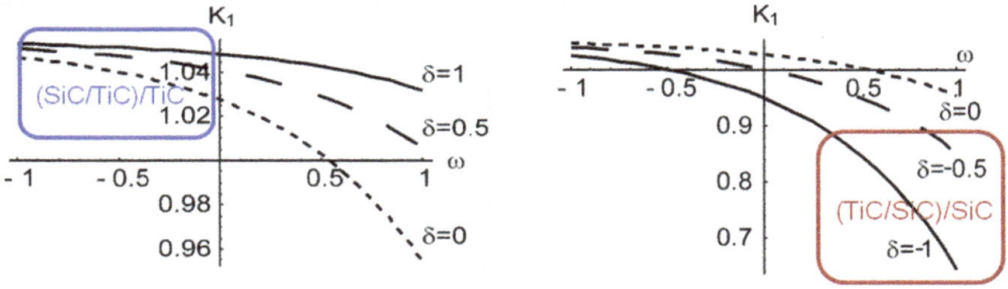

(a) (b)

Figure 5. The case of a heat flux and a tensile load. Influence of a non-symmetrical system of microcracks (with θ=0) above the interface crack (Fig. 2) on SIFs K_I at the right interface crack for (a) δ>0 and (b) δ<0.

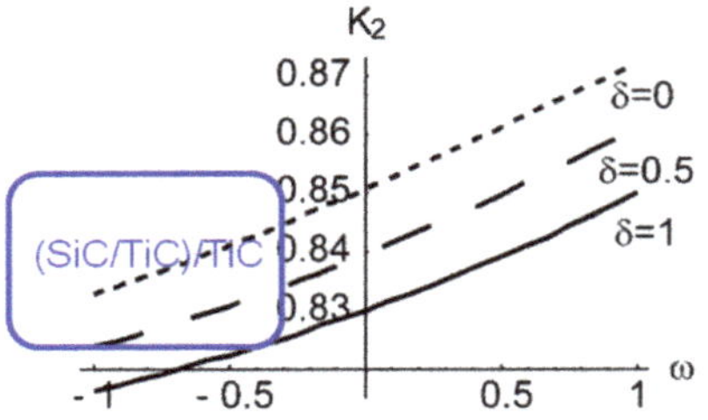

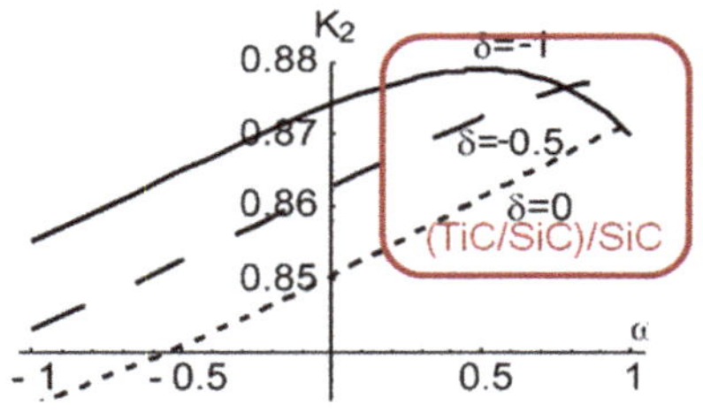

Figure 6. The case of a heat flux and a tensile load. Influence of a non-symmetrical system of microcracks (with θ=0) above the interface crack (Fig. 2) on SIFs K_2 at the right interface crack for (a) δ>0 and (b) δ<0.

Now we will use full formulae (12) – (19) for both thermal and mechanical loadings. Figs. 5 and 6 show SIFs k_1 and k_2 at the right interface crack tip as functions of the inhomogeneity parameter w and for different δ for the non-symmetrically disposed microcrack system ahead of the interface crack (Fig. 2) for $\theta = 0$. Because the mechanical part of the SIFs does not depend on inhomogeneity parameters, the curves of k_1 and k_2 are similar as in Figs. 3 and 4, but the values of k_1 and k_2 are different for these two cases. The difference in values of k_1 due to change of δ is up to 35% (maximum is reached for $w = 1$) and the difference in values of k_2 due to changes of δ is up to 17%.

In Fig. 5 the value of k_1 is positive for all parameters. A part of k_1 is larger than 1 and a part is less than 1, $k_1 = 1$ corresponds to the value for a single crack. The example of an FGM/homogeneous bimaterial is (SiC/TiC)/TiC (Fig. 5a), and in this case a system with cracks increases k_1, but for (TiC/ SiC)/ SiC (Fig. 5b) the same system of cracks decreases k_1 (at the same loading conditions).

In the case of a heat flux the SIF k_2 (Fig. 4) is higher than in the case of thermo-mechanical loading (Fig. 6) for all parameters δ and w, and is also higher than for a single crack, because of normalized $k_2 > 1$ ($k_2 = 1$ corresponds to the value for a single crack). In the case of both a heat flux and tensile load k_2 is lower than for a single crack, $k_2 < 1$ (Fig. 6).

Table 1. Thermal properties of some FGM/homogeneous materials and values of the parameters δ and ω.

FGM/H ($MoSi_2$/Al_2O_3)/ Al_2O_3 (Molybdenum disilicide $MoSi_2$, alumina Al_2O_3)						
	Thermal expansion coeff. [*10^{-6} K^{-1}]			Thermal conductivity $Wm^{-1}K^{-1}$		
$MoSi_2$	α_{t1}	5	$\alpha_{t1}/\alpha_{t2}=1$	k_1	52	$k_1/k_2>1$
Al_2O_3	α_{t2}	5	$\omega=0$	k_2	25	$\delta>0$
FGM/H (Al_2O_3/ $MoSi_2$)/ $MoSi_2$			$\omega=0$			$\delta<0$
FGM/H (SiC/ $MoSi_2$)/ $MoSi_2$ (Silicium carbide SiC, molibdenum desilicide $MoSi_2$)						
	Thermal expansion coeff. *10^{-6} K^{-1}			Thermal conductivity $Wm^{-1}K^{-1}$		
SiC	α_{t1}	4	$\alpha_{t1}/\alpha_{t2}<1$	k_1	60	$k_1/k_2>1$
$MoSi_2$	α_{t2}	5	$\omega<0$	k_2	52	$\delta>0$
FGM/H ($MoSi_2$/ SiC)/ SiC			$\omega>0$			$\delta<0$
FGM/H (SiC/ TiC)/ TiC (Silicium carbide SiC, titanium carbide TiC)						
	Thermal expansion coeff. *10^{-6} K^{-1}			Thermal conductivity $Wm^{-1}K^{-1}$		
SiC	α_{t1}	4	$\alpha_{t1}/\alpha_{t2}<1$	k_1	60	$k_1/k_2>1$
TiC	α_{t2}	7	$\omega<0$	k_2	20	$\delta>0$
FGM/H (TiC/ SiC)/ SiC			$\omega>0$			$\delta<0$

The results show that the inhomogeneity parameters δ and ω of thermo-conductivity and thermal expansion coefficient notably affect the SIFs of the interface crack. The parametric analysis reveals the dependence of the SIFs at the interface crack tips on the location and orientation of the cracks in the FGM. The SIFs can be amplified or shielded by the system of microcracks. It was also shown, that if only the thermal load is applied for some crack arrangements (Figs. 3) the SIF k_I at the interface crack tips can be negative, i.e. the interface crack surfaces could close. If both the thermal flux and the tensile load are applied the SIF k_I is always positive. The results are applicable to some kinds of FGM/homogeneous bimaterials examples of which are presented in Table 1.

Acknowledgement

V. Petrova acknowledges the support of the German Research Foundation under grants Schm 746/106-1 and Schm 746/113-1.

References

[1] V. Tamuzs, N. Romalis, V. Petrova, Fracture of Solids with Microdefects, NOVA Science Publishers Inc., New York, 2000.

[2] V. Petrova, V. Tamuzs, N. Romalis, A survey of macro-microcrack interaction problems, ASME Appl. Mech. Rev. 53 (2000) 117-146. https://doi.org/10.1115/1.3097344

[3] V. Birman, L.W. Byrd, Modeling and analysis of functionally graded materials and structures, ASME Appl. Mech. Rev. 60 (2007) 195-216. https://doi.org/10.1115/1.2777164

[4] Y. Ootao, Transient Thermoelastic and Piezothermoelastic Problems of Functionally Graded Materials, J. Therm. Stresses 32 (2009) 656–697. https://doi.org/10.1080/01495730902850918

[5] N.I. Shbeeb, W.K. Binieda, K.L. Kreider, Analysis of the driving forces for multiple cracks in an infinite nonhomogeneous plate. Part I: Theoretical analysis. Part II: Numerical solutions, ASME J. Appl. Mech. 66 (1999) 492-506. https://doi.org/10.1115/1.2791075

[6] N. Noda, B.L. Wang, Transient thermoelastic responses of functionally graded materials containing collinear cracks, Eng. Fract. Mech. 69 (2002) 1791–1809. https://doi.org/10.1016/S0013-7944(02)00055-3

[7] B.L. Wang, Y.-W. Mai, A periodic array of cracks in functionally graded materials subjected to thermo-mechanical loading, Int. J. Eng. Science 43 (2005) 432–446. https://doi.org/10.1016/j.ijengsci.2004.10.004

[8] T. Sadowski, A. Neubrand, Estimation of the crack length after thermal shock in FGM strip, Int. J. Fract. 127 (2004) L135–L140. https://doi.org/10.1023/B:FRAC.0000035087.34082.88

[9] V. Petrova, S. Schmauder, Thermal fracture of a functionally graded/homogeneous bimaterial with a system of cracks, Theor. Appl. Fract. Mech.55 (2011), 148–157. https://doi.org/10.1016/j.tafmec.2011.04.005

[10] V. Petrova, S. Schmauder, Crack – interface crack interactions in functionally graded/ homogeneous composite bimaterials subjected to a heat flux, Mech. Comp. Mater. 47 (2011), 125-136. https://doi.org/10.1007/s11029-011-9191-0

[11] V. Petrova, S. Schmauder, Mathematical modelling and thermal stress intensity factors evaluation for an interface crack in the presence of a system of cracks in functionally graded/ homogeneous bimaterials, Comp. Mater. Sci. 52 (1) (2012) 171-177. https://doi.org/10.1016/j.commatsci.2011.02.028

[12] V.V. Panasyuk, M.P. Savruk, A.P. Datsyshin, Stress Distribution near Cracks in Plates and Shells (in Russian), Naukova Dumka, Kiev, 1976.

[13] J.F. Shackelford, W. Alexander, CRC Materials Science and Engineering Handbook, CRC Press, Boca Raton, 2001. https://doi.org/10.1201/9781420038408

[14] O.P. Chakrabarti, P.K. Das, S. Mondal, Study of indentation induced cracks in MoSi2-reaction bonded SiC ceramics, Bull. Mater. Sci. 24 (2001) 181-184. https://doi.org/10.1007/BF02710098

CHAPTER 3

FGM/Homogeneous Bimaterials with Systems of Cracks under Thermo-Mechanical Loading: Analysis by Fracture Criteria

Vera Petrova [1,2]*, Siegfried Schmauder [1]

[1] IMWF, University of Stuttgart, Pfaffenwaldring 32, D-70569 Stuttgart, Germany

[2] Voronezh State University, University Sq.1, Voronezh 394006, Russia

veraep@gmail.com *, Siegfried.Schmauder@imwf.uni-stuttgart.de

Abstract

Fracture criteria for prediction of extension of the interface crack and of the crack growth direction in a bimaterial consisting of a homogeneous and a functionally graded material (FGM) with systems of internal defects are studied. The bimaterial is subjected to a heat flux and a tensile load applied at infinity. It is assumed that the thermal properties of the FGM have exponential form. The Young's modulus and Poisson's ratio are assumed to be constant. In the previous papers [1-4] asymptotic analytical formulas for the stress intensity factors (SIFs) at the interface crack tips were obtained as a series of a small parameter (the ratio between sizes of the internal and interface cracks). These SIFs are used in fracture criteria to obtain the possible direction of crack propagation and critical loads. The maximum circumferential stress criterion is used and some results for the fracture angles is obtained by the minimum strain energy density in order to compare the predictions for the fracture angles by two fracture criteria. The influence of geometry of the problem (location and orientation of cracks) and the parameters of non-homogeneity of FGMs on the main fracture characteristics is investigated.

Keywords

Functionally Graded Materials (Fgms), Interface, Cracks, Crack Kinking, Fracture Criteria, Stress Intensity Factor

1. Introduction

The paper is devoted to the problem of the thermal fracture of a bimaterial consisting of a homogeneous and a functionally graded material (FGM) with an interface crack and internal defects subjected to a heat flux and a tensile load applied at infinity. It is assumed that the thermal properties of the FGM have exponential form. The Young's modulus and Poisson's ratio are assumed to be constant. Thus, the material is elastically homogeneous, but thermally non-homogeneous. This kind of FGMs include some ceramic/ceramic and ceramic/metal FGMs. The problem is studied in the case when an interface crack is much larger than internal cracks in the FGM. In the previous papers [1-4] asymptotic analytical formulas for the stress intensity factors

(SIFs) at the interface crack tips were obtained as a series of a small parameter (the small parameter is equal to the ratio between sizes of the internal and interface cracks). These SIF's functions contain parameters of geometry of the problem and parameters of the non-homogeneity of the FGM.

The present work is devoted to analysis of the fracture criteria for prediction of extension of the interface crack and of the crack growth direction in FGM/homogeneous bimaterial. From experimental and theoretical investigations it is known that the crack deflection from initial crack propagation occurs under mixed mode loading and this deflection depends on the details of Modes I and II loadings. For FGMs the near-tip mixity can arise by virtue of the property variation in the material. Besides, the interaction of cracks, defects and interfaces adds additional near tip mixity. In this connection the following fracture criteria are used: the maximum circumferential stress criterion [5, 6, 7] and the minimum strain energy density [8, 9]. The parametric analysis shows the dependence of the initial interface crack propagation on the location and orientation of the crack systems. It is also shown that the non-homogeneity parameter of thermo-conductivity and thermal expansion coefficient notably affect the interface crack deflection angle. The influence of these parameters on the main fracture characteristics is investigated. Comparison of these results for considered criteria is performed.

2. Formulation of the problem

2.1 Geometry of the problem and assumptions

Fig. 1a shows the geometry of the problem. A bimaterial is composed of a FGM (denoted by number 1) and a homogeneous material (denoted by number 2). The bimaterial is perfectly bonded with the exception of an interface crack of length $2a_0$. It is assumed that the FGM contains N cracks of length $2a_k$. Cartesian coordinates (x, y) are centered at the midpoint of the interface crack; the x-axis lies along the interface line. Local coordinate systems (x_k, y_k) are attached to each internal crack. The crack position is determined by the defect midpoint coordinate and an inclination angle θ_k to the interface, i.e. to the x-axis (Fig. 1a). The bimaterial is subjected to a heat flux of intensity q and a tensile stress P applied at infinity. The cracks are thermally isolated and traction free.

It is assumed that the thermal conductivity coefficient and the thermal expansion coefficient are

$$k_1(y) = k_0 e^{\delta y}, \quad \alpha_1(y) = \alpha_0 e^{\omega y} \tag{1}$$

where the constant k_0 is the thermal conductivity, α_0 is the thermal expansion coefficient of the interface and of material 2, δ and ω are non-homogeneity parameters for the FGM. The Young's modulus and Poisson's ratio are assumed to be constant, $E_j = const$, $\nu_j = const$ (j = 1, 2).

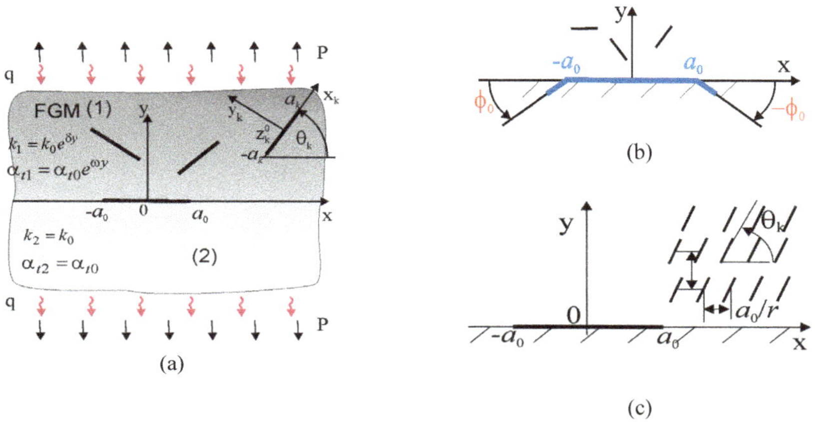

Figure 1. (a) The geometry of the problem; (b) The angle φ_0 of crack deflection; (c) The scheme of locations of the interface crack and microcracks.

The uncoupled, quasi-static thermoelastic theory is applicable to this problem that is the temperature distribution is independent of the mechanical field, and the solution consists of the determination of the temperature field and the determination of the thermal stresses.

2.2 Thermal and thermoelastic problem formulations

It is supposed that the cracks are thermally isolated and continuity conditions for thermal fluxes and temperatures are fulfilled at the interface. Using the superposition principle the temperature field in the bimaterial with cracks is presented as a sum of two terms: the temperature distribution in a bimaterial in the absence of cracks and the temperature perturbation caused by the cracks. For the crack problem the temperature perturbation should be determined.

The mechanical boundary conditions are: the cracks are traction-free and the continuity conditions at the interface are assumed, i.e. the stresses are equal and displacements are equal. Using the superposition principle the problem is transformed to the problem with boundary conditions on the crack lines. Because we suppose that the material is elastically homogeneous we can use directly the method presented in [7, 10]. The mathematical analysis of these problems leads to singular integral equations of the Cauchy-type [11]. Following the basic principle of linear elasticity the solution of purely thermal and purely mechanical loading can simply be added to give the solution for the combined loading case. This problem for combined loading was discussed in [12] with respect to stress intensity factors (SIFs) for an r-crack interacting with an inclusion.

2.3 Solution by small parameter method

The solution is derived for a special case where an interface crack is significantly larger in size than internal cracks in the FGM. The asymptotic analytical solution of the problem is obtained as a series of a small parameter which is equal to the ratio between sizes of the internal and interface cracks. The method was first suggested by Romalis and Tamuzs at 1984 and then was used for different macro-microcrack interaction problems for homogeneous materials [7, 10].

It is assumed that all internal cracks in the FGM have the same size $2a_k = 2a$ ($k = 1, 2, \ldots, N$), for example, they have the characteristic size of a grain size of the material. Suppose also that $2a << 2a_0$. In this case the small parameter is $\lambda = a/a_0$ and $\lambda << 1$. The solution is obtained in closed form up to λ^2 (see [1] for details)

$$f_0(x) = f_{00}(x) + f_{02}(x)\lambda^2. \tag{2}$$

Here the function f_0 is the solution of thermal and thermoelastic problems. The zero-th approximation f_{00} in Eq. (2) corresponds to an isolated interface crack and the second one f_{02} is taking into account the influence of each microcrack on the interface crack. Using this solution the SIFs are obtained. The detailed formulation and solution of these problems can be found in [1-4].

2.4 Stress intensity factors

In this work we consider elastically homogeneous materials so that we can use the classical definition of the stress intensity factor. It should be noted, that the crack tip singular field in FGMs has the same form as in homogeneous media [13] and the concept of the stress intensity factors can be also applied directly to cracks in FGMs. Besides, the interface crack between the FGM and the homogeneous material with smooth transition between these materials is also a classical crack with square-root singularities at the crack tips.

For a uniform heat flux and a tension load P acting at infinity the stress intensity factors at the interface crack tips are obtained up to λ^2 as [1, 3, 4]

$$k_{I0}^{\pm} - ik_{II0}^{\pm} = Q\sqrt{a_0}\left[\mp i + \lambda^2 \sum_{k=1}^{N} F_1(w_k, \theta_k, \delta, \omega)\right] + P\sqrt{a_0}\left[1 + \lambda^2 \sum_{k=1}^{N} F_2(w_k, \theta_k)\right] \tag{3}$$

Here,

$$Q = q\delta_t^h k_0 a_0, \quad \delta_t^h = \delta_t^h(\alpha_{t0}, E, \nu), \tag{4}$$

or writing full expressions we have for the thermal part of SIFs

$$\begin{aligned} k_{I0}^{q\pm} &= \pm\lambda^2 Q\sqrt{a_0}\frac{1}{2}\sum_{k=1}^{N} \{\mathrm{Re}(I_{k0}^T)\,\mathrm{Im}[m_{k1} - n_{k1}] \\ &\quad + \mathrm{Im}(I_{k0}^T)\,\mathrm{Re}[m_{k1} + n_{k1}] + 2\exp(\omega a_0\,\mathrm{Im}(w_k))J_k^T(\delta)\,\mathrm{Im}[m_{k0} - n_{k0}]\}, \end{aligned} \tag{5}$$

$$k_{II0}^{q\pm} = \pm Q\sqrt{a_0}\{1+\frac{\lambda^2}{2}\sum_{k=1}^{N}\{2J_k^T(\delta)\mathrm{Re}[e^{i\theta_k}(w_k/(w_k^2-1)^{1/2}-1)]$$
$$+\mathrm{Re}[I_{k0}^T]\mathrm{Re}[m_{k1}-n_{k1}]+\mathrm{Im}(I_{k0}^T)\mathrm{Im}[m_{k1}-n_{k1}]$$
$$-2\exp(\omega a_0\,\mathrm{Im}(w_k))J_k^T(\delta)\mathrm{Re}[m_{k0}-n_{k0}]\}\}, \tag{6}$$

and the mechanical part of SIFs is

$$k_{I0}^{P\pm} = P\sqrt{a_0}\left(1+\lambda^2\frac{1}{2}\sum_{k=1}^{N}\ \mathrm{Re}(J_k m_{k1}+\bar{J}_k n_{k1})\right) \tag{7}$$

$$k_{II0}^{P\pm} = P\sqrt{a_0}\lambda^2\frac{1}{2}\sum_{k=1}^{N}\ \mathrm{Im}(J_k m_{k1}+\bar{J}_k n_{k1}), \tag{8}$$

where $I_{k0}{}^T$, $J_k{}^T$ and J_k are

$$I_{k0}^T = \frac{1}{2}\left[\frac{e^{-2i\theta_k}-1}{\sqrt{\bar{w}_k^2-1}}+\frac{1}{\sqrt{w_k^2-1}}-\frac{e^{-2i\theta_k}(1-w_k\bar{w}_k)}{(\bar{w}_k^2-1)^{3/2}}\right], \tag{9}$$

$$J_k^T(\delta) = \exp(-\delta a_0\,\mathrm{Im}\,w_k)\cos\theta_k - \mathrm{Re}\left[e^{i\theta_k}\left(1-\frac{w_k}{\sqrt{w_k^2-1}}\right)\right], \tag{10}$$

$$J_k = \frac{1}{2}\left[e^{-2i\theta_k}-1+e^{-2i\theta_k}\frac{\bar{w}_k-w_k}{(\bar{w}_k^2-1)^{3/2}}+\frac{w_k}{\sqrt{w_k^2-1}}+\frac{\bar{w}_k}{\sqrt{\bar{w}_k^2-1}}\right] \tag{11}$$

and m_{k1}, n_{k1}, m_{k0} and n_{k0} are

$$m_{k1} = e^{i\theta_k}\,\mathrm{Re}\left[\frac{e^{i\theta_k}}{(\bar{w}_k\mp1)\sqrt{w_k^2-1}}\right],\ n_{k1} = \frac{e^{-2i\theta_k}}{2(\overline{w_k}\mp1)\sqrt{\overline{w_k}^2-1}}\left[(w_k-\bar{w}_k)\frac{2\bar{w}_k\pm1}{\bar{w}_k^2-1}+e^{2i\theta_k}-1\right],$$

$$m_{k0} = e^{i\theta_k}\left[1-\mathrm{Re}\left(\frac{\sqrt{w_k^2-1}}{\bar{w}_k\mp1}\right)\right],\ \ n_{k0} = \frac{e^{-i\theta_k}}{2}\frac{w_k-\bar{w}_k}{(\bar{w}_k\mp1)\sqrt{\bar{w}_k^2-1}}. \tag{12}$$

Besides $w_k = z_k^0/a_0$ is the non-dimensional complex coordinate of the midpoint of the cracks. In Eqs. (5)–(8) "Re" and "Im" denote the real and imaginary parts of complex numbers correspondingly and the upper part of the "$\pm$" or "$\mp$" signs refers to the right tip and the lower part to the left tip of the crack.

In Eq. (4) the material constant δ_t^h is $\delta_t^h = \beta_t/(\kappa+1)$. The following applies to the plane strain case $\kappa = 3\text{-}4\nu$ and $\beta_t = \alpha_{t0}E$, while the plane stress condition corresponds to $\kappa = (3 - \nu)/(1 + \nu)$, $\beta_t = \alpha_{t0}E/(1 + \nu)$ and $\mu = E/(2(1 + \nu))$ is the shear modulus.

For a single crack (without microcracks) in a homogeneous material subjected to a heat flux normal to the crack surfaces the thermal stress intensity factors are given as [5]

$$k_I^{\pm} = 0, \quad k_{II}^{\pm} = \pm Q\sqrt{a_0}, \tag{13}$$

and for the crack under the tensile load P remotely applied normal to the crack the stress intensity factors are

$$k_I^{\pm} = P\sqrt{a_0}, \quad k_{II}^{\pm} = 0. \tag{14}$$

The interaction of cracks leads to mixed mode conditions in the interface crack surfaces, i.e. $k_I \neq 0$ in the first case and $k_{II} \neq 0$ in the second one. The influence of both thermal and mechanical loading results in mixed-mode conditions near the interface crack.

For parametric analyses of the obtained results the SIFs are represented in a non-dimensional form. The SIFs k_I and k_{II} are normalized by k^0

$$k^0 = |k_{II}| = |Q\sqrt{a_0}| = P\sqrt{a_0} \tag{15}$$

and denoted as K_1 and K_2:

$$\begin{aligned} K_1^{\pm} - iK_2^{\pm} &= (k_I^{\pm} - ik_{II}^{\pm})/k^0 = [\mp i + \lambda^2 \sum\nolimits_{k=1}^{N} F_1(w_k, \theta_k, \delta, \omega)] \\ &+ A[1 + \lambda^2 \sum\nolimits_{k=1}^{N} F_2(w_k, \theta_k)] \end{aligned} \tag{16}$$

Here $A = P/Q$ is the ratio of the tension to the thermal loading. We suppose that both loads possess the same value, Eq. (15), and in this case A=1. It should be noted, that A is an important parameter and the influence of this parameter on fracture characteristics will be considered later.

3. Fracture criteria and direction of crack propagation

From experimental and theoretical investigations of cracks under mixed-mode loading, it is known that the cracks deviate from their initial propagation direction. For prediction of the crack growth and direction of this growth a fracture criterion should be applied. Two criteria will be considered: the maximum circumferential stress criterion [5, 6, 7] and the criterion based on the strain energy density function [8, 9].

3.1 Maximum circumferential stress criterion

Using the maximum circumferential stress criterion (see for references [5, 6, 7]) the direction of the initial crack propagation (Fig. 1b) is evaluated as

Materials Research Foundations **159** (2024) https://doi.org/10.21741/9781644902950

$$\phi_0 = 2\arctan\left[\left(k_I - \sqrt{k_I^2 + 8k_{II}^2}\right)\Big/4k_{II}\right] \tag{17}$$

and the critical stresses can be calculated from the expression

$$\cos^3(\varphi_0 / 2)\left(k_I - 3k_{II}\tan(\varphi_0 / 2)\right) = K_{Ic} / \sqrt{\pi}\,. \tag{18}$$

Here K_{Ic} is the fracture toughness of the material.

For a single crack in a homogeneous material under a heat flux the SIF factor k_I is equal to zero and Eq. (17) gives the fracture angle $\varphi_0 \approx \mp 70.5^\circ$ (the upper sign is for the right crack tip, the lower – for the left one). The initial direction of crack propagation in the general case is determined from Eq. (17) by substitution of the SIFs $k_I^\pm$ and $k_{II}^\pm$ from Eqs. (3) - (8). For the case of a heat flux some results for the fracture angle at the interface crack tips in FGM/homogeneous bimaterials were presented in [2].

3.2 Strain energy density criterion

Now the strain energy density criterion is used for fracture interpretation of the results. In [8,9] the minimum strain energy density factor criterion was introduced. The local strain energy density is given by $dW / dV = S / r$. Based on the stress intensity factor solutions k_I and k_{II}, the strain energy density (SED) factor $S(k_I, k_{II})$ is defined as

$$S(k_I, k_{II}) = a_{11}k_I^2 + 2a_{12}k_I k_{II} + a_{22}k_{II}^2, \tag{19}$$

where

$$\begin{aligned} a_{11} &= \frac{1}{16\mu}(1+\cos\varphi)(\kappa - \cos\varphi), \quad a_{12} = \frac{1}{16\mu}\sin\varphi[2\cos\varphi - (\kappa - 1)], \\ a_{22} &= \frac{1}{16\mu}[(\kappa+1)(1-\cos\varphi) + (1+\cos\varphi)(3\cos\varphi - 1)] \end{aligned} \tag{20}$$

and $\kappa = 3 - 4\nu$ is for plain strain, μ is the shear modulus and ν is Poisson's coefficient. In Eq. (20) φ is the polar angle of the polar coordinate system (r, φ) with the origin at the crack tip.

SED factor determines the mixed mode effects, i.e., the direction of crack initiation as well as the critical condition under which the crack would initiate. The criterion can be expressed mathematically as

$$\frac{\partial S}{\partial \varphi} = 0, \quad \frac{\partial^2 S}{\partial \varphi^2} > 0\,. \tag{21}$$

The crack growth occurs when the SED factor reaches critical value, i.e. $S = S_{cr}$ for $\varphi = \varphi_0$

In the SED criterion the angle φ_0 depends on Poisson's ratio. For Mode II cracks the maximum stress criterion predicts a fixed angle $\varphi_0 \approx \mp 70.5^0$, which corresponds to a material with zero Poisson's ratio in SED criterion. For $\nu = 0.3$ the angle of crack propagation is $\varphi_0 \approx \mp 82.3^0$.

4. Parameters of materials

The formulae for SIFs Eqs. (3)-(8) and hence the formulae for other fracture characteristics Eqs. (17)-(21) contain geometrical parameters of the problem, such as, length of cracks, coordinates of the centers of cracks w_k and inclination angles θ_k of small cracks to the interface, and parameters of materials, the main of them are inhomogeneity parameters of thermal conductivity and of the thermal expansion coefficient. The influence of these parameters on the fracture characteristics at the interface crack tips can be investigated.

The values of the inhomogeneity parameters are estimated based on the following considerations. From exponential form of the thermal conductivity Eq. (1) the inhomogeneity parameter δ is $\delta = (1/y)\ln(k_1/k_2)$. That means, the value depends on the ratio of material properties and the value of y. We consider an infinite domain and it is supposed that the value of δ changes slowly, we take $-1.0 \le \delta \le 1.0$. The same concerns the inhomogeneity parameter of the thermal expansion coefficient ω ($\omega = (1/y)\ln(\alpha_{t1}/\alpha_{t2})$.

Tables 1 and 2 give the thermal properties [14, 15] of some FGM/homogeneous material combinations and corresponding values of the inhomogeneity parameters δ and ω. The Young's moduli of these materials are similar; it means that these FGMs are elastically homogeneous.

FGMs are used in thermal barrier coatings to protect details from high temperatures as well as from wear and corrosion. The materials for protecting from high temperatures should have a low thermal conductivity and at the same time they are desired to have a thermal expansion coefficient close to that of the material for the protected substrate. Some appropriate examples of FGMs are ($MoSi_2$/ SiC)/ SiC and (TiC/ SiC)/ SiC in Table 1 and (ZrO_2/ Ni)/ Ni and (ZrO_2/ Steel)/ Steel in Table 2.

Table 1. Thermal properties for some ceramic/ceramic FGMs

	Thermal expansion coeff. [$*10^{-6}$ K^{-1}]					Thermal conductivity $Wm^{-1}K^{-1}$		
			FGM/H ($MoSi_2$/ SiC)/ SiC (Molybdenum disilicide $MoSi_2$, silicon carbide SiC)					
$MoSi_2$	α_{t1}	5	$\alpha_{t1}/\alpha_{t2} > 1$			k_1	52	$k_1/k_2 <$
SiC	α_{t2}	4	$\omega > 0$			k_2	60	1
								$\delta < 0$
FGM/H (SiC/ $MoSi_2$)/ $MoSi_2$					$\omega < 0$			$\delta > 0$
			FGM/H (TiC/ SiC)/ SiC (Titanium carbide TiC, silicon carbide SiC)					
TiC	α_{t1}	7	$\alpha_{t1}/\alpha_{t2} > 1$			k_1	20	$k_1/k_2 <$
SiC	α_{t2}	4	$\omega > 0$			k_2	60	1
								$\delta < 0$
FGM/H (SiC/ TiC)/ TiC					$\omega < 0$			$\delta > 0$

Table 2. Thermal properties for some ceramic/metal FGMs

	Thermal expansion coeff. $[*10^{-6}\ K^{-1}]$			Thermal conductivity $Wm^{-1}K^{-1}$		
FGM/H (ZrO_2/ Ni)/ Ni (Zirconia ZrO_2, nickel Ni)						
ZrO_2	α_{t1}	10	$\alpha_{t1}/\alpha_{t2}<1$ $\omega<0$	k_1	2	$k_1/k_2<1$ $\delta<0$
Ni	α_{t2}	18		k_2	90	
FGM/H (Ni/ ZrO_2)/ ZrO_2			$\omega>0$			$\delta>0$
FGM/H (ZrO_2/ Steel)/ Steel						
ZrO_2	α_{t1}	10	$\alpha_{t1}/\alpha_{t2}<1$	k_1	2	$k_1/k_2<1$
Steel	α_{t2}	12	$\omega<0$	k_2	20	$\delta<0$
FGM/H (Steel/ ZrO_2)/ ZrO_2			$\omega>0$			$\delta>0$

5. Numerical results

The influence of different arrays of microcracks on the thermal SIFs at the interface crack was investigated in the case of a heat flux in [1-3] and in the case of thermo-mechanical loading some results for SIFs can be found in [4]. In the present work, using formulae (3) – (12), the case of both heat flux and tensile load is considered for the geometry shown in Fig. 1c. In the following, we will consider SIFs, fracture angles and critical loads calculated according to relevant fracture criteria.

Assume that all microcracks have the same angle of inclination θ to the x-axis. The microcrack centers were presented by $x_n = a_0 n / r, y_n = a_0 m / s$ (n, m = ±1, ±2,...), with $r = s = 5$, $w_n = (x_n + i\, y_n)/ a_0$ (Fig. 1c). SIFs $k_{I,II}$ are normalized by k^0 and denoted by K_1 and K_2 in the figures, see Eqs. (15) and (16). The calculations were performed with $\lambda = 0.1$. The non-dimensional inhomogeneity parameters of thermal conductivity and of thermal expansion are δa_0 and ωa_0, but in the figures the designation δ and ω is used.

Fig. 2 shows SIF K_1 at the interface crack tips as functions of the inhomogeneity parameter ω and for different δ (the angle $\theta = 0$) for both thermal and mechanical loadings. Because the mechanical part of the SIFs (Eqs. (7) and (8)) does not depend on inhomogeneity parameters, the curves of K_1 are similar to the curves for the case of a heat flux in Fig. 6a in [3], but the values of K_1 are different for these two cases. The difference in values of K_1 due to change of δ is up to 35% (maximum is reached for $\omega = 1$). In Fig. 2 the value of K_1 is positive for all parameters. At the right crack tip a part of K_1 is larger than 1 and a part is less than 1, $K_1 = 1$ corresponds to the value for a single crack. At the left tip most of K_1 is larger than 1. Besides, K_1 at the right tip is less than K_1 at the left tip (the difference is up to 30%), that is the presence of system of cracks near the right crack tip suppress the crack propagation. Examples of FGM/homogeneous bimaterials are (TiC/ SiC)/ SiC and (SiC/TiC)/TiC (Fig. 2).

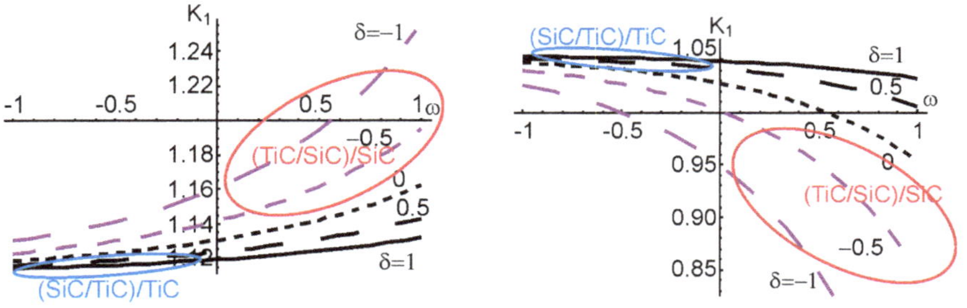

(a) (b)

Figure 2. The case of a heat flux and a tensile load. Influence of system of microcracks (Fig. 1c with θ=0) on SIFs K_1: (a) at the left interface crack tip and (b) at the right tip for different inhomogeneity parameters ω (of thermal expansion) and δ (of thermal conductivity).

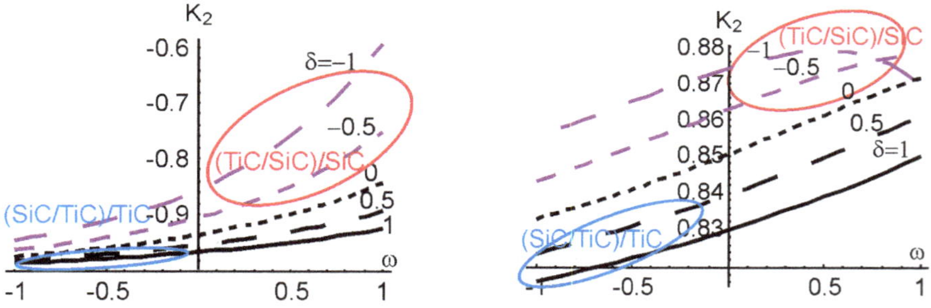

(a) (b)

Figure 3. The case of a heat flux and a tensile load. Influence of system of microcracks (Fig. 1c with θ=0) on SIFs K_2 : (a) at the left interface crack tip and (b) at the right tip for different inhomogeneity parameters ω (of thermal expansion) and δ (of thermal conductivity).

The normalized SIF K_2 is shown in Fig. 3. K_2 at the right interface crack tip is positive and at the left tip is negative, as it is expected and following from Eq. (3). The influence of parameters ω and δ is strong at the left tip (up to 40%) and is minimal at the right tip (below 7%). It should be noted, that K_2 at the right crack tip is less than 1 and less than the values of K_2 for the case of only thermal loading in Fig. 6a in [3] where Mode II is the dominant load.

Materials Research Foundations **159** (2024) https://doi.org/10.21741/9781644902950

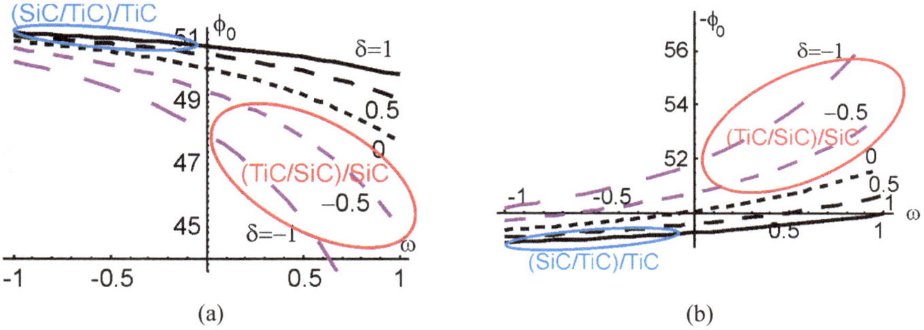

Figure. 4. The case of a heat flux and a tensile load. Influence of system of microcracks (Fig. 1c with θ=0) on the fracture angle ϕ_0: (a) at the left interface crack tip and (b) at the right tip for different ω and δ.

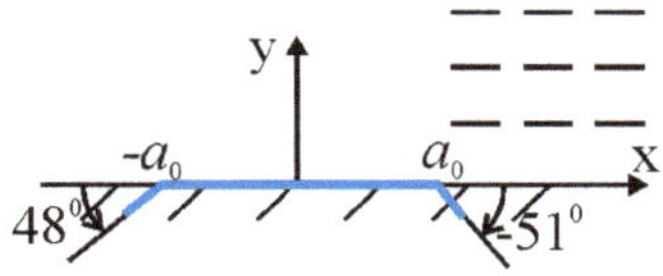

Figure. 5. Schematic representation of fracture angles for ω=0.5 and δ=-0.5.

The fracture angles ϕ_0, calculated by the maximum stress criterion Eq. (17), are presented in Fig. 4 for the case of the combination of heat flux and tensile load. The scheme of the direction of crack propagation is shown in Fig. 5 for ω=0.5 and δ=-0.5. For a crack in a homogeneous material subjected both to a heat flux and tension load normal to the crack line the formula (17) gives the fracture angle $\phi_0 \approx \mp 53^0$. As shown in Fig. 4 the absolute value of ϕ_0 is less than 53^0 for most parameters ω and δ at the left and right interface crack tips and is larger than 53^0 at the right crack tip for some combinations of ω and δ.

The influence of a system of microcracks (Fig. 1c with θ=0) on the critical load P_{cr} at the left and right interface crack tips is shown in Fig. 6 for different values of ω and δ. The maximum hoop stress criterion is applied. The critical load is normalized by P_0, and calculated by Eq. (18) for a single crack under a heat flux and tensile load, i.e. $P_{cr} = P^* / P_0$.

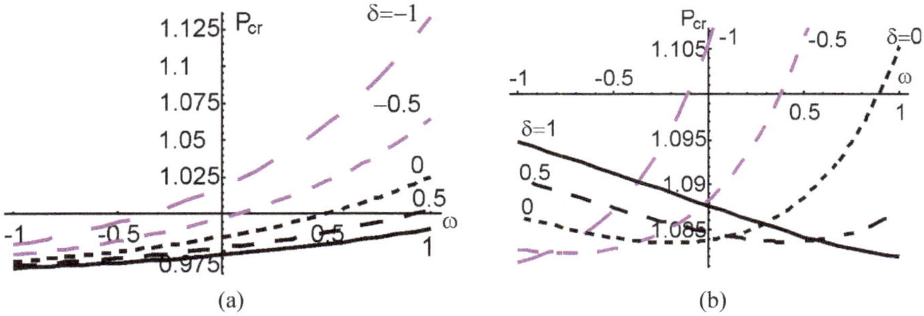

Figure. 6. The case of a heat flux and a tensile load: Influence of system of microcracks (Fig. 1c with θ=0) on the critical load P_{cr}: (a) at the left interface crack tip and (b) at the right tip for different ω and δ.

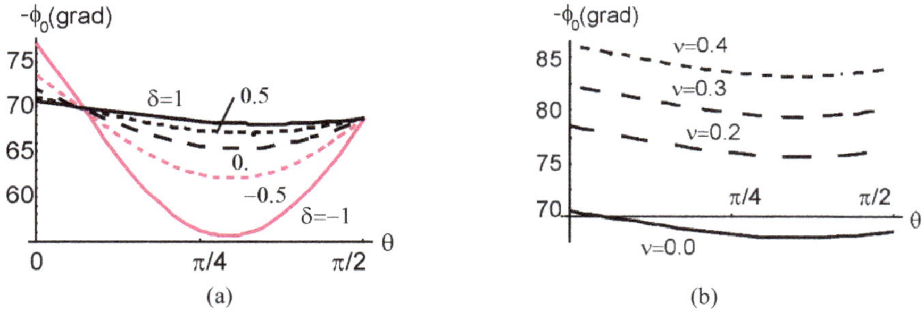

Figure. 7. The case of a heat flux: The fracture angle ϕ_0 as function of inclination angle θ at the right interface crack tip: (a) the maximum stress criterion, ω=1; (b) SED for different ν, δ=1 and ω=1.

In order to compare predictions for the fracture angles by two fracture criteria (maximum hoop stress and minimum strain energy density) selected results for the fracture angles ϕ_0 are depicted in Fig. 7 for the case of a heat flux. Fig. 7a presents results for the fracture angle ϕ_0 calculated by Eq. (17) for ω=1 and different δ and Fig. 6b – by Eq. (19)-(21) with ω=1 and δ=1. The case $\nu = 0$ in Fig. 7b corresponds to the maximum stress criterion prediction.

Fig. 8 shows non-dimensional critical loads (Eq. (15)) for different δ and for ω=0, 1.

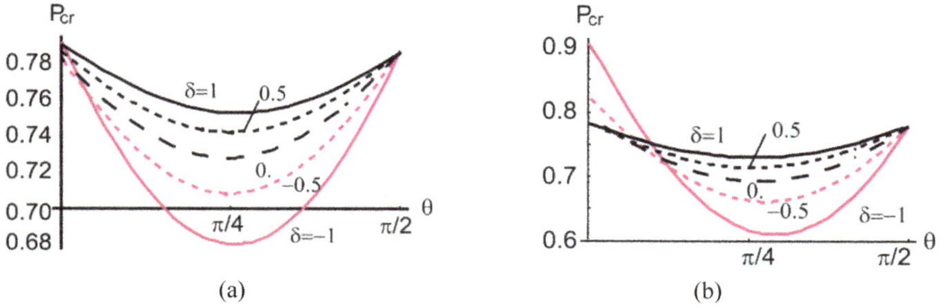

Figure. 8. The case of a heat flux: The critical load P_{cr} as function of inclination angle θ at the right interface crack tip: (a) $\omega=0$ and (b) $\omega=1$ (maximum stress criterion).

Conclusions

Mathematical modeling of the fracture processes in the vicinity of an interface crack in functionally graded/ homogeneous bimaterials with internal defects subjected to tensile loading and a thermal flux remotely applied normal to the interface surface was performed. Asymptotic analytical solution for a special case when an interface crack is significantly larger in size than internal cracks in the FGM was used in two fracture criteria for determination of the direction of the initial crack propagation and critical loads. The main fracture characteristics were obtained as functions of geometry of the problem and non-homogeneity parameters of FGMs. Examples of actual material combinations were presented, e.g. ceramic/ceramic TiC/SiC, $MoSi_2/Al_2O_3$ and $MoSi_2$/SiC, and also ceramic/metal FGMs, e.g., zirconia/nickel and zirconia/steel. Optimal crack configurations can be determined at which the stress intensity factors at the interface crack tips possess the minimal value or at which the critical loads are maximal and, accordingly, the interface crack failure will be minimal.

Acknowledgement

V. Petrova acknowledges the support of the German Research Foundation under grant Schm 746/113-1 and 746/131-1.

References

[1] V. Petrova, S. Schmauder, Thermal fracture of a functionally graded/homogeneous bimaterial with a system of cracks, Theor. Appl. Fract. Mech. 55 (2011)148-157. https://doi.org/10.1016/j.tafmec.2011.04.005

[2] V. Petrova, S. Schmauder, Crack – interface crack interactions in functionally graded/ homogeneous composite bimaterials subjected to a heat flux, Mech. Comp. Mater. 47(1) (2011)125-136. https://doi.org/10.1007/s11029-011-9191-0

[3] V. Petrova, S. Schmauder, Mathematical modelling and thermal stress intensity factors evaluation for an interface crack in the presence of a system of cracks in functionally graded/ homogeneous bimaterials, Comp. Mater. Sci. 52 (2012) 171-177.

https://doi.org/10.1016/j.commatsci.2011.02.028

[4] V. Petrova, S. Schmauder, Interaction of a System Cracks with an Interface Crack in Functionally Graded/Homogeneous Bimaterials under Thermo-Mechanical Loading, Comp. Mater. Sci. 64 (2012) 229-233. https://doi.org/10.1016/j.commatsci.2012.04.032

[5] V.V. Panasyuk, M.P. Savruk, A.P. Datsyshin, Stress Distribution near Cracks in Plates and Shells (in Russian), Naukova Dumka, Kiev, 1976.

[6] F. Erdogan, G.C. Sih, On the crack extension in plates under plane loading and transverse shear, J. Basic. Eng. 85 (1963)519-527. https://doi.org/10.1115/1.3656897

[7] V. Petrova, V. Tamuzs, N. Romalis, A survey of macro- microcrack interaction problems, ASME Appl. Mech. Rev. 53(5) (2000) 117-146. https://doi.org/10.1115/1.3097344

[8] G.C. Sih, Strain-energy-density factor applied to mixed-mode crack problems, Int. J. Fract. 10 (1974) 305–321. https://doi.org/10.1007/BF00035493

[9] G.C. Sih, Mechanics of Fracture Initiation and Propagation: Surface and volume energy density applied as failure criterion. Kluwer Academic Publishers, Dordrecht, the Netherlands, 1991.

[10] V. Tamuzs, N. Romalis, V. Petrova, Fracture of Solids with Microdefects. NOVA Science Publishers Inc., New York, 2000.

[11] N.I. Muskhelishvili, Singular Integral Equations. Dover Publ. Inc., Mineola, New York, 2011.

[12] W.H. Müller, S. Schmauder, Stress-intensity factors of r-crack in fiber-reinforced composites under thermal and mechanical loading, Int. J. Fract. 59 (1993) 307-343. https://doi.org/10.1007/BF00034562

[13] J.W. Eischen, Fracture of nonhomogeneous materials, Int. J. Fract. 34 (1987) 3-22. https://doi.org/10.1007/BF00042121

[14] J.F. Shackelford, W. Alexander, CRC Materials Science and Engineering Handbook. CRC Press, Boca Raton, 2001. https://doi.org/10.1201/9781420038408

[15] O.P. Chakrabarti, P.K. Das, S. Mondal, Study of indentation induced cracks in $MoSi_2$-reaction bonded SiC ceramics, Bull. Mater. Sci. 24 (2001)181-184. https://doi.org/10.1007/BF02710098

Materials Research Forum LLC
https://doi.org/10.21741/9781644902950

CHAPTER 4

Mathematical Modelling and Thermal Stress Intensity Factors Evaluation for an Interface Crack in the Presence of a System of Cracks in Functionally Graded/ Homogeneous Bimaterials

Vera Petrova [1,2]*, Siegfried Schmauder [1]
[1] IMWF, University of Stuttgart, Pfaffenwaldring 32, D-70569 Stuttgart, Germany
[2] Voronezh State University, University Sq.1, Voronezh 394006, Russia
veraep@gmail.com *, Siegfried.Schmauder@imwf.uni-stuttgart.de

Abstract

The work is devoted to mathematical modeling of the fracture processes in the vicinity of an interface crack in functionally graded/homogeneous bimaterials with internal defects subjected to a thermal flux. A previously obtained solution (Petrova, Schmauder, 2009, 2011) is used and supplemented with the additional possibility to take into account the crack closure. The solution is based on the integral equation method and it is assumed that thermal properties of the functionally graded material (FGM) have exponential form. For a special case where an interface crack length is much larger than the internal cracks in the FGM an asymptotic analytical solution of the problem is obtained as series of a small parameter (the ratio between sizes of the internal and interface cracks). Analyses of the effects of the location and orientation of the cracks, the material non-homogeneity parameters and the crack closure effects on the thermal stress intensity factors of the interface crack in FGM/homogeneous bimaterials are performed. Examples of some FGM/homogeneous bimaterial combinations (i.e., metal/metal, ceramic/metal) are discussed.

Keywords

Functionally Graded Materials, Interface, Cracks, Crack Closure, Thermal Fracture, Thermal Stress Intensity Factor

1. Introduction

Bimaterials and functionally graded materials (FGMs) are widely used in engineering structures, which are subject to different loads: mechanical, thermal and combinations of them. FGMs are designed so that to decrease bimaterial mismatch and residual stresses at the interface and prevent debonding along the interface. However, the interaction of defects causes additional stresses near interfaces and can lead to enhance or suppress crack propagation, so that crack interaction problems in FGMs and bimaterials are important for the investigation of fracture strength of materials.

Great progress was achieved in computational modelling of FGMs and cracks in FGMs (see a review [1]). Finite element method (FEM) modelling, boundary integral analysis and their different modifications are widely used for these purposes [2]. Boundary integral methods have advantages because of only boundary discretization is need in comparison with volume discretization in FEMs. Many integral equation methods, however, are based on knowing fundamental solutions of the corresponding partial differential equations with variable coefficients and have therefore limitations. One of the fundamental solutions, Green's functions are known only for the simplest cases, for example, the Green's function for an FGM wherein the material properties vary exponentially through the solid has been obtained in closed form for heat conduction problems and elasticity problems [3], but it is difficult to use this solution for crack interaction problems in FGMs. Most of these methods allow studying complicated boundary value problems for FGMs, but they require large efforts to develop special programs or using commercial programs and it consumes much computational time.

Another way is to derive approximate analytical solutions. It is evident, that these solutions can be obtained for special cases and with some assumptions and, hence, they have limitations. Meanwhile, these solutions are necessary for a first estimation of the mechanical and thermal parameters of FGMs with cracks, for better understanding the processes and possibly to show the direction of further investigations by numerical methods. Besides, the analytical solutions can check correctness of numerical solutions.

Crack interaction problems in homogeneous materials have been extensively investigated and large number of solutions have been obtained for different crack system configurations and different thermal and mechanical loadings [4]. Many papers also devoted to different models and semi-analytical solutions for cracks in FGMs, but crack interaction problems were obtained only for special crack arrangements (see, for example [5, 6]).

The present work is devoted to modelling of the fracture processes in the vicinity of an interface crack in functionally graded/homogeneous bimaterials with internal defects subjected to a thermal flux applied at infinity. It is assumed that thermal properties of FGMs possess exponential form. In previous works [7, 8] an approximate analytical solution was obtained for a special case where the interface crack length is larger than nearby internal cracks in the FGM. The thermal stress intensity factors (TSIFs) for the interface crack were derived as series of a small parameter (the ratio between sizes of the internal and interface cracks). Parametric analyses of the effects of the location and orientation of the cracks and the material non-homogeneity parameters on the thermal stress intensity factors of an interface crack in FGM/homogeneous bimaterials were performed in [8]. This solution is used in the present investigation and supplemented with the additional possibility to take into account the crack closure.

2. Formulation of the problem

2.1 Geometry of the problem and assumptions

The geometry of the problem is shown in Fig. 1. A bimaterial is composed of a functionally graded material (denoted by number 1) and a homogeneous material (denoted by number 2). The bimaterial is perfectly bonded with exception of an interface crack of length $2a_0$. It is assumed that the FGM contains N cracks of length $2a_k$. Cartesian coordinates (x, y) are centered at the midpoint of the interface crack; the x-axis lies along the interface line. Local coordinate systems (x_k, y_k) are attached to each internal crack and to the interface crack. The crack position is determined by the defect midpoint coordinate (x_k^0, y_k^0) and an inclination angle θ_k to the interface, i.e. to the x-axis (Fig. 1).

The bimaterial is subjected to a heat flux of intensity q applied at infinity. The cracks are thermally isolated and traction free.

It is assumed that properties of the functionally graded material depend only on the coordinate y. The thermal conductivity coefficient and the thermal expansion coefficient are

$$k_1(y) = k_0 e^{\delta y}, \quad \alpha_{t1} = \alpha_0 e^{\omega y} \tag{1}$$

where the constant k_0 is the thermal conductivity and α_0 is the thermal expansion coefficient of the interface and of material 2, while δ and ω are inhomogeneity parameters of the FGM. The Young's modulus and Poisson's ratio are assumed to be constant, $E_j = const$, $\nu_j = const$ (j = 1, 2). Thus, the material is elastically homogeneous, but thermally non-homogeneous.

The relation between global coordinates (x,y) and the local coordinate systems (x_k,y_k) can be written in complex form as follows $z = z_k^0 + z_k e^{i\theta_k}$, where $z_k = x_k + iy_k$ and $i = \sqrt{-1}$. $z_k^0 = x_k^0 + iy_k^0$ is the origin coordinate of the system (x_k,y_k) in the global system. In the local coordinate system connected with each arbitrarily oriented crack the coefficients k_1 and α_{t1} possess the form

$$k_1(x_k, y_k) = k_0 e^{\delta y_k^0} e^{\delta_1 x_k + \delta_2 y_k}, \ \delta_1 = \delta \sin\theta_k, \ \delta_2 = \delta\cos\theta_k,$$

$$\alpha_{t1}(x_k, y_k) = \alpha_{tk} = \alpha_0 e^{\omega y_k^0} e^{\omega_1 x_k + \omega_2 y_k}, \ \omega_1 = \omega\sin\theta_k, \ \omega_2 = \omega\sin\theta_k. \tag{2}$$

The uncoupled, quasi-static thermoelastic theory is applicable to this problem that is the temperature distribution is independent of the mechanical field, and the solution consists of the determination of the temperature field and the determination of the thermal stresses.

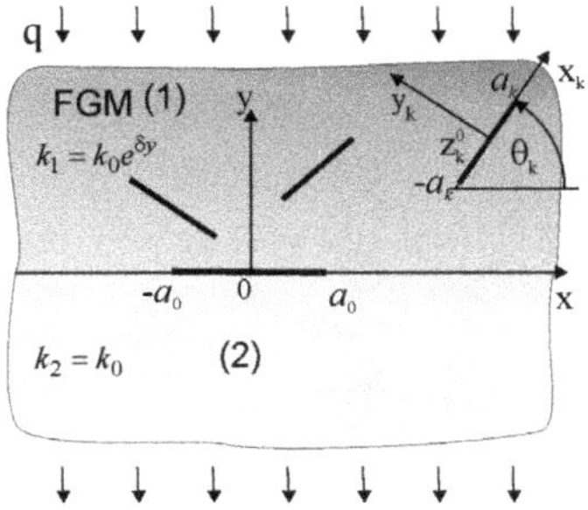

Figure. 1. The geometry of the problem.

2.2 Thermal problem

For the solution of the problem a superposition principle is used. Due to this principle the temperature field T_j^* (j=1, 2) in the bimaterial with cracks is presented as $T_j^*(x,y)=T_j^0(x,y)+T_j(x,y)(j=1,2)$, where $T_j^0(x, y)$ – the temperature distribution in a bimaterial in the absence of cracks, $T_j(x, y)$ – the temperature perturbation caused by the cracks.

The thermal boundary conditions for thermally isolated cracks and continuity conditions for the temperature perturbation $T_j(x, y)$ read as follows:

$$k_1\frac{\partial T_1(x,+0)}{\partial y}=k_2\frac{\partial T_2(x,-0)}{\partial y}=q_0(x)(|\,x\,|\le a_0),\ k_j\frac{\partial T_{jn}(x,\pm 0)}{\partial y_n}=q_n(x_n)(|\,x_n\,|\le a_n),$$

$$k_1\frac{\partial T_1(x,+0)}{\partial y}=k_2\frac{\partial T_2(x,-0)}{\partial y},\ \ T_1(x,+0)=T_2(x,-0)\ \ |\,x\,|\ge a_0, y=0, \tag{3}$$

$$T_1(\pm a_0,+0)=T_2(\pm a_0,-0),\ T_{jn}(\pm a_n,+0)=T_{jn}(\pm a_n,-0), \tag{4}$$

and the temperature perturbation vanishes at infinity. Here

$$q_0=-k_j\left(\partial T_j^0/\partial y\right)\Big|_{y=0},\qquad q_n=-k_j\left(\partial T_j^0/\partial y_n\right)\Big|_{y_n=0}. \tag{5}$$

The signs '+' and '–' denote the limiting values of the functions on the upper and lower surfaces of the crack or the interface, respectively.

The heat conduction equation for the steady state temperature in FGMs with thermal conductivity coefficient Eq. (1) is given by

$$\nabla^2 T_1+\delta\,\partial T_1/\partial y=0, \tag{6}$$

and for material 2 with $\delta=0$ we have the Laplace equation $\nabla^2 T_2=0$, where $\nabla^2=\partial^2/\partial x^2+\partial^2/\partial y^2$. Solving the temperature problem for an undamaged FGM/homogeneous bimaterial (Eq. (6) for T_j^0) the thermal fluxes on the crack lines, Eq. (5), are obtained as

$$q_0=-k_0 q,\quad q_n=-k_0 q\exp(-\delta y_n^0)\exp(-x_n\delta\sin\theta_n)\cos\theta_n,\ \ (n=1,2\ldots N), \tag{7}$$

Supposing that the non-homogeneity of the FGM is revealed only in non- homogeneous thermal fluxes (Eq. (7)) on crack surfaces, the perturbation problem can be solved by the method presented in [4] using similar integral equations. The system of $N+1$ singular integral equations for the unknown functions γ_k' is written as

$$\int_{-a_n}^{a_n} \frac{\gamma_n'(t)}{t-x} dt + \sum_{\substack{k=0 \\ k \neq n}}^{N} \int_{-a_k}^{a_k} \gamma_k'(t) P_{nk}(t,x) dt = \pi q_n(x), \quad |x| < a_n, \tag{8}$$

$$\int_{-a_n}^{a_n} \gamma_n'(t) dt = 0, \quad n = 0,1,...,N. \tag{9}$$

The Eq. (9) is the condition of the temperature continuity at the crack endpoints and represents Eq. (4) rewritten in the integral form. The functions $q_n(x)$ at the right part of equations (8) are determined by Eq. (7). The functions γ_k' are the derivatives of temperature jumps on the crack lines

$$2\gamma_k(t) = T_k^+ - T_k^-. \tag{10}$$

The regular kernels P_{nk} (n, k = 0, 1,..., N) contain the geometry of the problem and are given by the expressions:

$$P_{nk} = \mathrm{Re}[e^{i\theta_n} / (te^{i\theta_k} + z_k^0 - xe^{i\theta_n} - z_k^0)]. \tag{11}$$

In Eq. (11) and in all expressions below Re denotes the real part of complex numbers.

2.3 Thermoelastic problem

The mechanical boundary conditions for the stress-free cracks are

$$(\sigma_{1y} - i\tau_{1xy})^+ = (\sigma_{2y} - i\tau_{2xy})^- = 0, \ |x| \leq a_0, y = 0, \tag{12}$$

$$(\sigma_{jy} - i\tau_{jxy})^+ = (\sigma_{jy} - i\tau_{jxy})^- = 0, \ |x_k| \leq a_k, y_k = 0,$$

the continuity conditions at the interface are satisfied, i.e. the stresses are equal and displacements are equal. The condition at infinity is $\sigma_{ij} \to 0, \quad x^2 + y^2 \to \infty$. Here σ_{jy} and τ_{jxy} are normal and shear stresses, v_j and u_j are normal and shear displacements, j = 1, 2 for materials 1 and 2.

Because of we suppose that the material is elastically homogeneous we can use directly the method presented in [4]. The system of singular integral equations for the traction-free cracks is written as [9]:

$$\int_{-a_n}^{a_n} \frac{G_n(t)}{t-x} dt + \sum_{\substack{k=0 \\ k \neq n}}^{N} \int_{-a_k}^{a_k} [G_k(t) K_{nk}(t,x) + \overline{G_k(t)} L_{nk}(t,x)] dt = 0, \quad |x| < a_n, \tag{13}$$

$$\int_{-a_n}^{a_n} G_n(t) dt = iA_n = -2i \int_{-a_n}^{a_n} \beta_t t\gamma'(t) dt_n, \ n = 0,1,...,N. \tag{14}$$

$\overline{(...)}$ is the complex conjugate. The function G_k consists of two parts as

$$G_k(x) = g_k'(x) + 2i\beta_t\gamma_k(x), \tag{15}$$

where functions γ_k' (10) were found from the thermo-conductivity problem, while unknown functions $g_n'(x)$ are the derivatives of the displacement jumps on the crack lines

$$g_n'(x) = (2\mu / i(\kappa+1))\partial[(u_n^+ - u_n^-) + i(v_n^+ - v_n^-)] / \partial x. \tag{16}$$

In Eqs. (13) – (15) β_t is $\beta_t = \alpha_t E^*$. The following applies to the plane strain case $E^* = E/(1+\kappa)$ and $\kappa = 3\text{-}4\nu$, while the plane stress condition corresponds to $E^* = E/(1+\kappa)/(1+\nu)$ and $\kappa = (3-\nu)/(1+\nu)$; $\mu = E/(2(1+\nu))$ is the shear modulus. Taking into account Eq. (2) the parameter β_t is written as

$$\beta_t = \alpha_{tk}E^* = \beta_t^h e^{\omega y_k^0} e^{\omega_1 x_k + \omega_2 y_k}, \quad \beta_t^h = \alpha_0 E^* \tag{17}$$

The condition of Eq. (14) provides that displacements are single-valued at the end points of the cracks. The condition of temperature continuity at the crack tips $\gamma_n(\pm a_n) = 0$ is also taken into account in Eq. (14). The regular kernels K_{nk} and L_{nk} contain the geometry of the problem and can be found in [4, 8] or [9].

3. Solution of the problem

The solution is derived for a special case when an interface crack is significantly larger in size than internal cracks in the FGM. The asymptotic analytical solution of the problem is obtained as a series of the small parameter which is equal to the ratio of the size of small internal cracks to the interface crack size. The method was first suggested by Romalis and Tamuzs at 1984 and then was used for different macro- microcrack interaction problems for homogeneous materials [4]. In the interest of brevity the solution will not be repeated here in details. For the macro – microcrack interactions in a homogeneous material under thermal flux the full solution can be found in [4].

3.1 Solution by small parameter method

Let us assume that all internal cracks in the FGM have the same size $2a_k = 2a$ ($k = 1, 2,..., N$), for example, they have the characteristic size of a grain size of the material. Suppose also that $2a << 2a_0$. In this case the small parameter is $\lambda = a/a_0$ and $\lambda << 1$. Introducing non-dimensional coordinates χ and τ by $x = a_n\chi$ and $t = a_k\tau$, the Eqs. (8) are rewritten in dimensionless variables. The unknown functions γ_k' (10) are sought as a power series with respect to λ

$$\gamma_n'(\chi) = \sum_{p=0}^{\infty} \gamma_{np}'(\chi)\lambda^p, \quad n = 0,1,...,N, \quad \lambda = a / a_0 << 1. \tag{18}$$

The regular kernels P_{nk} (11) are expanded in series of λ, too. For convergence of these series the condition $\left|\lambda / (\chi - z_k^0 / a_0)\right| < 1$, $|\chi| < 1$ must be satisfied. It is fulfilled if cracks are not intersected.

By substituting the series (18) and kernels into Eqs. (8), and equating the coefficients of corresponding powers of λ, a recurrent system of equations for coefficients γ'_{np} in series (18) is formed (see [4] for details). After solving the recurrent equations, the solution can be represented as a Maclaurien series over even powers of the small parameter λ (all coefficients with odd second index are equal to zero). The second approximation for the temperature jump γ'_0 on the interface crack line is derived in closed form as

$$\gamma'_0(\chi) = \gamma'_{00}(\chi) + \lambda^2 \gamma'_{02}(\chi). \tag{19}$$

The full expression of this function can be found in [8]. The zero-th approximation γ'_{00} in Eq. (19) corresponds to an isolated interface crack and the second γ'_{02} is taking into account the influence of each microcrack on the interface crack.

The solution (19) is used in the thermoelastic problem. The scheme of the solution of the thermoelastic problem is the same as for the solution of the thermal problem. The second approximation for the function G_0 is given as [7]

$$G_0(\chi) = G_{00}(\chi) + G_{02}(\chi)\lambda^2. \tag{20}$$

Using this function the thermal stress intensity factors are obtained.

3.2 Stress intensity factors

In this work we consider elastically homogeneous materials so that we can use the classical definition of the stress intensity factor. It should be noted, that the crack tip singular field in FGMs has the same form as in homogeneous media [10] and the concept of the stress intensity factors can be also applied directly to cracks in FGMs. Besides, the interface crack between the FGM and the homogeneous material with smooth transition between these materials is also classical crack with square-root singularities at the crack tips. Therefore, in all these cases the thermal stress intensity factors (TSIFs) are found by [4, 9]

$$k_{In}^{\pm} - ik_{IIn}^{\pm} = \mp \lim_{\chi \to \pm 1} \sqrt{a_n(1-\chi^2)} G_n(\chi) \ (n = 0,1,2,\ldots,N), \tag{21}$$

where the upper part of the "$\pm$" or "$\mp$" signs refers to the right tip and the lower part to the left tip of cracks. Substituting Eq. (20) into Eq. (21), the thermal stress intensity factors at the interface crack tips are obtained up to λ^2 as

$$\begin{aligned} k_{I0}^{\pm} = \pm\lambda^2 \beta_t^h q k_0 a_0 \sqrt{a_0} \frac{1}{2} \sum_{k=1}^{N} \{ &\mathrm{Re}(I_{k0}^T) \mathrm{Im}[m_{k1} - n_{k1}] \\ &+ \mathrm{Im}(I_{k0}^T) \mathrm{Re}[m_{k1} + n_{k1}] + 2\exp(\omega a_0 \,\mathrm{Im}(w_k)) J_k^T(\delta) \,\mathrm{Im}[m_{k0} - n_{k0}] \}, \end{aligned} \tag{22}$$

$$k_{II0}^{\pm} = \pm\beta_t^h q k_0 a_0 \sqrt{a_0}\{1 + \frac{\lambda^2}{2}\sum_{k=1}^{N}\{2J_k^T(\delta)\mathrm{Re}[e^{i\theta_k}(w_k/(w_k^2-1)^{1/2}-1)]$$
$$+\mathrm{Re}[I_{k0}^T]\mathrm{Re}[m_{k1}-n_{k1}]+\mathrm{Im}(I_{k0}^T)\mathrm{Im}[m_{k1}-n_{k1}]$$
$$-2\exp(\omega a_0\,\mathrm{Im}(w_k))J_k^T(\delta)\mathrm{Re}[m_{k0}-n_{k0}]\}\},\tag{23}$$

where $I_{k0}{}^T$ and $J_k{}^T$ are

$$I_{k0}^T = \frac{1}{2}\left[\frac{e^{-2i\theta_k}-1}{\sqrt{\overline{w}_k^2-1}}+\frac{1}{\sqrt{w_k^2-1}}-\frac{e^{-2i\theta_k}(1-w_k\overline{w}_k)}{(\overline{w}_k^2-1)^{3/2}}\right]$$

$$J_k^T(\delta) = \exp(-\delta a_0\,\mathrm{Im}\,w_k)\cos\theta_k - \mathrm{Re}\left[e^{i\theta_k}\left(1-\frac{w_k}{\sqrt{w_k^2-1}}\right)\right]\tag{24}$$

and m_{k1}, n_{k1}, m_{k0} and n_{k0} are

$$m_{k1} = e^{i\theta_k}\,\mathrm{Re}\left[\frac{e^{i\theta_k}}{(\overline{w}_k\mp 1)\sqrt{w_k^2-1}}\right],\quad n_{k1} = \frac{e^{-2i\theta_k}}{2(\overline{w}_k\mp 1)\sqrt{\overline{w}_k^2-1}}\left[(w_k-\overline{w}_k)\frac{2\overline{w}_k\pm 1}{\overline{w}_k^2-1}+e^{2i\theta_k}-1\right],$$

$$m_{k0} = e^{i\theta_k}\left[1-\mathrm{Re}\left(\frac{\sqrt{w_k^2-1}}{\overline{w}_k\mp 1}\right)\right],\quad n_{k0} = \frac{e^{-i\theta_k}}{2}\frac{w_k-\overline{w}_k}{(\overline{w}_k\mp 1)\sqrt{\overline{w}_k^2-1}},\quad w_k = z_k^0/a_0\tag{25}$$

Here w_k is the non-dimentional complex coordinate of the midpoint of cracks. In Eqs (22) – (25) Re and Im denote the real and imaginary parts of complex numbers correspondingly, the parameter β_t^h is defined in Eq. (17).

The TSIFs for a single crack (without microcracks) is given as [4, 9]

$$k_{II}^{\pm} = \pm\beta_t^h q k_0 a_0\sqrt{a_0},\quad k_I^{\pm} = 0.\tag{26}$$

The interaction of cracks leads to mixed mode conditions in the interface crack surfaces, i.e. $k_I \neq 0$.

4. Cracks with closure

As follows from the formula (22), the TSIF k_{I0} may be negative. It means that the crack surfaces may overlap. Therefore, it might be necessary to account for the closure of the crack surfaces for some crack configurations. The thermal macro- microcrack interaction problem in homogeneous materials taking into account the crack closure was studied in Refs. [4]. We can apply this scheme for the solution in the present investigation.

The asymptotic solution to the problem of the interaction between an interface crack and internal microcracks in the FGM under the thermal loading is obtained with the following assumptions [4]:

- A solitary contact free portion of the crack with the unknown length $2c$ is located on the open part of the interface macrocrack, a local coordinate system (x_0^c, y_0^c) is attached to this part of the interface crack (Fig. 2);

- The temperature distribution does not depend on the crack closure. It means that on the closed portion of the crack the heat flux remains negligible;

- A smooth contact is assumed on the closed portion of the crack.

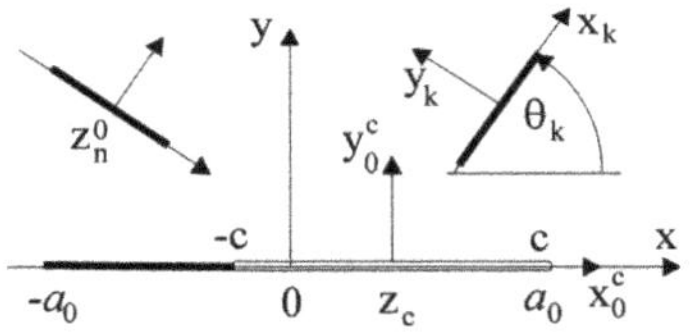

Figure. 2. An interface crack with a contact zone and microcracks.

Due to the second assumption, the temperature distribution is the same as was obtained in previous section. Therefore, only the problem of thermoelasticity needs to be considered. To account for the crack closure, it is necessary to reformulate boundary conditions. For completely open cracks, the boundary conditions (12) are valid. On a closed portion of the interface crack or on the closed edges of microcracks, the shear stresses $\tau_n^{\pm}$ and the transverse displacement jumps $[v_n]$ are zero:

$$\tau_n^{\pm}(x,0) = 0, \quad [v_n] = 0. \tag{27}$$

For the case of open cracks, the unknowns are the shear and transverse displacement jumps $[u_n]$ and $[v_n]$ Eq. (16). For closed cracks, the shear displacement jumps and normal tractions are unknown. The boundary value problem consisting of the system of equations (13) with boundary conditions (12) and (27) decouples into two problems: one for the real part and another for the imaginary one. Separating the real and imaginary parts the system of equations are obtained [4]. The second approximation of the TSIFs at the interface crack tips are obtained as

$$\begin{aligned} k_{I0}^{\pm} &= \pm\lambda^2 \beta_t^h q k_0 a_0 \sqrt{a_0} \frac{1}{2\hat{c}^2} \sum_{k=1}^{N} \{\mathrm{Re}(I_{k0}^T)\,\mathrm{Im}[m_{k1}^c - n_{k1}^c] \\ &\quad + V\,\mathrm{Im}(I_{k0}^T)\,\mathrm{Re}[m_{k1}^c + n_{k1}^c] + 2\exp(\omega a_0\,\mathrm{Im}(w_k)) J_k^T(\delta)\,\mathrm{Im}[m_{k0}^c - n_{k0}^c]\}, \end{aligned} \tag{28}$$

and the TSIFs at the microcracks tips are

$$k_{In}^{\pm} = \lambda \beta_t^h k_0 a_0 \sqrt{a}\, \mathrm{Im}(I_{k0}^T) \tag{29}$$

The expression for k_{II0} is defined by Eq. (23) in which the term with $\mathrm{Im}(I_{k0}^T)$ should be multiplied by V.

In Eq. (28) $m_{k1}^c, n_{k1}^c, m_{k0}^c$ and n_{k0}^c are obtained from Eqs. (25) by setting $(w_k - d_0)/\hat{c}$ instead of w_k, $\hat{c}$ and d_0 are nondimentional parameters $\hat{c} = c/a_0$ and $d_0 = z_c/a_0$, z_c is the center of the open portion of the interface crack (Fig. 2).

By virtue of the boundary condition (27) the parameter V is introduced to Eq. (28) as

$$V = \begin{cases} 0 & \text{if } k_{In} \le 0 \\ 1 & \text{if } k_{In} > 0 \end{cases} \tag{30}$$

which allows to take into account the microcrack closure. k_{In} is determined by Eq. (29).

The function of TSIF k_{I0} (28) contains the unknown parameters of the open portion of the interface crack. The crack is closed smoothly and unknown coordinates of the points of separation of the closed and open parts of the interface crack are found from the condition that the singularity vanishes at these points (see [4, 11])

$$k_{I0}(\pm c) = 0. \tag{31}$$

This condition is written in the local system (x_0^c, y_0^c), Fig. 2.

An iterative procedure is used for finding the length of non-contacting regions of the interface crack. The first approximation of the length of the overlapping portion is replaced by every next approximation until the equality (31) is satisfied with sufficient accuracy. The same crack closure algorithm was also applied in [12] to treat the problem of having negative k_I for a partially insulated interface crack between a functionally graded coating and a homogeneous substrate subjected to both thermal and mechanical loading.

In the frame of this approximation it has been established that small cracks are existed in only two modes - fully open or fully closed. Microcracks depending on their location and orientation can cause either full or partial closure of the interface crack. The value of k_{II} at the interface macrocrack tip can be evaluated disregarding the closed portion of the interface crack, however, for a correct determination of k_I the presence of the closed portion should be taken into account.

5. Results and discussion

The obtained asymptotic analytical formulae for TSIFs Eqs. (22), (23), (28) and (29) allow investigating the influence of different arrays of microcracks and parameters of inhomogeneity of FGMs on the interface crack. Assume that all microcracks have the same angle of inclination θ to the x-axis. The microcrack centers are presented by $x_n = a_0 n/r, y_n = a_0 m/s$ (n, m = ±1, ±2,...), with $r = s = 5$, $w_n = (x_n + i\, y_n)/a_0$ (Fig. 3). The schemes of different arrangements of multiple microcracks in the FGM are shown in Fig. 3. TSIFs $k_{I,II}$ are normalized by $|\ k_{II}\ |$ (26) and denoted by K_1 and K_2 in the figures. The calculations were performed with $\lambda = 0.1$. The non-dimensional inhomogeneity parameters of thermal conductivity and of thermal expansion are δa_0 and ωa_0, but in the figures we will leave the designation δ and ω. Besides, we can put a_0 equals to 1 without reducing generality.

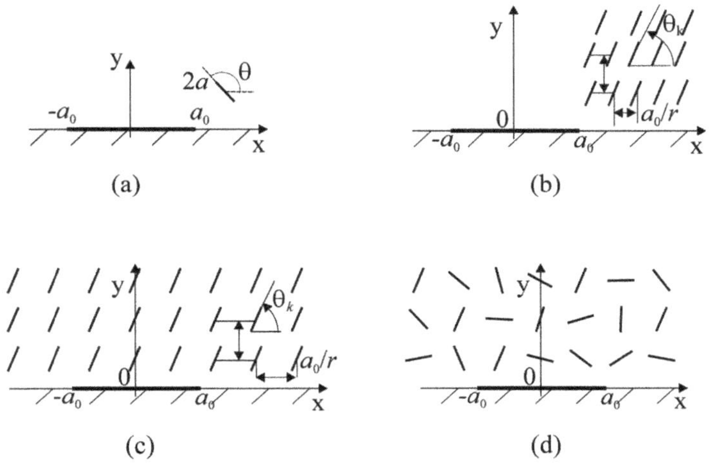

Figure. 3. Schemes of locations of the interface crack and microcracks: (a) one crack with midpoint coordinate w_k=1.2+i0.2; (b) non-symmetrically disposed microcrack system ahead of the interface crack; (c) a symmetric system of microcracks above the interface and (e) randomly orientated system of cracks.

The values of the inhomogeneity parameters are estimated on the following reasons. From exponential form of the thermal conductivity Eq. (1) the inhomogeneity parameter δ is $\delta = (1/y^*)\ln(k_1/k_2)$. That is, the value of δ depends on the ratio of material properties and the value of y^* the region where these properties of the FGM vary. We consider infinite domain and so that we suppose that the value of δ is changed slowly, we take $-1.0 \le \delta \le 1.0$. Besides, if $k_1 > k_2$ then $\delta > 0$, and if $k_1 < k_2$ – $\delta < 0$. The same concerns the inhomogeneity parameter of the thermal expansion coefficient ω.

At first we will use the formulae (22) – (25). Normalized TSIFs K_1 and K_2 as functions of inclination angle θ of a microcrack with midpoint coordinate $w_k = 1.2 + i0.2$ (Fig. 3-a) are shown in Fig. 4 for different values of the non-homogeneity parameter δ: $\delta = 0$ – large dashed line, 0.5 – small dashed line, 1 – solid line. The curves of TSIFs at the right interface crack tip are denoted by $K_{1,2}^{+}$ and at the left tip by $K_{1,2}^{-}$. The figures indicate that the TSIFs at the right interface crack tip are higher than at the left tip. The δ is slightly influencing on the TSIFs and if the microcrack inclination angle approaches $\pi/2$ this influence disappears.

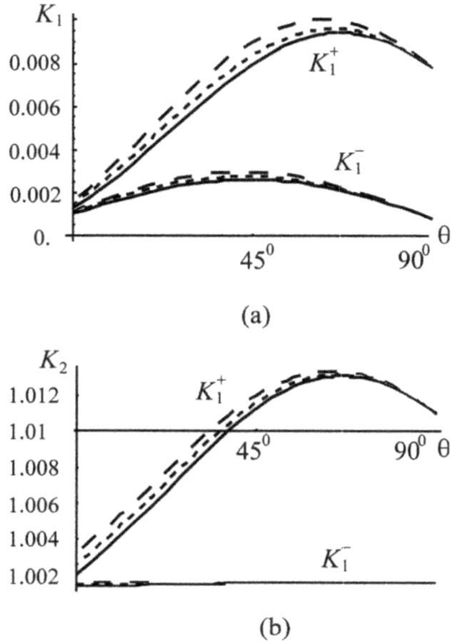

Figure. 4. Normalized stress intensity factors K_1 and K_2 vs inclination angle θ at the right interface crack tip $K_{1,2}^+$ and at the left tip $K_{1,2}^-$, one crack ahead of the interface crack (w_k=1.2+i0.2, Fig. 3-a) for different values of the non-homogeneity parameter δ : large dashed line – δ = 0, small dashed line – 0.5, solid line – 1.

Fig. 5 show the influence of the non-homogeneity parameter δ of thermo-conductivity and the microcrack inclination angle θ on the thermal stress intensity factors K_1 and K_2 at the interface crack tips for the uniform distributed system of microcracks (Fig. 3c). We can see that K_1 at the right interface crack tip is greater than at the left tip for all inclination angles θ of microcracks and for all δ values.

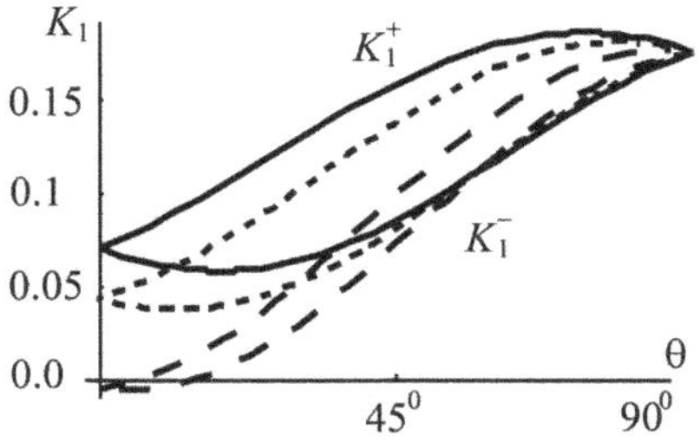

(a)

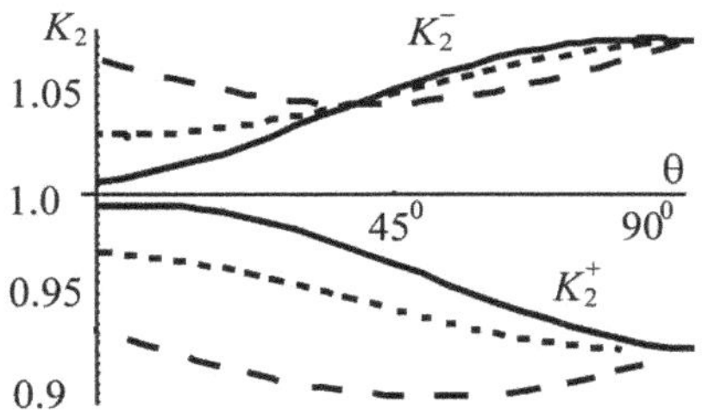

(b)

Figure. 5. Influence of system of microcracks above the interface crack (Fig. 3-c) on TSIFs K_1 (a) and K_2 (b) at the interface crack tips, for different values of non-homogeneity parameter δ. The designations are the same as in the previous figure.

The presented in Figs. 4 and 5 results correspond to the case when the materials have similar thermal expansion coefficients, that is $\omega = 0$, and the thermal conductivity k is increased with increasing the coordinate y, i.e. $\delta > 0$. It could be, for example, for an FGM/homogeneous bimaterial $(MoSi_2/Al_2O_3)/Al_2O_3$ that is in the lower part in Fig. 1 the material is alumina Al_2O_3 with $k^{Al_2O_3} = 25$ Wm^{-1}K^{-1} and in the upper part the material is an FGM $MoSi_2/Al_2O_3$ gradually varying from molibdenum disilicide $MoSi_2$ with $k^{MoSi_2} = 52$ Wm^{-1}K^{-1} to Al_2O_3. At the same time $MoSi_2$ and Al_2O_3 have similar thermal expansion coefficients [13, 14].

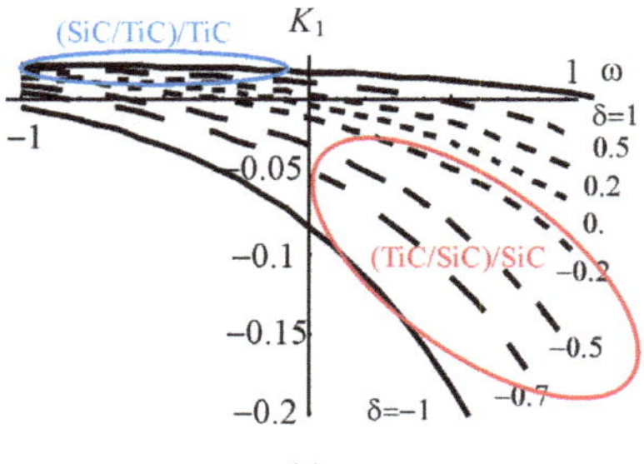

(a)

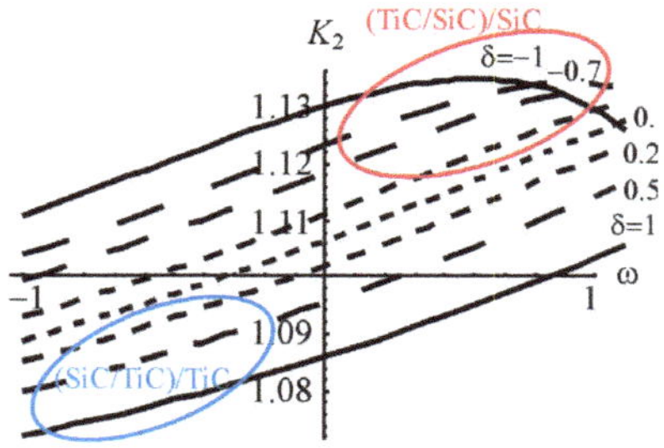

(b)

Figure. 6. Influence of non-symmetrical system of microcracks (with $\theta = 0$) above the interface crack (Fig. 3-b) on TSIFs (a) K_1 and (b) K_2 at the right interface crack for $1 < \delta < -1$.

TSIFs K_1 and K_2 at the right interface crack tip as functions of the inhomogeneity parameter ω of thermal expansion coefficient and for different δ is presented in the Fig. 6 for the non-symmetrically disposed microcrack system ahead of the interface crack (Fig. 3-b) for $\theta = 0$ and in the Fig. 7 for $\theta = \pi/4$. The same system of cracks but with different inclination angles θ produces different influence on the TSIFs. The microcracks with $\theta = 0$ cause the interface crack closure – K_1 is negative for most parameters of δ and ω. At the same time for $\theta = \pi/4$ K_1 is positive for all inhomogeneity parameters. The examples of the material combinations are the following: parameters $\delta > 0$ and $\omega < 0$ correspond to the FGM/homogeneous material combinations (SiC/TiC)/TiC and (SiC/$MoSi_2$)/$MoSi_2$; and parameters $\delta < 0$ and $\omega > 0$ correspond to (TiC/SiC)/SiC and ($MoSi_2$/SiC)/SiC/. These regions are indicated in Figs. 6 and 7. The thermal properties of these materials are: $k^{\mathrm{SiC}} = 60$ $\mathrm{Wm^{-1}K^{-1}}$, $k^{\mathrm{TiC}} = 20$ $\mathrm{Wm^{-1}K^{-1}}$, $\alpha_t^{\mathrm{SiC}} = 4*10^{-6}$ $\mathrm{K^{-1}}$, $\alpha_t^{\mathrm{TiC}} = 7*10^{-6}$ $\mathrm{K^{-1}}$, $\alpha_t^{\mathrm{MoSi_2}} = 5*10^{-6}$ $\mathrm{K^{-1}}$ [13, 14]. The elastic modulus, Young's modulus and Poisson's ratio, of these materials are similar.

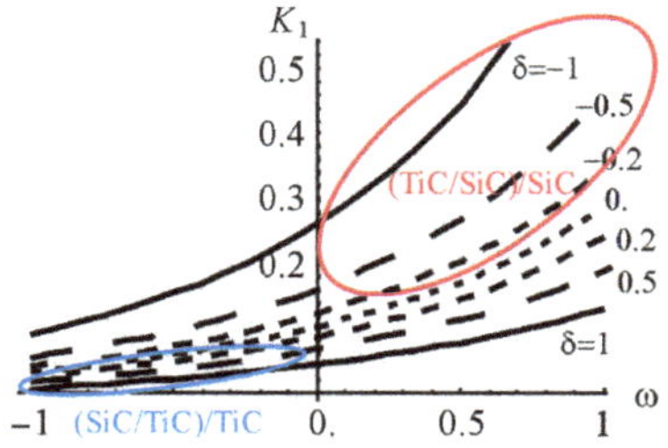

(a)

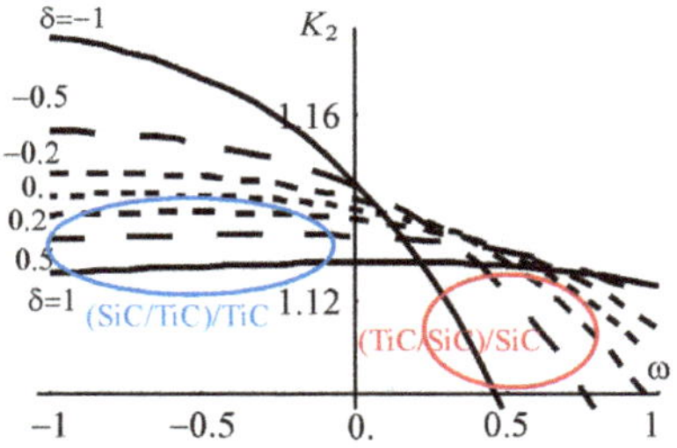

(b)

Figure. 7. Influence of non-symmetrical system of microcracks (with $\theta=\pi/4$) above the interface crack (Fig. 3-b) on TSIFs (a) K_1 and (b) K_2 at the right interface crack for $1 < \delta < -1$.

These examples of material combinations show that in the study of the influence of inhomogeneity parameters on the mechanical characteristics of FGMs with cracks the parameters δ and ω are usually not identical and it is not typical to put these parameters to the same value as it is performed in many studies (see, for example, [12]). In each case of actual FGMs the values of their inhomogeneity parameters should be estimated. Ref. [15] gives a number of examples of material properties of FGMs and on this basis different combinations of the non homogeneity parameters, which control the variation of the heat conductivity, the shear modulus and the thermal expansion in the graded coating, were studied.

Another example of the influence of the system of cracks on the TSIF of the interface crack is given in Table 1. The TSIFs at the right interface crack tip are presented for randomly oriented system of cracks in Fig. 3 (d), and for comparison for the crack system shown in Fig. 3 (c) with $\theta = 0$, $\pi/4$ and $\pi/2$. The calculations were performed for $\delta = 1$ and $\omega = 0$. We can see the difference in the TSIFs.

Table 1. TSIFs at the right interface crack tip as influenced by crack systems in Fig. 3 (c,d).

θ	Random	0	$\pi/4$	$\pi/2$
K_1	0.1688	0.071	0.159	0.1762
K_2	0.9658	0.9937	0.9645	0.922

Table 2. TSIF accounting for interface crack closure.

θ	K_1	K_1^+	K_1^-	c/a_0	$\lvert K_1 - K_1^c \rvert / K_1$
$\pi/4$	1	−0.0169	0.00267		
	2	0	0.00336	0.927	25 %
$\pi/3$	1	−0.0185	0.00354		
	2	0	0.00398	0.950	12.3 %

Table 2 presents a calculation accounting for interface crack closure for the TSIF K_1 for the interface crack as influenced by a crack in the FGM with midpoint coordinate $w_k = 1.0 + i0.2$ ($x_k = 1.0$, $y_k = 0.2$), inclination angles $\theta = \pi/4$ and $\pi/3$, and inhomogeneity parameters $\delta = 1$ and $\omega = 0$. The non dimensional half-length c/a_0 of the open portion of the interface crack is also given in the table (see Fig. 2). The case 1 corresponds the TSIFs without taking into account the interface crack closure (Eq. 22), the case 2 corresponds the TSIFs accounting for the crack closure. The calculation was performed using the Eqs. (28) and (31) and the procedure described in Sec. 4 is applied. The parameter V (30) is equal to 1, i.e. the microcrack is open. The small crack near the right interface crack tip causes the interface crack shielding, K_1^+ is negative. A small contact zone near the right crack tip appears, the half-length of this zone is equal to $(a_0 - c)/a_0 = 0.073$ for $\pi/4$ and 0.05 for $\pi/3$. The recalculation of the TSIFs accounting for interface crack closure gives the difference in 25% between case 1 and case 2 for $\pi/4$ and in 12.3% for $\pi/3$.

Conclusions

An approximate model for investigating the influence of microcracks on an interface crack in FGM/homogeneous bimaterials under a heat flux is presented and based on the asymptotic analytical formulas for the thermal stress intensity factors at the interface crack tips. The parametric analysis shows the dependence of the TSIFs at the interface crack tips on the location and orientation of the cracks in the FGM. It is also shown that the inhomogeneity parameters δ and ω of thermo-conductivity and thermal expansion coefficient notably affect the TSIFs of the interface crack. The TSIFs can be amplified (K_2>1) or shielded (K_2< 1) by the system of microcracks. It is also shown, that for some crack arrangements (Figs. 6-a) the TSIF k_I at the interface crack tips can be negative, i.e. the interface crack surfaces could close. The solution taking into account the crack closure is derived and an example of this calculation is given. These results are applicable to such kinds of FGMs as: ceramic/ceramic FGMs, i.e. TiC/SiC,

$MoSi_2/Al_2O_3$ and $MoSi_2/SiC$, and also some ceramic/metal FGMs, i.e. zirconia/nickel and zirconia/steel. Examples of some FGM/homogeneous bimaterials are discussed.

Acknowledgement

V. Petrova acknowledges the support of the German Research Foundation under grants Schm 746/92-1 and Schm 746/106-1.

References

[1] V. Birman, L.W. Byrd, Modeling and analysis of functionally graded materials and structures, ASME Appl. Mech. Rev. 60 (2007) 195-216. https://doi.org/10.1115/1.2777164

[2] J. Sladek, V. Sladek, Ch. Zhang, An advanced numerical method for computing elastodynamic fracture parameters in functionally graded materials, Comp. Mater. Sci. 32 (2005) 532-543. https://doi.org/10.1016/j.commatsci.2004.09.011

[3] P. A. Martin, J. D. Richardson, L. J. Gray, and J. Berger, Green's functions for an Exponentially Graded Elastic Material, Proc. Royal Soc. 458 (2002) 1931-1948. https://doi.org/10.1098/rspa.2001.0952

[4] V. Tamuzs, N. Romalis, V. Petrova, Fracture of Solids with Microdefects, NOVA Science Publishers Inc., New York, 2000.

[5] N.I. Shbeeb, W.K. Binieda, K.L. Kreider, Analysis of the driving forces for multiple cracks in an infinite nonhomogeneous plate. Part I: Theoretical analysis. Part II: Numerical solutions, ASME J. Appl. Mech. 66 (1999) 492-506. https://doi.org/10.1115/1.2791075

[6] N. Noda, B.L. Wang, Transient thermoelastic responses of functionally graded materials containing collinear cracks, Eng. Fract. Mech. 69 (2002) 1791-1809. https://doi.org/10.1016/S0013-7944(02)00055-3

[7] V. Petrova, S. Schmauder, Crack interaction problems in functionally graded/ homogeneous bimaterials subjected to a heat flux, Proc. ICF12, Ottawa, Canada (2009), Proceedings on CD ROM, 10 p.

[8] V. Petrova, S. Schmauder, Crack - interface crack interactions in functionally graded/ homogeneous composite bimaterials subjected to a heat flux, Mech. Compos. Mater. 47 (1) (2011) 125-136. https://doi.org/10.1007/s11029-011-9191-0

[9] V.V. Panasyuk, M.P. Savruk, A.P. Datsyshin, Stress Distribution near Cracks in Plates and Shells (in Russian), Naukova Dumka, Kiev,1976.

[10] J.W. Eischen, Fracture of nonhomogeneous materials, Int. J. Fracture 34 (1987) 3-22. https://doi.org/10.1007/BF00042121

[11] R. Goldstein, Yu. Zhitnikov, T. Morozova, Equilibrium of the cuts system with forming the closing and opening regions on them (Russian), Prikl. Mat. Mekh. **55**(4) (1991) 672-678. https://doi.org/10.1016/0021-8928(91)90019-Q

[12] S. El-Borgi, F. Erdogan, F. Ben Hatira, Stress intensity factors for an interface crack

between a functionally graded coating and a homogeneous substrate, Int. J. Fract. 123 (2003) 139-162. https://doi.org/10.1023/B:FRAC.0000007373.29142.57

[13] J.F. Shackelford, W. Alexander, CRC Materials Science and Engineering Handbook, CRC Press, Boca Raton, 2001. https://doi.org/10.1201/9781420038408

[14] O.P. Chakrabarti, P.K. Das, S. Mondal, Study of indentation induced cracks in $MoSi_2$-reaction bonded SiC ceramics, Bull. Mater. Sci. 24 (2001) 181-184. https://doi.org/10.1007/BF02710098

[15] S. El-Borgi1, M. F. Djemel1, and R. Abdelmoula, A surface crack in a graded coating bonded to a homogeneous substrate under thermal loading, J. Therm. Stress. 31 (2008) 176-194. https://doi.org/10.1080/01495730701737886

CHAPTER 5

Crack Closure Effects in Thermal Fracture of Functionally Graded/ Homogeneous Bimaterials with Systems of Cracks

Vera Petrova [1,2]*, Siegfried Schmauder [1]

[1] IMWF, University of Stuttgart, Pfaffenwaldring 32, D-70569 Stuttgart, Germany

[2] Voronezh State University, University Sq.1, Voronezh 394006, Russia

veraep@gmail.com *, Siegfried.Schmauder@imwf.uni-stuttgart.de

Abstract

In the presented model the microstructure of a functionally graded material (FGM) is accounted in two ways: via FGM properties and by distributed small cracks which could be on grain boundaries of the material. The contribution to the solution of material gradation of FGM and of material structure reflected via geometry of distributed cracks is investigated. Besides, some additional effects arising due to possible crack closure and contact of crack surfaces is taken into account in the model and its contribution is also assessed. The goal of this investigation is to show these affects and to estimate the contribution on the SIFs of each of these, if they are strong or can be neglected.

Keywords

Thermal Fracture, Crack Closure, Functionally Graded Materials, Stress Intensity Factors.

1. Introduction

Functionally graded materials (FGMs) are a special group of composites with continuously varying properties in one space direction [1 - 3]. These graded properties are achieved by special technologies, e.g., plasma spraying, powder metallurgy, physical vapor deposition (PVD), centrifugal casting method [4]. The principal motivation for FGM development is to decrease bimaterial mismatch, to reduce residual stresses at interfaces and to prevent delamination along the interfaces [3]. FGMs have attracted growing interest because of their wide application in numerous branches of modern technology. In particular, FGMs are used in thermal barrier coating to protect details from high temperatures as well as from wear and corrosion. The materials for protecting from high temperatures should have a low thermal conductivity and at the same time they are desired to have a thermal expansion coefficient close to that of the material for the protected substrate. Thermo-elastic fracture analysis of FGMs is important for improving fracture characteristics of FGMs.

In the present work a mathematical modeling of the fracture processes in the vicinity of an interface crack in FGM/homogeneous bimaterials with internal defects subjected to a thermal flux or a combination of thermal flux and shear loading is performed with emphasis on crack closure effects. The solution is based on the integral equation method and it is assumed that thermal properties of the FGM possess exponential form. It should be noted that this work is motivated by early works by Tamuzs and co-authors [5-11] where the problems of macro-microcrack interaction in a homogeneous material were solved with taking into account the crack closure. At first the shear loading was considered [5, 8] and then the thermal fracture problem [5]. These ideas were used in the works [12, 13].

Crack-closure mechanisms and their influence on fracture characteristics of materials were discussed by many authors [14, 15]. It was revealed that a diversity of mechanisms may produce the crack closure, among them are the following. *Loading-induced* crack closure is observed when external loads (shear, tension, compression, thermal and combination of them) cause the mixed-mode loading conditions near a crack with compression and shear stresses. *Geometry-induced* closure is due to a specific shape of cracks, e.g., zig-zag and kinked cracks [16-18], curvilinear cracks [19], or due to interaction among cracks, defects and boundaries of elastic body with cracks, see, e.g., [5-12]. Crack closure can be due to *material inhomogeneity* such as well-known interface cracks between two different materials [21]. Non-homogeneity of FGMs can influence on crack propagation path and as a result compression zones near the kinked crack may appear [22, 23]. It should also be mentioned *environmentally induced* crack closure such as oxide-induced in hydrothermal environment. In [24] it was shown that crack closure may occur in the moisture absorption process in glass-fiber/epoxy laminates with transverse matrix cracks.

In the present study we consider the interaction between an interface crack and internal cracks in FGM/homogeneous bimaterials under the influence of combination of Mode II load and thermal flux, so we will have three types of crack closure mechanisms, such as load, geometry and material inhomogeneity, and the influence of these mechanisms on the possible crack closure is investigated.

2. Formulation of the problem and solution

2.1 Geometry of the problem and assumptions

Fig. 1 (a) shows the geometry of the problem. A bimaterial is composed of a functionally graded material (denoted by number 1) and a homogeneous material (denoted by number 2). The bimaterial is perfectly bonded with the exception of an interface crack of length $2a_0$. It is assumed that the FGM contains N cracks of length $2a_k$. Cartesian coordinates (x, y) are centered at the midpoint of the interface crack; the x-axis lies along the interface line. Local coordinate systems (x_k, y_k) are attached to each internal crack. The crack position is determined by the defect midpoint coordinate (x_k^0, y_k^0) and an inclination angle θ_k to the interface, i.e. to the x-axis (Fig. 1 (a)).

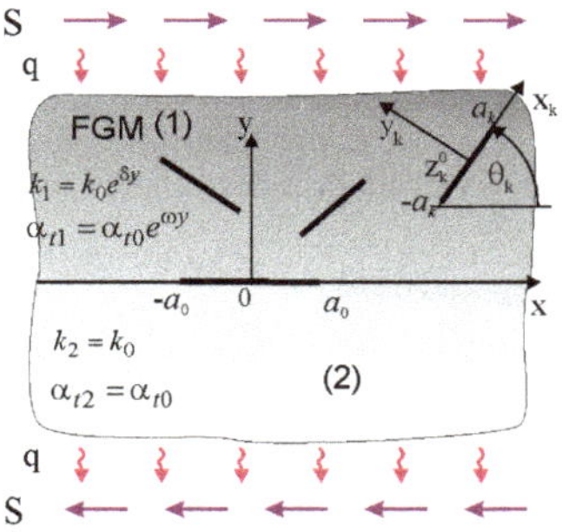

Figure 1. The geometry of the problem: an interface crack and internal cracks in a FGM/ homogeneous bimaterial.

The bimaterial is subjected to a heat flux of intensity q and a shear stress S applied at infinity. It is supposed, that the cracks are thermally isolated and traction free.

As in the previous works [12, 13] it is assumed that the properties of the functionally graded material depend only on the coordinate y and are given in exponential form

$$k_1(y) = k_0 e^{\delta y}, \ \alpha_1 = \alpha_0 e^{\omega y}, \tag{1}$$

where k_1 and α_1 are the thermal conductivity and the thermal expansion coefficients for an FGM, the constant k_0 is the thermal conductivity and α_0 is the thermal expansion coefficient of the interface and of material 2, while δ and ω are non-homogeneity parameters for the FGM. The Young's modulus and Poisson's ratio are assumed to be constant, $E_j = const$, $\nu_j = const$ (j = 1, 2). For this special case, the material is elastically homogeneous, but thermally non-homogeneous. Besides, it was shown [25] that Poisson's ratio has negligible effect on the crack-tip stress intensity factors.

The solution of the problem consists of the determination of the temperature field and the determination of the thermal stresses since the uncoupled, quasi-static thermoelastic theory is used and the temperature distribution is independent of the mechanical field according to this theory.

2.2 Formulation of the thermal and thermoelastic problems

The thermal and mechanical boundary conditions are based on the following assumptions: the cracks (both internal and interface) are thermally isolated and traction free. At the interface the continuity conditions are fulfilled, i.e. temperature, thermal fluxes, stresses and displacements are equal.

Using the superposition principle the temperature field in the bimaterial with cracks is presented as a sum of two terms: the temperature distribution in a bimaterial in the absence of cracks and the temperature perturbation caused by the cracks. For the crack problem the temperature perturbation should be determined. Supposing that the non-homogeneity of the FGM is revealed only in non-homogeneous thermal fluxes on crack surfaces, the perturbation problem can be solved by the method presented in [13] using similar integral equations. The details of this solution can be found in [12, 13].

For the thermoelastic problem we also use the superposition principle according to which the problem is transformed to the problem with boundary conditions on the crack surfaces:

$$\sigma_{ny}^{\pm} - i\tau_{nxy}^{\pm} = p_n(x), \quad p_n = -iSe^{-2i\theta_n}, |x| \leq a_n, n = 0,1,...,N, \tag{2}$$

and with zero stresses at infinity $\sigma_{ij} \to 0, \quad x^2 + y^2 \to \infty$.

Because we suppose that the material is elastically homogeneous we can use directly the method presented in [11]. The mathematical analysis of these problems leads to singular integral equations of the Cauchy-type [26, 27]. Following the basic principle of linear elasticity the solution of purely thermal and purely mechanical loading can simply be added to give the solution for the combined loading case (superposition principle). This problem for combined loading was discussed in [28] with respect to stress intensity factors (SIFs) for an r-crack interacting with an inclusion.

2.3 Solution by small parameter method

In previous works [12, 13] the solution was obtained asymptotically by the method of small parameter. It was done for a special case where an interface is significantly larger in size than internal cracks in the FGM. The method was first suggested by Romalis and Tamuzs in 1984 and then was used for different macro-microcrack interaction problems for homogeneous materials [5-11]. In the interest of brevity the solution will not be repeated here in details

It is assumed that all internal cracks in the FGM have the same size $2a_k = 2a$ ($k = 1, 2, ..., N$), for example, they have the characteristic size of a grain size of the material. Suppose also that $2a << 2a_0$ and in this case the small parameter is $\lambda = a/a_0$ and $\lambda << 1$, namely, the small parameter is equal to the ratio of the size of small internal cracks to the interface crack size. For a uniform heat flux q and a shear load S acting at infinity the stress intensity factors at the interface crack tips are obtained up to λ^2 as

$$k_I^{\pm} - ik_{II}^{\pm} = Q\sqrt{a_0}[\mp i + \lambda^2 \sum_{k=1}^{N} F_1(w_k, \theta_k, \delta, \omega)] + S\sqrt{a_0}[-i + \lambda^2 \sum_{k=1}^{N} F_2(w_k, \theta_k)],$$
$$w_k = z_k^0 / a_0 \tag{3}$$

$$Q = q\delta_t^h k_0 a_0, \; \delta_t^h = \delta_t^h(\alpha_{t0}, E, \nu) \tag{4}$$

In Eq. (3) the upper part of the "$\pm$" or "$\mp$" signs refers to the right tip and the lower part to the left tip of the crack. '*i*' is the imaginary unity.

The full expression for Eq. (3) is cited in the Appendix.

For a single crack (without microcracks) subjected to a heat flux normal to the crack surfaces the thermal stress intensity factors are given as [11, 29]

$$k_I^{\pm} = 0, \quad k_{II}^{\pm} = \pm Q\sqrt{a_0}, \; Q = \delta_t^h q k_0 a_0 \tag{5}$$

and for the crack under shear load S remotely applied parallel to the crack the stress intensity factors are given as

$$k_I^{\pm} = 0, \quad k_{II}^{\pm} = S\sqrt{a_0}\,. \tag{6}$$

In both cases, we have pure Mode II conditions near the interface crack and for combined thermal and mechanical loading the SIFs are written as

$$k_I^{\pm} = 0, \quad k_{II}^{+} = (S+Q)\sqrt{a_0}, \quad k_{II}^{-} = (S-Q)\sqrt{a_0}\,. \tag{7}$$

The magnitude of k_{II} (7) at the right crack tip is much larger than at the left one. The interaction of cracks results in mixed-mode conditions near the interface crack, i.e. $k_I \neq 0$, with dominant Mode II contribution as we can see from Eqs. (3), (A.1), (A.2) and (A.3).

3. Cracks with closure

Analysis of the formulae (3), (A.1), (A.2) and (A.3) shows that the SIF k_{I0} takes negative values for certain crack configurations. It means that the crack surfaces may overlap. Therefore, it might be necessary to account for the closure of the cracks. A detailed solution to the macro-microcrack interaction problem taking into account a crack closure was derived in [5, 11] for a homogeneous material under shear loading and in [6, 11] for the thermal problem. Later a similar approach was used for an interface crack - microcracks interaction problem for FGM/homogeneous bimaterials under thermal load [12]. In the present investigation we apply this scheme for the solution for the combined thermal and shear (Mode II) mechanical loading.

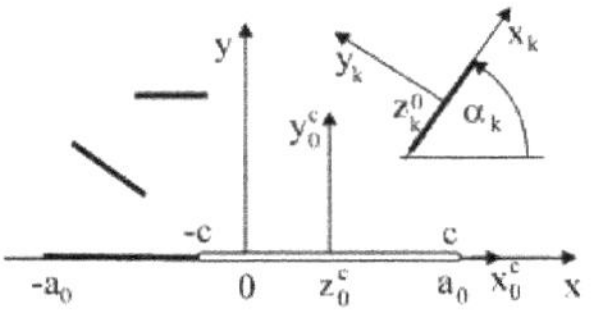

Figure 2. An interface crack with a contact zone and microcracks.

3.1 Formulation of the problem and solution

The problem is solved by the method stated in [12] under similar assumptions on the presence of an open region of length 2c at the interface crack and on smooth contact within the closed regions (Fig. 2) and under the additional assumption that the temperature distribution does not depend on the contact of the crack faces, i.e. the heat-insulation conditions are still satisfied on the closed regions of the interface crack or on closed microcracks (microcracks are fully closed or opened.)

Since the solution for the thermal conductivity problem is the same by virtue of the latter assumption, it remains to solve the thermoelastic problem with regard for the possible crack closure.

To account for the crack closure the boundary conditions are reformulated. For completely open cracks, the previous traction-free boundary conditions are valid. On the closed portion of the interface crack or on the closed surfaces of microcracks, the shear stresses are given as

$$\tau_n^{\pm}(x,0) = 0 \tag{8}$$

and the transverse displacement jumps $[v_n]$ are required to vanish:

$$[v_n] = 0. \tag{9}$$

For the case of open cracks, the unknowns are the shear and transverse displacement jumps $[u_n]$ and $[v_n]$. For closed cracks, the shear displacement jumps and normal tractions are unknown.

The second approximation of the SIFs at the interface crack tips is obtained as

$$\begin{aligned} k_{I0}^{q\pm} &= \pm\lambda^2 Q\sqrt{a_0}\frac{1}{2\hat{c}^2}\sum_{k=1}^{N}\{\mathrm{Re}(I_{k0}^T)\,\mathrm{Im}[m_{k1}^c - n_{k1}^c] \\ &\quad + V\,\mathrm{Im}(I_{k0}^T)\,\mathrm{Re}[m_{k1}^c + n_{k1}^c] + 2\exp(\omega a_0\,\mathrm{Im}(w_k))J_k^T(\delta)\,\mathrm{Im}[m_{k0}^c - n_{k0}^c]\}, \end{aligned} \tag{10}$$

$$\begin{aligned} k_{II0}^{q\pm} &= \pm Q\sqrt{a_0}\{1 + \frac{\lambda^2}{2}\sum_{k=1}^{N}\{2J_k^T(\delta)\,\mathrm{Re}[e^{i\theta_k}(w_k/(w_k^2-1)^{1/2}-1)] + \mathrm{Re}[I_{k0}^T]\,\mathrm{Re}[m_{k1} - n_{k1}] \\ &\quad + V\,\mathrm{Im}(I_{k0}^T)\,\mathrm{Im}[m_{k1} - n_{k1}] - 2\exp(\omega a_0\,\mathrm{Im}(w_k))J_k^T(\delta)\,\mathrm{Re}[m_{k0} - n_{k0}]\}\}, \end{aligned} \tag{11}$$

$$k_{I0}^{S\pm} = \mp\lambda^2 S\sqrt{a_0}\frac{1}{2\hat{c}^2}\sum_{k=1}^{N}\{-\mathrm{Re}(J_k)\,\mathrm{Im}[m_{k1}^c + n_{k1}^c] + V\,\mathrm{Im}(J_k)\,\mathrm{Re}[m_{k1}^c + n_{k1}^c]\}, \tag{12}$$

$$k_{II0}^{S\pm} = S\sqrt{a_0}\{1 + \frac{\lambda^2}{2}\sum_{k=1}^{N}[\mathrm{Re}(J_k)\,\mathrm{Im}(m_{k1} - n_{k1}) + V\,\mathrm{Im}(J_k)\,\mathrm{Re}(m_{k1} - n_{k1})]\} \tag{13}$$

and the SIFs at the microcrack tips are derived

$$k_{In}^{\pm} = \lambda[Q\sqrt{a_0}\,\mathrm{Im}(I_{k0}^T) + S\sqrt{a_0}\,\mathrm{Im}(J_k)] \tag{14}$$

In Eqs. (10) - (13) $m_{k1}^c, n_{k1}^c, m_{k0}^c$ and n_{k0}^c are obtained from Eqs. (A.7) by setting $(w_k - d_0)/\hat{c}$ instead of w_k, here $\hat{c}$ and d_0 are nondimentional parameters $\hat{c} = c/a_0$ and $d_0 = z_0^c/a_0$ for the open portion of the interface crack (Fig. 2).

By virtue of the boundary condition (9) the parameter V is introduced into Eqs. (10) - (13) as

$$V = \{0, \text{if } k_{In} \le 0;\ 1, \text{if } k_{In} > 0\} \tag{15}$$

which allows to take into account the microcrack closure. Here k_{In} is Eq. (14).

The functions of SIF k_I (10) and (12) contain the unknown parameters of the open portion of the interface crack. Unknown coordinates of the points of separation of the closed and open parts of the interface crack are found from the condition that the crack is closed smoothly and the singularity vanishes at these points (see [19, 20])

$$k_{I0}(\pm c) = 0. \tag{16}$$

This condition has been written in the local system (x_0^c, y_0^c), Fig. 2. In Eq. (16) k_{I0} is either Eq. (10) for only thermal loading, or Eq. (12) for pure shear loading, or the sum of Eqs. (10) and (12) for combined thermo-mechanical loading.

3.2 Analysis of the solution

The analysis of Eq. (14) reveals that in the frame of this approximation the small cracks exist in only two modes - fully open or fully closed. Eqs. (10) and (12) show that microcracks (depending on their location and orientation) can cause either full or partial closure of the interface crack and, hence, for a correct determination of k_I the presence of the closed portion should be taken into account. Since the frictionless contact for cracks is considered, these closed zones do not influence on shear displacements and correspondingly on SIFs k_{II}. The contact zones with friction will influence on both SIFs k_I and k_{II}.

The microcrack closure influences on the SIFs k_I at the microcrack tips, and for simplicity we suppose that for closed microcracks $k_I = 0$, while for the opened one k_I is determined by Eq. (14). The parameter V in Eq. (15) is introduced in order to take into account microcrack closure. Similar as for the interface crack, due to the assumption of frictionless contact the microcrack closure does not affect the SIF k_{II} at the microcrack tips. Besides, the contribution of the terms with V into the SIFs k_{I0} and k_{II0} is not large as can be seen in Eqs. (10) – (14) where it is a part of the expression multiplied by the small parameter λ^2.

The crack contact problem is nonlinear, because the lengths of the contact zones are not known in advance and have to be solved as a part of the solution, Eq. (16). Generally a suitable iteration process must be employed. This procedure was suggested in [20] for crack interaction problems with frictionless contact zones and used in macro-microcrack interaction problems in [5, 6]. A similar crack closure algorithm was also applied in [30, 31] to treat the problem of having negative k_I stress intensity factors for a partially insulated interface crack between a functionally graded coating and a homogeneous substrate subjected to both thermal and mechanical loading.

The frictional contact problems are more complicated in comparison with frictionless problems for cracks. Three different zones may appear on the cracks, namely, open, sliding and sticking. Besides, the crack closure problem with friction makes the solution history dependent and an incremental solution procedure has to be employed as it was done, e.g., in [18] for zigzag crack problems. In [32] an iterative algorithm was adopted for modeling crack closure and sliding with the displacement discontinuity method. The cases of partially closed cracks with both frictionless and frictional contacts were considered and benchmark results for the SIFs for a subsurface crack were presented and compared with the results of other authors. An iterative hybrid boundary element method for solving non-linear closed crack problems can also be found in [33].

3.3 Influence of contact zones on SIFs

Let us consider some examples which will show the influence of the crack closure on SIFs. For the calculations it is convenient to represent SIFs in a non-dimensional form. The SIFs k_I and k_{II} are normalized by k_0

$$k_0 = |k_{II}| = |Q\sqrt{a_0}| = S\sqrt{a_0} \tag{17}$$

and denoted as K_1 and K_2:

$$K_1^{\pm} - iK_2^{\pm} = (k_I^{\pm} - ik_{II}^{\pm}) / k_0 . \tag{18}$$

We suppose that both loads, the thermal Q (Eq. 4) and the shear S, possess the same values. If we abandon this assumption, then we should introduce a new parameter, e.g., $A = S / Q$ – the ratio of the shear load to the thermal loading. For example, the nondimensional k_I, Eq. (14), at the microcrack tips is written as

$$K_{1n}^{\pm} = k_{In}^{\pm} / |Q\sqrt{a_0}| = \lambda[\mathrm{Im}(I_{k0}^T) + A\,\mathrm{Im}(J_k)] . \tag{19}$$

It should be noted, that A is an important parameter and the influence of this parameter on fracture characteristics will be considered later.

The functions of SIF k_I (10) and (12) contain the unknown parameters of the open portion of the interface crack such as the size $\hat{c} = c / a_0$ and the midpoint coordinate $d_0 = z_0^c / a_0$. From geometry (Fig. 2) we have the following relation between d_0 and $\hat{c}$:

$$d_0 = \pm(1 - \hat{c}) . \tag{20}$$

The upper sign refers to the open zone near the right crack tip and the lower – near the left tip.

In order to use the condition (16) with Eqs. (10) and (12) and calculate the size of the open portion of the crack $\hat{c}$ we should substitute d_0 by Eq. (20) and accordingly in Eq. (A.7) w_k is substituted by

$$(w_k - d_0) / \hat{c} = \begin{cases} (w_k - (1 - \hat{c})) / \hat{c} & \text{if } K_1^- < 0 \text{ (and } K_1^+ > 0) \\ (w_k - (-1 + \hat{c})) / \hat{c} & \text{if } K_1^+ < 0 \text{ (and } K_1^- > 0) \end{cases} \tag{21}$$

Algorithm of the calculation is following.

Step 1. Determine the presence of the closed zone on the interface crack. Using Eqs. (3), (A.1) and (A.3) we calculate SIFs K_1 at the left crack tip (K_1^-) and at the right tip (K_1^+) and check the sign of K_1. If $K_1^- < 0$ or $K_1^+ < 0$, then the interface crack has closed zones, and we should determine K_1 using Eqs. (10) and (12), which are taking into account the crack closure.

Step 2. Obtain the size of the open portion of the crack, $2\hat{c}$. We can solve the nonlinear equation (16) applying an iterative procedure or we can find the root of the equation (16) using the computational software program "Mathematica", where this procedure is available for functions. As if the $\hat{c}$ has been found $\hat{c} = c_0$, the corrected K_1 is calculated by substituting the obtained $\hat{c} = c_0$ in Eqs. (10) and (12). For example, if the closed zone is near the left tip of the interface crack, then $K_1^- = 0$ and the corrected $K_1^+ = K_1^+(c_0)$.

4.   Results

4.1   Heat flux

A calculation accounting for interface crack closure for the thermal SIF K_1 for the interface crack as influenced by a crack in the FGM with midpoint coordinate $w_k = 1.0 + i0.2$ ($x_k = 1.0$, $y_k = 0.2$) and inclination angle $\theta = \pi/4$ is performed for the case of a heat flux. The parameter V is equal to 1, i.e. the microcrack is open, the small parameter λ is taken 0.1.

In the Table 1 the results present for K_1 for different inhomogeneity parameters δ and ω. *Case a* – corresponds to the SIFs without taking into account the interface crack closure and the *case b* – with accounting for the closure effects (K_1^c). The small crack near the right interface crack tip causes interface crack shielding, where K_1^+ is negative. A small contact zone near the right crack tip appears, the length of this zone is equal to $(a_0 - c)/a_0 = 0.0702$ for $\pi/4$ and for other materials are smaller. For all cases recalculation of the SIFs gives the difference from 20.8% for a homogeneous material to 13, 42% for FGM with $(\delta, \omega) = (-1, 0.5)$ between the results accounting for crack closure and disregarding this zone.

The influence of parameters inhomogeneity (δ, ω) on SIF K_1 can be also seen from the Table 1. The corrected value of K_1 (b) for the case 2 for $(\delta, \omega) = (-0.5, 0)$ is differed from case 1 on 6.4%, the case 3 for $(\delta, \omega) = (-1, 0)$ is differed on 13.5% and the case 4 for $(\delta, \omega) = (-1, 0.5)$ – on 29.6%.

Table 1.   *SIF accounting for interface crack closure. Heat flux.* $w_k = 1.0 + i0.2$, $\theta = 45^o$

N	δ, ω	K_1	$K_1^- \times 10^{-2}$	$K_1^+ \times 10^{-2}$	c/a_0	$\| K_1 - K_1^c \| / K_1$ $\times 100\%$
1	0, 0	a	0.312	-1.634	not considered	
		b	0.377	0	0.9298	20.8
2	-0.5, 0	a	0.338	-1.601	not considered	
		b	0.401	0	0.932	18.64
3	-1, 0	a	0.3667	-1.56	not considered	
		b	0.428	0	0.9335	16.72
4	-1, 0.5	a	0.4307	-1.4846	not considered	
		b	0.4885	0	0.9375	13.42

4.2 Thermal and shear loading

Table 2. SIF accounting for interface crack closure. Thermo-mechanical loading. $w_k = 1.0 + i0.2$, θ = *45°.*

N	δ, ω	K_1	$K_1^- \times 10^{-2}$	$K_1^+ \times 10^{-2}$	c/a_0	$\lvert K_1 - K_1^c \rvert / K_1 \times 100\%$
1	0, 0	a	0.918	-0.2586	not considered	
		b	1.00631	0	0.9377	9.62
2	-0.5, 0	a	0.944	-0.226	not considered	
		b	1.031	0	0.9384	9.2
3	-1, 0	a	0.973	-0.1898	not considered	
		b	1.057	0	0.9392	8.6
4	-1, 0.5	a	1.0369	-0.1092	not considered	
		b	1.1182	0	0.941	7.8

The results for K_1 for different inhomogeneity parameters δ and ω for the case of combined thermal and shear loading are presented in Table 2. The parameters are the same as in the previous case. We can see that the influence of the crack closure on SIFs is smaller in comparison to a pure heat flux. The maximum length of the closed zone is equal to $(a_0 - c)/a_0 = 0.0623$ for case 1 (a homogeneous material). For all cases the recalculation of the SIFs gives the difference from 9.62% for a homogeneous material to 7.8% for a FGM with $(\delta, \omega) = (-1, 0.5)$ between the results accounting for crack closure and disregarding this zone.

The influence of inhomogeneity parameters (δ, ω) on the SIF $K_1$are the following. The corrected value of K_1 (b) for the case 2 for (δ, ω) = (–0.5, 0) is differed from case 1 by 2.45%, the case 3 for (δ, ω) = (–1, 0) is differed by 5% and the case 4 for (δ, ω) = (–1, 0.5) – by 11.1%.

Conclusions

Theoretical modelling of the fracture process in FGM/homogeneous bimaterials with an interface crack and internal defects subjected to a heat flux and Mode II shear loading is presented. Asymptotic analytical formulas for SIFs at the interface crack tips are derived for a special case where the interface crack size is much larger than the size of internal cracks in the FGM. A version of these formulas includes the possibility to account for the crack closure. With the assumption of the frictionless contact between the crack faces the crack closure affects the SIFs k_I at the interface crack tips and SIFs k_{II} are calculated disregarding these contacts. Microcracks can be fully closed or opened and the microcrack closure influences on both k_I and k_{II}, but not much.

Strong impact of the non-homogeneity parameters δ and ω of thermo-conductivity and thermal expansion coefficients of FGM on stress intensity factors k_I and on the length of the contact zones is observed. This effect is smaller on about 50% for combined thermal and shear loading in comparison with the pure thermal loading. The observed crack closure is caused by thermo-mechanical loading, the geometry of the problem and the non-homogeneity of functionally graded materials.

The influence of the non-homogeneity of the material due to the presence of microcracks and due to the non-homogeneity of FGMs on the dominant SIFs k_{II} and other fracture characteristics can be investigated on the basis of the obtained explicit formulas for SIFs and will be analyzed in a future work.

Acknowledgement

The support of the German Research Foundation under the grant Schm 746/139-1 is greatly acknowledged.

References

[1] Y. Miyamoto, W. A. Kaysser, B. H. Rabin, A. Kawasaki, R. G. Ford, Functionally graded materials: Design, Processing and Applications, Kluwer Academic, Dordrecht, 1999. https://doi.org/10.1007/978-1-4615-5301-4

[2] V. Birman, L.W. Byrd, Modeling and analysis of functionally graded materials and structures, ASME Appl. Mech. Rev. 60 (2007) 195-216. https://doi.org/10.1115/1.2777164

[3] M. Tilbrook, R. Moon, M. Hoffman, Crack propagation in graded composites, Comp. Sci. Techn. 65 (2005) 201-220. https://doi.org/10.1016/j.compscitech.2004.07.004

[4] Y. Watanabe, H. Sato, J. Cuppoletti, review fabrication of functionally graded materials under a centrifugal force, nanocomposites with unique properties and applications in medicine and industry, InTech, Rijeka, Croatia, Shanghai, China, 2011. https://doi.org/10.5772/20988

[5] V. Tamuzs, V. Petrova, Main crack in the field of microdefects in transverse shear conditions, Materials Science 29(3) (1993) 322-331. https://doi.org/10.1007/BF00558978

[6] V. Tamuzs, V. Petrova, N. Romalis, Thermal fracture of a macrocrack with closure as influenced by microcracks, Theor. Appl. Fract. Mech. 21 (1994) 207-218. https://doi.org/10.1016/0167-8442(94)90034-5

[7] V. Petrova, Shear fracture of the main crack at the presence of rigid inclusions and an accompanying crack closure, Mech. Comp. Mater. 30 (5) (1994) 609-618. https://doi.org/10.1007/BF00616772

[8] V. Tamuzs, V. Petrova, N. Romalis, Plane problem of crack-microcrack interaction with taking into account a crack closure, Eng. Fract. Mech. 55 (1996) 957-967. https://doi.org/10.1016/S0013-7944(96)00073-2

[9] V. Tamuzs, V. Petrova, S. Tarasovs, Interaction of micro-cracks with a macro-crack yielded in a narrow strip, Theor. Appli. Fract. Mech. 41 (2004) 291-299. https://doi.org/10.1016/j.tafmec.2003.11.016

[10] V. Petrova, V. Tamuzs, N. Romalis, A survey of macro- microcrack interaction problems, ASME Appl. Mech. Rev. 53(5) (2000) 117-146. https://doi.org/10.1115/1.3097344

[11] V. Tamuzs, N. Romalis, V. Petrova, Fracture of Solids with Microdefects, NOVA Science Publishers Inc., New York, 2000.

[12] V. Petrova, S. Schmauder, Mathematical modelling and thermal stress intensity factors evaluation for an interface crack in the presence of a system of cracks in functionally graded/ homogeneous bimaterials, Comp. Mater. Sci. 52 (2012) 171-177. https://doi.org/10.1016/j.commatsci.2011.02.028

[13] V. Petrova, S. Schmauder, Thermal fracture of a functionally graded/homogeneous bimaterial with a system of cracks, Theor. App. Fract. Mech. 55 (2011) 148-157. https://doi.org/10.1016/j.tafmec.2011.04.005

[14] R. Ritchie, The conflicts between strength and toughness, Nat. Mater. 10 (2011) 817-821. https://doi.org/10.1038/nmat3115

[15] R.C. McClung, The influence of applied stress, crack length, and stress intensity factor on crack closure, Metallurgical Trans. A 22 (1991) 1559-1571. https://doi.org/10.1007/BF02667369

[16] Xian-Fang Li, L. Roy Xu, T-Stress across static crack kinking, J. Appl. Mech. 74 (2007) 181-190. https://doi.org/10.1115/1.2188016

[17] W. L. Zang, P. Gudmundson, Contact problems of kinked cracks modelled by a boundary integral method, Int. J. Numer. Methods Eng. 29 (4) (1990) 847-860. https://doi.org/10.1002/nme.1620290412

[18] W. L. Zang, P. Gudmundson, Frictional contact problems of kinked cracks modelled by a boundary integral method, Int. J. Numer. Methods Eng. 31 (3) (1991) 427-446. https://doi.org/10.1002/nme.1620310303

[19] A. V. Andreev, R. V. Goldstein, Y. V. Zhitnikov, Equilibrium of curvilinear cracks taking into account formation of zones of contact, slipping and sticking the crack surfaces, Izv. AN SSSR. Mechanics of Solids (3) (2000) 137-148.

[20] R. V. Goldstein, Y. V. Zhitnikov, T. M. Morozova, Equilibrium of a system of cracks with contact and opening regions, J. Appl. Math. Mech. 55 (4) (1991) 539-544. https://doi.org/10.1016/0021-8928(91)90019-Q

[21] M. Comninou, An overview of interface cracks, Eng. Fract. Mech. 37(1) (1990) 197-208. https://doi.org/10.1016/0013-7944(90)90343-F

[22] J. Abanto-Bueno, J. Lambros, An experimental study of mixed mode crack initiation and growth in functionally graded materials, Exp. Mech. 46 (2006) 179-196. https://doi.org/10.1007/s11340-006-6416-6

[23] A. Oral, J. Lambros, G. Anlas, Crack initiation in functionally graded materials under mixed mode loading: Experiments and Simulations, Trans. ASME J. Appl. Mech. 75 (2008) 0511101-0511108. https://doi.org/10.1115/1.2936238

[24] J. Lundgren, P. Gudmundson, Moisture absorption in glass-fiber/epoxy laminates with transverse matrix cracks, Compos. Sci. Technol. 59 (1999) 1983-1991. https://doi.org/10.1016/S0266-3538(99)00055-X

[25] F. Erdogan F, B.H. Wu, The surface crack problem for a plate with functionally graded

properties. ASME J. Appl. Mech. 64 (1997) 449-456. https://doi.org/10.1115/1.2788914

[26] N. I. Muskhelishvili, Singular Integral Equations, Noordhoff, Groningen, 1958.

[27] N. I. Muskhelishvili, Some Basic Problems of Mathematical Theory of Elasticity, Noordhoff Press, Amsterdam, 1953.

[28] W.H. Müller, S. Schmauder, Stress-intensity factors of r-crack in fiber-reinforced composites under thermal and mechanical loading, Int. J. Fract. 59 (1993) 307-343. https://doi.org/10.1007/BF00034562

[29] V.V. Panasyuk, M. P. Savruk, A. P. Datsyshin, Stress Distribution near Cracks in Plates and Shells, (in Russian) Naukova Dumka, Kiev, 1976.

[30] S. El-Borgi, F. Erdogan, L. Hidri. Stress intensity factors for an interface crack between a functionally graded coating and a homogeneous substrate, Int. J. Fract. 123 (2003) 139-162. https://doi.org/10.1023/B:FRAC.0000007373.29142.57

[31] S. El-Borgi, F. Erdogan, L. Hidri, A partially insulated embedded crack in an infinite functionally graded medium under thermo-mechanical loading, Int. J. Eng. Sci. 42 (2004) 371-393. https://doi.org/10.1016/S0020-7225(03)00287-8

[32] J. Wang, S. Crouch, An Iterative algorithm for modeling crack closure and sliding, Eng. Fract. Mech. 75 (2008) 128-135. https://doi.org/10.1016/j.engfracmech.2007.03.030

[33] N. Elvin, Ch. Leung, A fast iterative boundary element method for solving closed crack problems, Eng. Fract. Mech. 63 (1999) 631-648. https://doi.org/10.1016/S0013-7944(99)00035-1

Appendix

The full expressions for SIFs are

$$k_{I0}^{q\pm} = \pm\lambda^2 Q\sqrt{a_0}\,\frac{1}{2}\sum_{k=1}^{N}\ \{\mathrm{Re}(I_{k0}^{T})\,\mathrm{Im}[m_{k1}-n_{k1}] + \mathrm{Im}(I_{k0}^{T})\,\mathrm{Re}[m_{k1}+n_{k1}] + 2\exp(\omega a_0\,\mathrm{Im}(w_k))J_k^T(\delta)\,\mathrm{Im}[m_{k0}-n_{k0}]\}, \tag{A.1}$$

$$k_{II0}^{q\pm} = \pm Q\sqrt{a_0}\{1+\frac{\lambda^2}{2}\sum_{k=1}^{N}\{2J_k^T(\delta)\,\mathrm{Re}[e^{i\theta_k}(w_k/(w_k^2-1)^{1/2}-1)] + \mathrm{Re}[I_{k0}^{T}]\,\mathrm{Re}[m_{k1}-n_{k1}] + \mathrm{Im}(I_{k0}^{T})\,\mathrm{Im}[m_{k1}-n_{k1}] - 2\exp(\omega a_0\,\mathrm{Im}(w_k))J_k^T(\delta)\,\mathrm{Re}[m_{k0}-n_{k0}]\}\}, \tag{A.2}$$

and the mechanical part of the SIFs is

$$k_{I0}^{S\pm} - ik_{II0}^{S\pm} = S\sqrt{a_0}\left(-i+\lambda^2\frac{1}{2}\sum_{k=1}^{N}\ (J_k m_{k1} - \bar{J}_k n_{k1})\right) \tag{A.3}$$

where $I_{k0}{}^{T}$, $J_k{}^{T}$ and J_k are

$$I_{k0}^{T} = \frac{1}{2}\left[\frac{e^{-2i\theta_k}-1}{\sqrt{\overline{w}_k^2-1}} + \frac{1}{\sqrt{w_k^2-1}} - \frac{e^{-2i\theta_k}(1-w_k\overline{w}_k)}{(\overline{w}_k^2-1)^{3/2}}\right] \tag{A.4}$$

$$J_k^T(\delta) = \exp(-\delta a_0 \operatorname{Im} w_k)\cos\theta_k - \operatorname{Re}\left[e^{i\theta_k}\left(1-\frac{w_k}{\sqrt{w_k^2-1}}\right)\right] \tag{A.5}$$

$$J_k = \frac{1}{2}\left[(2e^{-2i\theta_k}-1)\frac{\overline{w}_k}{\sqrt{\overline{w}_k^2-1}} + \frac{w_k}{\sqrt{w_k^2-1}} - e^{-2i\theta_k}\frac{\overline{w}_k - w_k}{(\overline{w}_k^2-1)^{3/2}}\right] \tag{A.6}$$

and m_{k1}, n_{k1}, m_{k0} and n_{k0} are

$$m_{k1} = e^{i\theta_k}\operatorname{Re}\left[\frac{e^{i\theta_k}}{(\overline{w}_k \mp 1)\sqrt{w_k^2-1}}\right],\ n_{k1} = \frac{e^{-2i\theta_k}}{2(\overline{w}_k \mp 1)\sqrt{\overline{w}_k^2-1}}\left[(w_k-\overline{w}_k)\frac{2\overline{w}_k \pm 1}{\overline{w}_k^2-1} + e^{2i\theta_k} - 1\right],$$

$$m_{k0} = e^{i\theta_k}\left[1-\operatorname{Re}\left(\frac{\sqrt{w_k^2-1}}{\overline{w}_k \mp 1}\right)\right],\ \ n_{k0} = \frac{e^{-i\theta_k}}{2}\frac{w_k-\overline{w}_k}{(\overline{w}_k \mp 1)\sqrt{\overline{w}_k^2-1}}. \tag{A.7}$$

Besides

$$w_k = z_k^0 / a_0$$

is the non-dimensional complex coordinate of the midpoint of each of the small cracks. In Eqs (23) – (29) "Re" and "Im" denote the real and imaginary parts of complex functions, respectively, 'i' is the imaginary unity.

CHAPTER 6

Revisit of Antiplane Shear Problems for an Interface Crack. Does the Stress Intensity Factor for the Interface Mode III Crack Depend on the Bimaterial Modulus?

Vera Petrova [1,2]*, Siegfried Schmauder [1], Mikael Ordyan [3], Alexander Shashkin [2]

[1] IMWF, University of Stuttgart, Pfaffenwaldring 32, D-70569 Stuttgart, Germany

[2]Voronezh State University, University Sq.1, Voronezh 394006, Russia

[3]Voronezh State Technical University, 20-Letiya Oktyabrya 84, 394006 Voronezh, Russia

veraep@gmail.com *, Siegfried.Schmauder@imwf.uni-stuttgart.de

Abstract

An overview of the results for Mode III interface crack problems available in the literature, as well as those obtained by the authors, is presented in the paper. The analytical and semi-analytical solutions for the main fracture characteristics (i.e. stress intensity factors and energy release rates) are discussed with respect to their dependency on bimaterial constants. Detailed formulation for the antiplane shear problem for an interface crack interacting with internal cracks is presented using the methods of complex potentials and singular integral equations. Some particular solutions (e.g. for the problem of a singularity interacting with an interface and for the Mode III interface crack) are obtained, and these can serve as benchmark solutions or as parts of other similar problems. The approximate closed-form solution is derived for a special case of the interaction between the interface crack and internal micro-cracks. The resulting form of the solution for the stress intensity factors is rather simple and its structure is easily analyzed. A theoretical analysis of the explicit form of expressions for the stress intensity factors shows the conditions under which the solution for the interface crack depends on the bimaterial mismatch, and those under which it does not depend. This formula also allows to define the condition, when a small internal crack does not influence the stress intensity factor of the interface crack (i.e. invisible for the interface crack).

Keywords

Interface, Mode III Crack, Stress Intensity Factors, Closed-Form Solution

1. Introduction

Interface crack problems are important in the investigation of strength and fracture resistance of composites, layered structures, and different bimaterial compounds, because of the quality of interface bonds influences the fracture resistance of whole structures [1]. Since the 60s a large number of publications have been devoted to interface crack problems. Many solutions for the stress intensity factors (SIFs) for interface cracks are included in the SIFs handbook by Murakami [2] with references to the relevant literature. In the recent years, the interface crack problems are being solved as a part of complex practical problems for modeling of composite materials under the influence of mechanical and physical fields (Talreja and Varna [3], Zhuang et al. [4]). The up-to-date reviews on the interface crack problems give a description of different aspects of the interface crack problems, e.g. see [5 - 7].

The plane problem of an interface crack in a bimaterial is well examined. It was shown that the stress intensity factors (SIFs) of interface crack tips depend on the elastic constants of bimaterials. Besides, for the plane deformation of ideally bonded isotropic materials, Dundurs has derived that these constants can be reduced to two parameters α and β, which are now called Dundurs parameters [8, 9]. Based on the physical limits of material properties, the admissible values of the parameters α and β are restricted to a parallelogram in the α-β-plane [9]. The distribution of the Dundurs parameters for typical material combinations of technologically important composites in the α-β diagram was derived by Suga et al. [10]. The examination of these parameters shows that the values of α and β are limited to a narrow range. These data collection can serve as reference data (a catalog) for α and β parameters for technologically important composites.

In contrast to the plane problem, the study of the interface crack problems for Mode III shear loading has received minor attention. Meanwhile, the Mode III problems are similar (up to a certain extent) to the torsion problems [11, 12]. Besides, the studies of these problems can be found in other topics of physics. For example, the mathematical formulation for the Mode III shear stress problem is similar to the stationary thermal conductivity problem with respective boundary conditions and with respective material parameters. The analogy between the steady-state heat transfer and antiplane shear problems was discussed, for example, in a review by Paggi and Carpinteri [11]. Moreover, similar equations describe other physical phenomena, such as electrostatic, electromagnetism, diffusion processes as well as fluid dynamics (steady-state case). Furthermore, the antiplane shear problem is a part of 3D interface crack problems in the linear fracture mechanics approach [13]. The effect of Mode III contribution on the interface crack growth in combined Modes I, II, and III was investigated by Tvergaard and Hutchinson [14]. The conditions of small-scale yielding and with the fracture process at the interface represented by a cohesive zone model were assumed in [14].

There are many examples of solutions to the thermal interface crack problems, including are the following. The heat flow distribution was studied for a partially insulated interface crack as a part of the thermoelastic solution in Lee and Park [15]. In Weiss and Keer [16] the thermal effects of a periodic array of partially or fully insulated interface cracks were investigated using the complex variable technique. Many investigations of the Mode III interface cracks are devoted to functional materials, i.e. piezo-electro and thermo-electric materials as studied by Govorukha et al. [7]. Li and Tang [17] investigated an antiplane interface crack between two bonded dissimilar layers for piezo-electric materials with intrinsic electro-mechanical coupling behaviors. A screw dislocation interacting with interface cracks in a piezoelectric bimaterial was studied by Wu et al. [18]. A problem for an interface crack with partially electrically conductive crack faces under antiplane mechanical and in-plane electric loadings was solved by Lapusta et al. [19]. A problem for interface cracks between dissimilar magneto-electro-elastic strips under out-of-plane mechanical and in-plane magneto-electrical impacts was investigated by Su et al.

[20] for magneto-electric permeable cracks and for an arbitrary number of interface cracks. The interface crack problem in a layered thermoelectric or metal/thermoelectric material subjected to thermoelectric loadings was studied by Zhang and Wang [21] for electric impermeable and heat semi-permeable crack boundary conditions.

The present paper is devoted to the analysis of solutions for Mode III interface crack problems. The solutions available in the literature for an antiplane shear interface crack are reviewed with respect to the dependence of the main fracture characteristics (i.e. SIFs and energy release rates) and the stresses and strains in the vicinity of the interface crack of the bimaterial mismatch. To demonstrate the presence (or absence) of the influence of bimaterial constants on the solution of the interface crack problem, a problem for an interface crack interacting with internal cracks in a bimaterial under anti-plane shear load is presented. For the formulation of singular integral equations, the method of complex potentials is used. As a part of this solution, the interface crack problem is formulated and an explicit closed-form expression for the SIFs is derived. The approximate closed-form solution is obtained for a special case of the interaction between the interface crack and small cracks. The resulting form of the solution is rather simple, therefore, its structure is easily analyzed. The problem contains the geometric parameters and the parameters of the constituent materials. The study of the influence of these parameters on the basic fracture characteristics (for example, SIFs) is important for predicting the fracture resistance of bimaterials. Theoretical analysis of the explicit form of expressions for the SIFs shows the conditions under which the solution for the interface crack depends on the bimaterial mismatch, and under which it does not depend. Besides, it clarifies what characteristics of the Mode III interface crack problem depend on the bimaterial parameters, and what characteristics do not depend on them. Moreover, a condition when a small internal crack does not influence the interface crack (with respect to the SIF) is derived in explicit form and discussed.

2. Review of available solutions for Mode III interface crack problems

The formulation of the basic equations for the antiplane shear problem will be done in the next section. Here we briefly remind that in this problem the unique non-zero displacement component w_j in the z-direction is a function of the coordinates x and y, i.e. $w_j(x,y)$. The stresses τ_{xzj} and τ_{yzj} are also functions of the coordinates (x,y) [22]. Index $j = 1, 2$ denotes the components for material 1 and 2, respectively.

At the present time, a huge amount of publications are devoted to different aspects of interface crack problems with different results. In the SIFs handbook by Murakami [2] the references [23, 24] state that the SIF Mode III for an interface crack in an infinite bimaterial compound does not depend on the bimaterial shear modulus, see Fig. 1a and b for the geometry. (It should be mentioned, that in other references in [2] the problems are more general, and this particular case for the Mode III interface crack is not evident.)

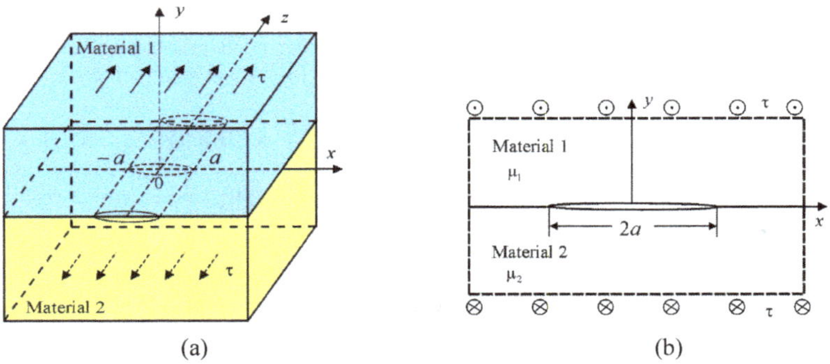

Figure 1. A bimaterial compound with an interface crack under antiplane load: a) three-dimensional representation and coordinate system; b) two-dimensional representation.

In Erdogan and Gupta [23] the closed-form solution was obtained for two bonded wedges with an interface crack, Fig. 2. The form of the singular integral equations obtained in [23] (Eqs. (24) and (26) in [23]) is the same as for a single interface crack in an infinite bimaterial. The unknown function is the derivative of displacement jumps on the interface crack $f(r) = \partial[w_1(r,+0) - w_2(r,-0)] / \partial r$ and this function depends on μ_1/μ_2, where μ_1 and μ_2 are the shear moduli of material 1 and 2, respectively. The SIFs are obtained for the common case of the geometry in Fig. 2. In a special case, when the wedges have equal angles $\theta_1 = \theta_2 = \pi/2$ and the interface crack is normal to the half-space boundary, the SIFs at the interface crack tips were obtained in an explicit form and it was shown that these SIFs are independent on the bimaterial parameters μ_1/μ_2.

In an earlier paper of Erdogan [25] the Mode III interface crack problem was solved for an infinite medium (Fig. 1b) and the variation of displacement jumps $w^+ - w^- = h(x)$ on the interface crack with changing the ratio μ_1/μ_2 was investigated. However, the SIF was not written in an explicit form (see Eq. (40) in [25]) and it was only mentioned that the SIF may depend on μ_1 and μ_2, but without any details for this dependency.

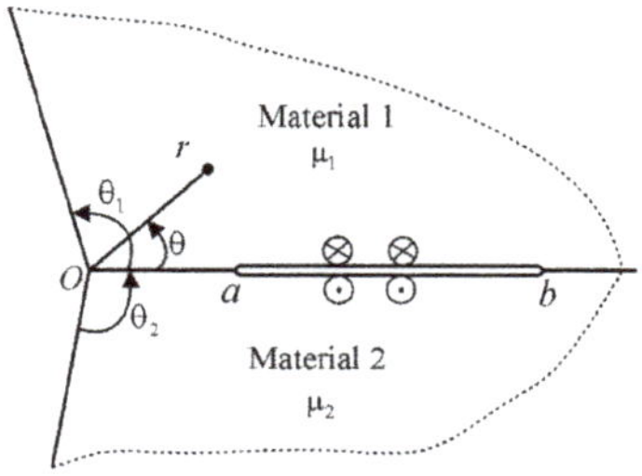

Figure 2. Geometry of two bonded wedges with an interface crack.

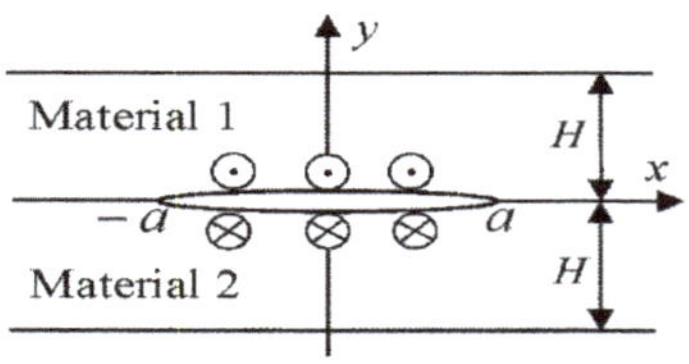

Figure 3. A Mode III crack at the interface between two bonded dissimilar layers.

In another paper by Zhang [24] (cited in [2]), a general solution for the stress intensity factor of a two-layered composite in a rectangular sheet containing a central crack under Mode III loading was obtained by means of the Fourier transform and Fourier series. It was found that the Mode III stress intensity factors at the interface crack tips are independent of the material constants of the rectangular plate. Regarding the full stresses field, it was noted that "the stresses in the dissimilar media will be dependent upon the shear moduli", [24] (page 713). The paper by Delale [26] studies the Mode III fracture of two bonded functionally graded materials with the exponential model for shear moduli. The Mode III SIFs and energy release rates were obtained and a parametric study was done for various values of the ratio of the shear moduli. A particular case – two bonded homogeneous half spaces was also investigated. The conclusion was made that non-homogeneity may have a significant effect on the fracture of bimaterials.

Let us consider further results on Mode III interface crack problems. Li [27] studied a Mode III crack at an interface between two bonded dissimilar elastic layers of equal finite thickness H, Fig. 3. The closed-form solution for a crack of length 2a has been analytically derived for two sets of boundary conditions on the outer surfaces of the layers: stress-free and clamped conditions. Furthermore, the analytical solution for a semi-infinite interface crack in two bonded dissimilar layers has been directly determined. The obtained formulas for SIFs and energy release rate (ERR) are Eqs. (22) and (23) in [27] for free boundaries and for clamped boundaries are Eqs. (37), (38). It was written in [27]: "It is seen from (37) that for the case of clamped boundaries, the SIF depends either on μ_1 or μ_2, whereas the SIF is independent of material constants for the case of stress-free boundaries." A more general case of this problem was considered by Wu and Dzenis [28], where a closed-form solution was obtained for the problem of a Mode III interfacial edge crack between two bonded semi-infinite dissimilar elastic strips ($0 \le x < \infty, -H \le y \le H$), Fig. 4.

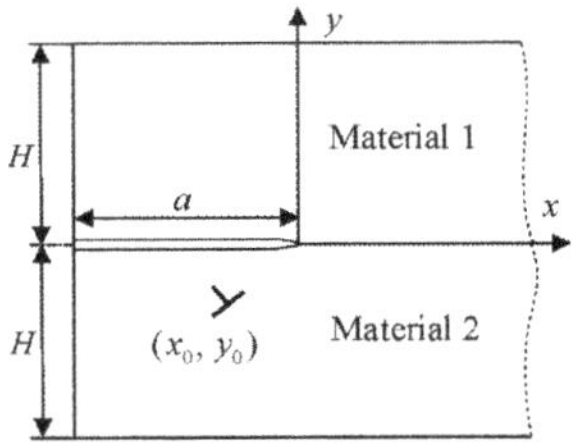

Figure 4. An edge-cracked bimaterial strip under anti-plane singularity located at point (x_0, y_0).

A general out-of-plane displacement potential for the crack interacting with a screw dislocation or a line force is constructed using conformal mapping technique and existing dislocation solutions. Based on this displacement potential, the SIF K_{III} and the energy release rate (ERR and G_{III}) for the interfacial edge crack were obtained explicitly. The solution [28] can be used as the Greens function to analyze interfacial edge cracks subjected to arbitrary anti-plane loadings applied on the crack or edge surfaces or in the interior of the strips. In the general case, the SIFs depend on the bimaterial mismatch. However, the SIF solutions for special cases, namely, for an edge crack loaded with uniform forces on the crack surfaces and for a semi-infinite crack between two bonded strips and loaded with a pair line forces acting on the crack surfaces, are independent on the bimaterial parameters. If a pair of line-forces P acts on the lower and upper crack surfaces at $z_0 = -b \pm 0i$ ($\pm 0i$ denotes the points on the upper and lower surfaces, respectively), the SIF is obtained as

$$K_{III} = \frac{\sqrt{2}P}{\sqrt{\pi a}}\sqrt{\frac{\pi a}{H}\sinh\left(\frac{\pi a}{H}\right)} \Big/ \sqrt{\cosh\left(\frac{\pi a}{H}\right) - \cosh\left(\frac{\pi(a-b)}{H}\right)}, \tag{1}$$

and for a semi-infinite crack ($a \to \infty$) between two bonded strips:

$$K_{III} = P\sqrt{\frac{2}{H}} \Big/ \sqrt{1 - \exp\left(-\frac{\pi b}{H}\right)}, \tag{2}$$

and for the edge crack loaded with the uniform forces P acting on the crack surfaces:

$$K_{III} = \tau\sqrt{\pi a}\frac{2}{\pi}\sqrt{\frac{2H}{\pi a}\tanh\left(\frac{\pi a}{2H}\right)}K\left[\tanh\left(\frac{\pi a}{2H}\right)\right] \tag{3}$$

where $K(\cdot)$ denotes the complete elliptic integral of the first kind.

These particular solutions (1) - (3) coincide with corresponding results obtained in [27].

The energy release rate is evaluated using the relationship

$$G_{III} = \frac{1}{4}\left(\frac{1}{\mu_1} + \frac{1}{\mu_2}\right)K_{III}^2. \tag{4}$$

Thus, this characteristic a-priory (by definition) contains bimaterial parameters.

In a problem for two bonded dissimilar infinite media with a semi-infinite interface crack under mixed-mode conditions (one part of the crack is loaded by stresses, and on the other crack surface the displacements are prescribed) the Mode III SIFs depend on the bimaterial compliance μ_1/μ_2 according to Silvestrov et al. [29]. The other antiplane shear problem of edge and embedded cracks located at the interface of orthotropic–isotropic half-layers under dynamic anti-plane loads was solved by Yousefi et al. [30]. Several examples of edge and embedded cracks were considered and the interaction between these cracks was studied in [30], but the influence of bimaterial combinations on the SIFs was not investigated. In Craciun et al. [31] a mathematical model for the antiplane interface crack in a pre-stressed fiber-reinforced elastic composite is presented, the complex potential for the problem is obtained in the explicit form. This complex potential contains bimaterial parameters, but, unfortunately, the SIFs are not derived in [31]. In Mishuris et al. [32] special models for imperfect interfaces and for interfaces with cracks were presented for the antiplane problem.

Another large class of problems, which is important for the fracture mechanics of composites, is the interaction of interface cracks with other cracks, dislocations and inclusions. Some fundamental solutions for singularities (such as a point force, an edge dislocation) interacting with interfaces and cracks were obtained by Suo for isotropic bimaterial problems [33] and for anisotropic bimaterials [34]. As a part of these studies, the Mode III interface crack problem has been considered. In particular, in [34] the interface crack interacting with a dislocation has been analyzed, and the Mode III solution has been constructed. In this solution, the dependence of the complex potential and SIF upon the bimaterial parameters has been obtained explicitly. The problems considered in Choi et al. [35] are close to the problems in [33, 34]. In [35] the solutions to the interaction problems for singularities, interfaces, and cracks in infinite anisotropic bimaterials were summarized, to be used for the cases of isotropic/isotropic and anisotropic/isotropic bimaterials. The general forms of complex potentials and SIFs were obtained for semi-infinite and finite interface cracks interacting with a dislocation for the plane and antiplane elasticity problems.

A generalized approach for the study of a finite interface crack interacting with near-interface small cracks under antiplane loading was presented by Wang and Meguid [36, 37]. The original problem was decomposed into a number of simpler subproblems each containing either a small crack or the interface crack. Using Fourier transforms, singular integral equations have been constructed for these subproblems. The final result has been obtained by applying a self-consistent iterative procedure to superimpose the different subproblems. Explicit expressions have been obtained for the stresses and SIFs of the interface crack in the presence of microcracks. The stresses and SIFs depend on the bimaterial constants. In Mishuris et al. [38] and Piccolroaz et al. [39], the interaction of an interface crack with small impurities under antiplane load was analyzed on the basis of an asymptotic formula derived by the authors. From the solution for the SIF it is not clear if the Mode III SIF for a single interface crack depends on the bimaterial constants or not because the result is formulated in terms of concentrated forces applied on the crack surfaces. In this non-symmetric case, the SIFs depend on bimaterial constants, as has been shown in other papers [26, 27, 28] and discussed above.

3. Interaction of an interface crack with a system of internal cracks

In this section, the problem for an interface crack interacting with systems of internal cracks is presented with an emphasis on the dependence of the strain-stress state and fracture characteristics on the bimaterial parameters. The formulation for the problem is given in terms of the complex potentials, and the essential parts of the solution are written. Full details are given in Appendix A.

3.1 Formulation of the problem

The geometry of the problem is shown in Fig. 5. The position of cracks (N cracks of length $2a_k$) are determined by their midpoint coordinates z_k^0 ($z_k^0 = x_k^0 + iy_k^0$) and inclination angles α_k in the global Cartesian coordinate system (x, y). A crack of length $2a_0$ is located at the interface. The local coordinate systems (x_k, y_k) are attached to each internal crack. The bimaterial is subjected to antiplane shear loading τ faraway of the interface. Out of the interface crack, $|x| > a_0, y = 0$, the materials are perfectly bonded, i.e. the tractions are equal, and the displacements are equal at the interface.

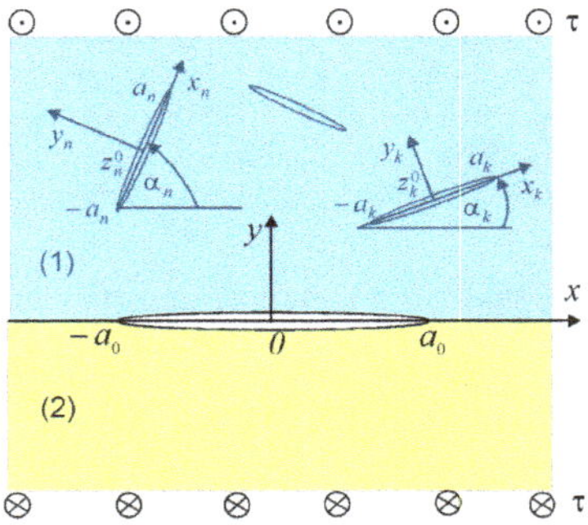

Figure 5. Arbitrarily located internal cracks near an interface crack in a bimaterial under antiplane shear.

In the anti-plane shear problem the constitutive and equilibrium equations are given by (see, e.g. [22, 40]):

$$\tau_{xzj} = \mu_j \frac{\partial w_j}{\partial x}, \quad \tau_{yzj} = \mu_j \frac{\partial w_j}{\partial y} \qquad (j = 1, 2); \tag{5}$$

$$\frac{\partial \tau_{xzj}}{\partial x} + \frac{\partial \tau_{yzj}}{\partial y} = 0, \qquad (j = 1, 2); \tag{6}$$

and the compatibility equations are identically satisfied. In Eqs. (5) and (6) μ_j is the shear modulus, $w_j(x, y)$ is the displacement component and τ_{xzj} and τ_{yzj} are the stresses; as previously, the index j = 1 and 2 refers to materials 1 (upper half-plane) and 2 (lower half-plane), respectively.

Substitution of Eq. (5) into (6) yields Laplace's equation

$$\mu_j \left(\frac{\partial^2 w_j}{\partial x^2} + \frac{\partial^2 w_j}{\partial y^2} \right) \equiv \Delta(\mu_j w_j) = 0. \tag{7}$$

Since a solution to Laplace's equation is a harmonic function, the unknown function $\mu_j w_j$ can be presented in terms of a single analytic function $f_j(z)$ of the complex variable $z = x + iy$ [41-43] as follows:

$$\mu_j w_j = \mathrm{Re}[f_j(z)], \ \tau_{xzj} - i\tau_{yzj} = F_j(z) \tag{8}$$

where

$$F_j(z) = f_j'(z). \tag{9}$$

is the complex potential of the problem. Re[...] denotes the real part of a complex function.

The displacement jumps on the crack faces are introduced as:

$$2\phi_0(x) = w_1(x,0^+) - w_2(x,0^-) \qquad (|x| < a_0)$$

$$2\phi_{nj}(x) = w_{nj}(x,0^+) - w_{nj}(x,0^-) \qquad (|x| < a_n,\, n = 1, 2, ..., N) \tag{10}$$

and their derivatives, respectively, for the interface crack and for the internal cracks are written as:

$$2\phi_0'(x) = \frac{\partial}{\partial x}[w_1(x,0^+) - w_2(x,0^-)] \qquad (|x| < a_0), \tag{11}$$

$$2\phi_{nj}'(x) = \frac{\partial}{\partial x}[w_{nj}(x,0^+) - w_{nj}(x,0^-)] \qquad (|x| < a_n,\, n = 1, 2, ..., N). \tag{12}$$

The physical antiplane shear problem is rewritten in terms of the complex potential F_j, and then the complex potential technique is used (see next sections and Appendix A, Eqs. (A.1)-(A.8)). As soon as the function F_j will be obtained, the displacements and stresses are derived from Eq. (8), and the stress intensity factors are calculated as

$$K_{IIIn}^{\pm} = \mp \lim_{x\to\pm a_n} \sqrt{\pi(a_n^2 - x_n^2)/a_n}\; \tau_{yzj}(x)$$

$$= \mp \lim_{x\to\pm a_n} \sqrt{\pi(a_n^2 - x_n^2)/a_n}\; [-\mathrm{Im}\, F_j(z)] \qquad (j = 1, 2;\, n = 0, 1,.., N) \tag{13}$$

The SIF K_{IIIn}^{+} stands for the right crack tip and K_{IIIn}^{-} – for the left one. Im[…] denotes the imaginary part of a complex function.

The superposition principle is used for the formulation of the basic relations for the problem of interaction between an interface crack and internal cracks. The examples of these studies for the crack-singularity interaction problems can be found in [33-35]. Due to this principle, the main problem can be decomposed into a sequence of sub-problems, each of which contains a single crack. Besides, the solution to the problem for a crack (internal or interface) in a bimaterial under an applied load is the sum of two sub-problems: (1) the infinite uncracked bimaterial under the influence of the far-loading, and (2) the bimaterial with the crack with the respective load on the crack surfaces. The solution for the first problem does not contain a singularity, meanwhile, the second problem contains a crack, i.e. the singularity. The second problem is important from the viewpoint of fracture mechanics. The solution to the first problem is used in the second one and is also useful for determining the full elastic field in the bimaterial with cracks. It should be noted that this decomposition does not mean that we can solve all these sub-problems independently (except some special cases). There is a coupling between the sub-problems, and this is shown below when constructing integral equations of the problem.

Due to the superposition principle, the load at infinity reduces to the load on the crack surfaces. The transmission conditions at the interface and the boundary conditions on cracks are written in terms of displacements using relation (1) as follows:

at the interface the displacements are equal and the tractions are equal, i.e.

$$w_1(x,0^+) = w_2(x,0^-),\ \mu_1 \frac{\partial w_1(x,0^+)}{\partial y} = \mu_2 \frac{\partial w_2(x,0^-)}{\partial y},\quad |x| \ge a_0,\ y=0; \tag{14}$$

on the interface crack

$$\mu_1 \frac{\partial w_1(x,0^+)}{\partial y} = \mu_2 \frac{\partial w_2(x,0^-)}{\partial y} = \tau_0(x),\quad |x| \le a_0,\ y=0; \tag{15}$$

on the internal cracks

$$\mu_j \frac{w_{jn}(x_n,0^\pm)}{\partial y_n} = \tau_n(x_n),\ |x_n| \le a_n,\ n=0,1,..,N; \tag{16}$$

at the crack tips

$$w_1(\pm a_0,0^+) = w_2(\pm a_0,0^-),\ w_{jn}(\pm a,0^+) = w_{jn}(\pm a,0^-); \tag{17}$$

and at infinity

$$\mu_1 \frac{\partial w_1(x,y)}{\partial y} = \mu_2 \frac{\partial w_2(x,y)}{\partial y} = 0,\ x^2+y^2 \to \infty. \tag{18}$$

Here τ_0 and τ_n are known tractions on the crack surfaces

$$\tau_0(x) = -\mu_j \frac{\partial w_j^0}{\partial y};\ \tau_n(x_n) = -\mu_j \left.\frac{\partial w_{nj}^0}{\partial y_n}\right|_{y_n=0} = -\mu_j\left(-\frac{\partial w_{nj}^0}{\partial x}\sin\alpha_n + \frac{\partial w_{nj}^0}{\partial y}\cos\alpha_n\right)\Bigg|_{y_n=0} \tag{19}$$

which are calculated for the bimaterial without cracks and are taken with the opposite sign. In Eqs. (14)-(19) the signs "+" and "–" denote the limiting values of the functions on the upper and lower surfaces of the crack or at the interface. w_j^0 and w_{nj}^0 are displacements on the crack lines, which are known from the solution of the uncracked bimaterial under the influence of the far-loading.

3.2 Interface crack

The procedure for obtaining the complex potential for an interface crack is similar as described by Suo [33] and also close to the technique used in [40] for homogeneous materials. This method was used to some extent in the stationary thermal conductivity problem for a bimaterial with cracks [44, 45]. The procedure for obtaining the solution is based on the use of the boundary conditions on the interface crack and the continuity conditions on the crack-free parts of the interface ($|x| > a_0, y=0$) rewritten via the complex potential F_j, the details are given in Appendix A, Eqs. (A.9)-(A.14).

The singular integral equation for the unknown ϕ_0' (11) is written as

$$\int_{-a_0}^{a_0} \frac{\phi_0'(t)}{t-x} dt = \frac{1}{2}\pi \frac{\mu_1+\mu_2}{\mu_1\mu_2}\tau_0(x),\quad (|x| \le a_0) \tag{20}$$

with the condition

$$\int_{-a_0}^{a_0} \phi_0'(t)dt = 0. \tag{21}$$

Eq. (21) is the integral form of single-valuedness condition for the displacements at the crack tips. Eq. (20) is necessary for further construction of the system of equations for a system of multiple cracks.

The complex potential F_0 for the interface crack problem is derived as

$$F_j(z) = F_0(z) = \frac{\mu_1\mu_2}{\mu_1+\mu_2}\frac{2}{\pi i}\int_{-a_0}^{a_0}\frac{\phi_0'(t)}{t-z}dt. \tag{22}$$

After determination of ϕ_0' the stresses are obtained by Eq. (8) and the SIFs by Eq. (13).

The solution of the singular integral equation (20) with condition (21) is [41, 42, 46]

$$\phi_0'(x) = -\frac{\mu_1+\mu_2}{2\mu_1\mu_2}\frac{1}{\pi\sqrt{a_0^2-x^2}}\int_{-a_0}^{a_0}\frac{\tau_0(t)\sqrt{a_0^2-t^2}}{t-x}dt. \tag{23}$$

The integral in Eq. (23) is calculated explicitly if τ_0 is a polynomial function. Thus, $F_j(z)$ can be derived in a closed-form as follows, Eqs. (22) and (23) yield

$$F_j(z) = F_0(z) = \frac{1}{\pi i\sqrt{z^2-a_0^2}}\int_{-a_0}^{a_0}\frac{\tau_0(t)\sqrt{a_0^2-t^2}}{t-z}dt,$$

and then for $\tau_0 = -\tau = const$ $F_j(z)$ is written as

$$F_j(z) = i\tau\left(1 - z/\sqrt{z^2-a_0^2}\right). \tag{24}$$

Using Eq. (8) the stress distribution is derived as

$$\tau_{xzj} = \mathrm{Re}[F_j(z)] = -\tau\,\mathrm{Im}\left[-z/\sqrt{z^2-a_0^2}\right], \tau_{yzj} = -\mathrm{Im}[F_j(z)] = -\tau\,\mathrm{Re}\left[1-z/\sqrt{z^2-a_0^2}\right] \tag{25}$$

The stress intensity factors at the interface crack tips are obtained as

$$K_{III0}^{\pm} = \mp\lim_{x_0\to\pm a_0}\sqrt{\pi(a_0^2-x_0^2)/a_0}\;\tau_{yz0}(x) = \sqrt{\pi a_0}\;\tau \tag{26}$$

As seen from Eq. (23) the displacement jumps on the interface crack depend on the bimaterial constants. The independence of the SIF of the bimaterial parameters is shown in Eq. (26). The expression for the SIF (26) is the same as for the SIF for a crack in a homogeneous medium under antiplane shear loading.

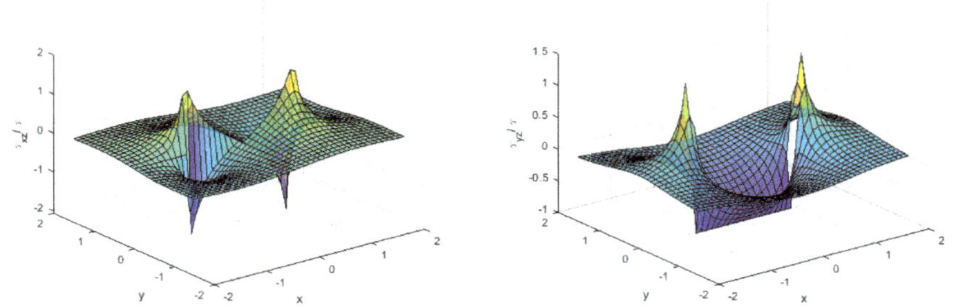

(a) Stresses τ_{xz}/τ (b) Stresses τ_{yz}/τ

Figure 6. Distribution of the shear stresses (a) τ_{xz}/τ and (b) τ_{yz}/τ near a crack located at x = [-1, 1] and y =0.

Fig. 6 shows the distribution of shear stresses (Eqs. (8) and (25)) in the vicinity of a crack with evident singularity behavior around the crack tips.

3.3 An internal crack in a bimaterial

The details of the formulation of the complex potential of this part of the solution are written in Appendix A, Eqs. (A.15)-(A.19). The solution to this subproblem is presented in Appendix, Eqs. (A.15)-(A.19). The complex potential for an internal crack in the bimaterial with the undamaged interface was obtained using the technique given in [33, 34] and is written as

$$F(z)=\begin{cases} F^0(z)-M\overline{F^0}(z), & z\in(1) \\ (1+M)F^0(z), & z\in(2) \end{cases}, \text{ where } M=\frac{\mu_2-\mu_1}{\mu_2+\mu_1}(-1\le M\le 1) \quad (27)$$

F^0 is the complex potential for the problem for the same crack in a homogeneous material. It is assumed that the crack is located in material 1, and in this case, F^0 contains the shear modulus μ_1. In Suo [33, 34] the potential for a singularity in an infinite homogeneous medium is used for the construction of the complex potential for the same singularity in a bimaterial compound. The singularity can be of any type, e.g. a dislocation, a point force or others.

The complex potential $F(z)$ (27) coincides with the potential in Eq. (3.10) [35] if we rewrite the potential for a singularity located in the material 2 and take the bimaterial parameter M with the negative sign.

3.4 Interaction of internal cracks with an interface crack

Based on the superposition principle and using the solutions given above (the interface crack problem and an internal crack in bimaterial), the system of singular integral equations for interacting cracks is obtained as

$$\int_{-a_0}^{a_0}\frac{\phi_0'(t)}{t-x}dt+\sum_{k=1}^{N}\int_{-a_k}^{a_k}P_{0k}(t,x)\phi_k'(t)dt=\frac{1}{2}\pi\tau_0(x)\frac{\mu_1+\mu_2}{\mu_1\mu_2}, \quad (|x|<a_0), \quad (28)$$

$$\mu_1 \int_{-a_n}^{a_n} \frac{\phi_n'(t)}{t-x} dt + \frac{2\mu_1\mu_2}{\mu_1+\mu_2} \int_{-a_0}^{a_0} P_{n0}(t,x)\phi_0'(t)dt + \mu_1 \sum_{\substack{k=1 \\ k\neq n}}^{N} \int_{-a_k}^{a_k} P_{nk}(t,x)\phi_k'(t)dt = \pi\tau_n(x)$$

$$|x| < a_n,\ n = 1, 2, \ldots, N), \tag{29}$$

with the condition

$$\int_{-a_n}^{a_n} \phi_n'(t)dt = 0 \qquad (n = 0, 1, 2, \ldots, N), \tag{30}$$

and with the regular kernel, containing the geometry of the problem,

$$P_{nk}(t,x) = \mathrm{Re}[\frac{e^{i\alpha_n}}{te^{i\alpha_k} + z_k^0 - xe^{i\alpha_n} - z_n^0}], \quad (n,\ k = 0,1, \ldots, N). \tag{31}$$

In Appendix A, Eqs. (A.20)-(A.30) provide the complete information for obtaining Eqs. (28)-(31).

In the general case, the equations can be solved numerically by a method based on the application of Chebyshev polynomials, see [40, 47]. In [48] the numerical solution to the problem was provided. The following section presents an approximate analytical solution for a particular case.

4. Parameters of materials

For a particular case, but important for applications where the interface crack is much larger than other internal cracks, the solution of the singular integral equations (28)-(30) has been obtained in a closed asymptotic form. Suppose that the sizes of internal cracks in the bimaterial is much smaller than the size of the interface crack, i.e. $2a_k << 2a_0$, and, besides, $a_k = a$ ($k = 1,2, \ldots, N$). In this case, we can choose $\lambda = a / a_0$ as a small parameter. Eqs. (28) and (29) with condition (30) are rewritten in dimensionless form. Non-dimensional coordinates χ and θ are defined as $x = \chi a_k$ and $t = \theta a_k$, respectively, and the non-dimensional coordinates of crack centers are $\delta_k = z_k^0 / a_0$. The solution to Eqs. (28) and (29) is sought as a power series with respect to λ as

$$\tilde{\phi}_0'(\chi) = \sum_{p=0}^{\infty} \tilde{\phi}_{0p}'(\chi)\lambda^p, \quad \tilde{\phi}_n'(\chi) = \sum_{p=0}^{\infty} \tilde{\phi}_{np}'(\chi)\lambda^p. \tag{32}$$

$\tilde{\phi}_n'$ and $\tilde{\phi}_{np}'$ denote the dimensionless functions for ϕ_n' and ϕ_{np}'.

The solution is obtained up to λ^2, which takes into account the interaction between the interface crack and each of the small internal cracks and does not include mutual interactions between small cracks. The description of this technique is given in Appendix B, Eqs. (B.1)-(B.14). For the functions $\tilde{\phi}_n'$, the following expressions are derived

$$\tilde{\phi}_0'(\chi) = \tilde{\phi}_{00}'(\chi) + \lambda^2\tilde{\phi}_{02}'(\chi) = \frac{\mu_1+\mu_2}{2\mu_1\mu_2}\frac{\chi}{\sqrt{1-\chi^2}}\tilde{\tau}_0 - \\ -\frac{\lambda^2}{2}\frac{1}{\mu_1}\frac{1}{\sqrt{1-\chi^2}}\sum_{k=1}^{N}\mathrm{Re}[e^{i\alpha_k}\frac{1-\chi\delta_k}{(\chi-\delta_k)^2\sqrt{\delta_k^2-1}}]\{\tilde{\tau}_k - \tilde{\tau}_0\,\mathrm{Re}[e^{i\alpha_k}(1-\frac{\delta_k}{\sqrt{\delta_k^2-1}})]\}, \tag{33}$$

$$\tilde{\phi}_n'(\chi) = \lambda\tilde{\phi}_{n1}'(\chi) + \lambda^2\tilde{\phi}_{n2}'(\chi) = \frac{1}{\mu_1}\left\{\lambda\frac{\chi}{\sqrt{1-\chi^2}}\{\tilde{\tau}_n - \tilde{\tau}_0\,\mathrm{Re}[e^{i\alpha_n}(1-\frac{\delta_n}{\sqrt{\delta_n^2-1}})]\}\right. \\ \left. +\lambda^2\frac{-\chi^2+\frac{1}{2}}{\sqrt{1-\chi^2}}\tilde{\tau}_0\,\mathrm{Re}[\frac{e^{2i\alpha_n}}{(\delta_n^2-1)^{3/2}}]\right\} \quad (n=1,2,\ldots,N) \tag{34}$$

and, then, using Eq. (13) with accounting (22) and (8), the stress intensity factors (SIFs) are obtained as

$$K_{III0}^{\pm} = \tau\sqrt{\pi a_0}\left\{1+\frac{\mu_1\mu_2}{\mu_1+\mu_2}\frac{\lambda^2}{\mu_1}\sum_{k=1}^{N}\mathrm{Re}\left[\frac{e^{i\alpha_k}}{(\delta_k\mp 1)\sqrt{\delta_k^2-1}}\right]\right. \\ \left.\left(\cos\alpha_k - \mathrm{Re}\left[e^{i\alpha_k}-\frac{\delta_k e^{i\alpha_k}}{\sqrt{\delta_k^2-1}}\right]\right)\right\} \tag{35}$$

or

$$K_{III0}^{\pm} = \tau\sqrt{\pi a_0}\left\{1+\frac{1}{1+\mu_1/\mu_2}\lambda^2\sum_{k=1}^{N}\mathrm{Re}\left[\frac{e^{i\alpha_k}}{(\delta_k\mp 1)\sqrt{\delta_k^2-1}}\right]\mathrm{Re}\left[\frac{\delta_k e^{i\alpha_k}}{\sqrt{\delta_k^2-1}}\right]\right\}. \tag{36}$$

Solution (35) or (36) consists of the term for a single interface crack and the second term that takes into account the influence of other internal small cracks on the interface crack; the interaction between small cracks is not taken into account [49, 50]. The influence term contains not only the parameters of geometry (the non-dimensional coordinate of the crack centers $\delta_k = z_k^0/a_0$ and inclination angles α_k), but also the bimaterial constants μ_1/μ_2. For a homogeneous material ($\mu_1=\mu_2$), Eqs. (35) and (36) coincide with the expression obtained earlier in [51].

For internal cracks, the SIF is written as

$$K_{IIIn}^{\pm} = \frac{\sqrt{\pi a_n}\tau}{\mu_1}\{\lambda\{\cos\alpha_n - \mathrm{Re}[e^{i\alpha_n}(1-\frac{\delta_n}{\sqrt{\delta_n^2-1}})]\}\mp\lambda^2\,\mathrm{Re}[\frac{e^{2i\alpha_n}}{(\delta_n^2-1)^{3/2}}]\}, \\ n=1,2,\ldots,N. \tag{37}$$

5. Results and discussion

5.1 Comparison with a closed-form solution

It is interesting to note that solution (36) has a form similar to the form of a closed analytical solution obtained by Wang and Schiavone [52] for the anti-plane problem for an arbitrary shape Eshelby inclusion embedded in a bimaterial with a semi-infinite interface crack. The bimaterial coefficient in the solution in [52] is the same as in Eq. (36). Besides, a condition was obtained when the Eshelby inclusion will not affect the SIF at the interface crack tip. Under this condition "the inclusion is essentially invisible to the crack" [52]. A similar condition in which a small crack has no influence on an interface crack is easily obtained from (28) and is written as

$$F = \mathrm{Re}\left[\delta_k e^{i\alpha_k} / \sqrt{\delta_k^2 - 1}\right], \quad \mathrm{Re}\left[\delta_k e^{i\alpha_k} / \sqrt{\delta_k^2 - 1}\right] = 0. \tag{38}$$

In the crack interaction problems, this phenomenon is called the "neutral" crack or the "neutral" position of the crack [39, 49]. From expression (38) it is evident that this "invisibility" of cracks depends only on the geometry (on the midpoint coordinates $\delta_k = z_k^0 / a_0$ of cracks and the inclination angles α_k) and does not depend on the bimaterial constants.

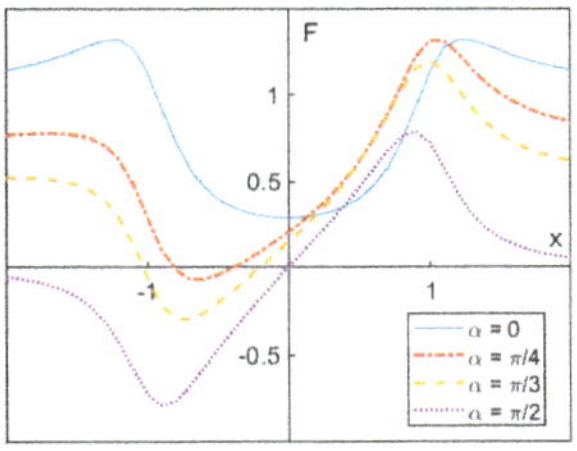

(a) y = 0.3

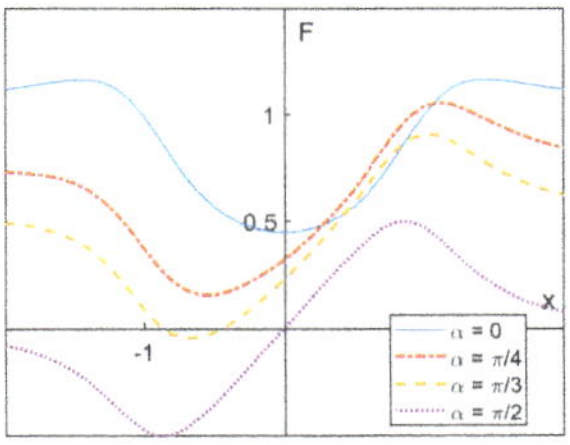

(b) y = 0.5

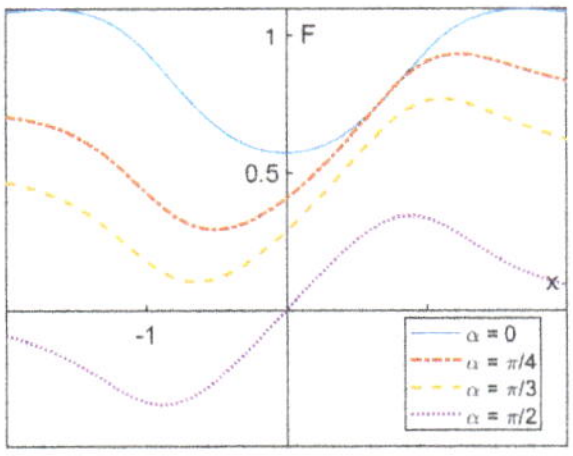

(c) y = 0.7

Figure 7. The influence function $F = \mathrm{Re}[\delta_k e^{i\alpha_k} / (\delta_k^2 - 1)^{1/2}]$ *vs. the coordinate x* (x_k^0/a_0 = Re[δ_k]) *for a fixed coordinate y* (y_k^0/a_0 = Im[δ_k]) *for several values of y =0.3, 0.5 and 0.7, figures a, b and c, respectively, and for the angles α = 0, π/4, π/3, π/2.*

Fig. 7 shows the dependence of the influence function $F = \text{Re}[\delta_k e^{i\alpha_k} / (\delta_k^2 - 1)^{1/2}]$ on the coordinate x for a fixed coordinate y (with values $y = 0.3$, 05 and 0.7) and for inclination angles $\alpha = 0, \pi/4, \pi/3, \pi/2$. For simplicity the variable x denotes $x_k^0/a_0 = \text{Re}[\delta_k] = \text{Re}[x_k^0 + iy_k^0]/a_0$, and y denotes $y_k^0/a_0 = \text{Im}[\delta_k] = \text{Im}[x_k^0 + iy_k^0]/a_0$ in Fig. 7. The plots graphically demonstrate how the zeros of the function F depend on α and the midpoint coordinates. For $\alpha = \pi/2$ condition (38) is satisfied for all cracks located on the imaginary axis, i.e. for $\delta_k = iy_k^0/a_0$, while for other inclination angles the zeros depend on δ_k, and for $\alpha = 0$ the influence function has no zeros.

5.2 Structure of the solution

The explicit formula (36) allows us to investigate the influence of the parameters of the problem on the stress intensity factors. Results for SIFs are presented in non-dimensional form

$$K_{III}^{*\pm} = K_{III0}^{\pm} / K^0, \quad K^0 = \tau\sqrt{\pi a_0} \tag{39}$$

It is known that cracks amplify or shield the main crack. For the normalized SIF $K_{III}^* > 1$ denotes the amplification effect, and $K_{III}^* < 1$ – the shielding effect and $K_{III}^* = 1$ – no influence of other cracks on the interface crack.

The first term in Eq. (36) stands for the interface crack, and the other term determines the influence of internal cracks. This influence function depends on the bimaterial parameter, which we denote by

$$F(\mu_1/\mu_2) = \frac{\mu_2}{\mu_1 + \mu_2} = \frac{1}{1 + \mu_1/\mu_2}, \tag{40}$$

and also depends on the geometry, i.e. the midpoint coordinates δ_k and the inclination angles α_k of the cracks. Denote this part of the influence function as

$$S_k(\delta_k, \alpha_k) = \text{Re}\left[\frac{e^{i\alpha_k}}{(\delta_k \mp 1)\sqrt{\delta_k^2 - 1}}\right] \text{Re}\left[\frac{\delta_k e^{i\alpha_k}}{\sqrt{\delta_k^2 - 1}}\right]\Bigg\}. \tag{41}$$

The values of $F(\mu_1/\mu_2)$, Eq. (40), are always positive, and, accordingly, $F(\mu_1/\mu_2)$, affects only the magnitude of amplification-shielding and does not affect this phenomenon itself (the transition from amplification to shielding). The values of the geometrical part, S_k, Eq. (41), may be positive or negative, and, therefore, they influence the transition from amplification to shielding. The behavior of the function $F(\mu_1/\mu_2) = 1/(1 + \mu_1/\mu_2)$ (Eq. 40) with respect to the variable μ_1/μ_2 (the ratio of shear moduli) is shown in Fig. 8. The positive function F decreases continuously with increasing μ_1/μ_2.

The dependence on coordinates δ_k means the dependence of the function S_k on the distances between the cracks. In the polar coordinate system (ρ_k, β_k) with the origin at the right tip of the interface crack, the non-dimensional coordinate δ_k can be written as

$$\delta_k = \rho_k e^{i\beta_k} + 1$$

and the distances between the interface crack tip and the midpoint coordinates of small cracks is determined by ρ_k. An analysis of the behavior of the function S_k (Eq. 41) with respect to ρ_k shows that this function S_k decreases with the distance ρ_k following the inverse-square law. A detailed analysis of the effect of ρ_k on the behavior of the stress intensity factors can be found in [53] for the plane problem for the macro-microcrack interaction in a homogeneous material.

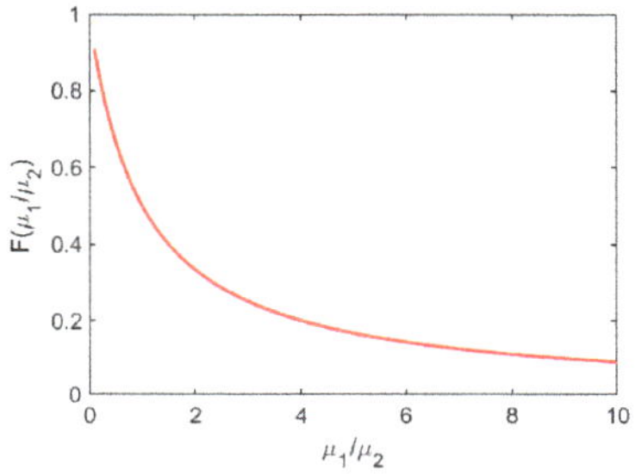

Figure 8. Function $F(\mu_1 / \mu_2) = 1/(1 + \mu_1 / \mu_2)$ of the variable $\mu_1 / \mu_2 = \{0.1 - 10\}$

5.3 Influence of systems of small cracks on an interface crack

Formula (36) can be used for any number of internal cracks of arbitrary location and orientation, see Fig. 9. The only limitation is that the cracks should not intersect. In addition, for better accuracy of the solution, the distance between the crack tips should be greater than the half-length of a small internal crack.

Consider the influence of systems of microcracks on the interface crack. As illustrative examples the following geometries are studied: a non-symmetrically located microcrack system ahead of the interface crack (Fig 10a), a column of microcracks directly above the interface crack (Fig. 10b) and microcracks above the interface crack (Fig. 10c). Fig. 10 shows both the microcrack location patterns and the midpoint coordinates for microcracks.

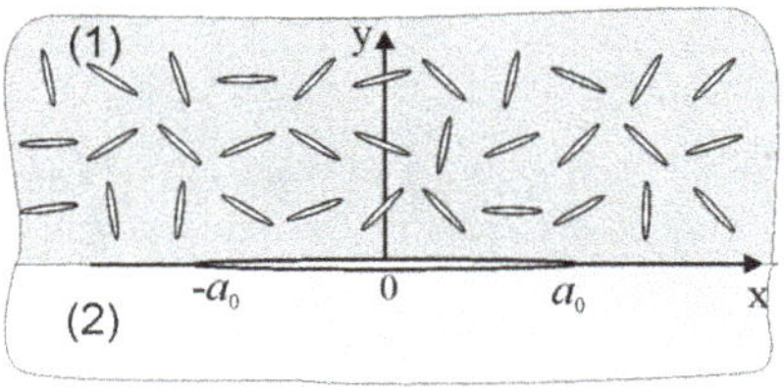

Figure 9. General scheme of the location of the interface crack and microcracks.

Materials Research Foundations **159** (2024) https://doi.org/10.21741/9781644902950

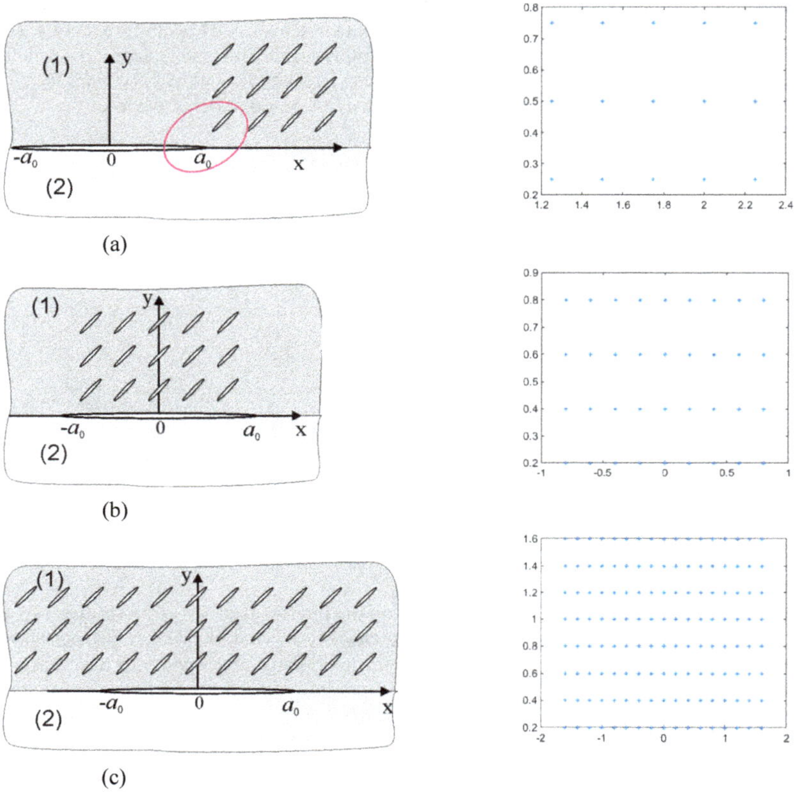

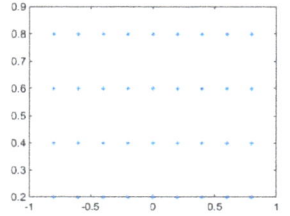

Figure 10. Schemes of locations of microcrack systems in material 1 and their midpoint coordinates: (a) a non-symmetrically located system ahead of the interface crack; (b) a system of microcracks located directly above the interface crack; (c) a system of microcracks located above the interface crack.

Materials Research Foundations **159** (2024) https://doi.org/10.21741/9781644902950

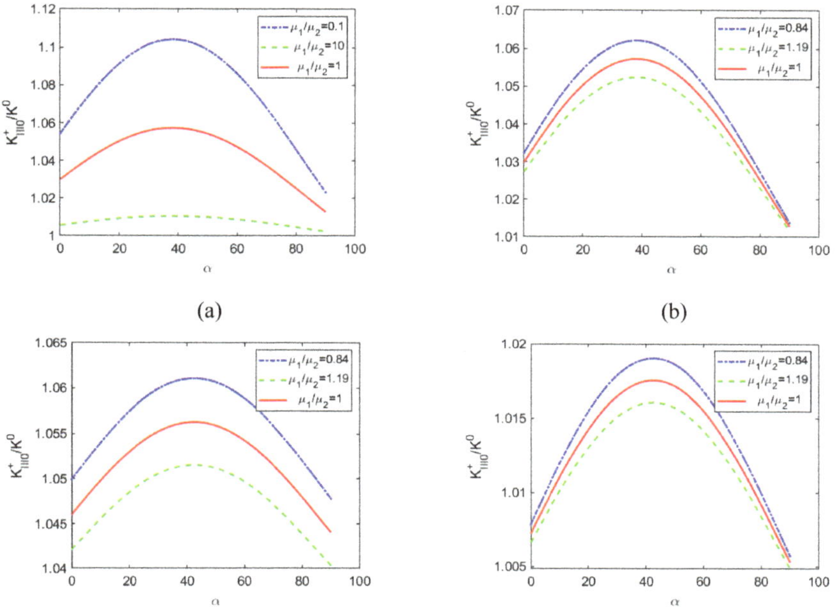

(a) (b)

(c) (d)

Figure 11. Nondimensional SIFs for the right interface crack tip vs. the inclination angle α for the system of microcracks depicted in Fig. 10a: (a) for $\mu_1/\mu_2 = 0.1$, 10 and 1.0; (b) for $\mu_1/\mu_2 = 0.844$, 1.19 and 1.0; (c) the angle α for one crack is varied (the crack is shown in a red circle in Fig. 10a) and fixed with the value $\alpha = 45°$ for other cracks, μ_1/μ_2 is the same as in (b); (d) for one microcrack with the midpoint coordinate (1.25, 0.25) (the crack is shown in a red circle in Fig. 10a), and μ_1/μ_2 is the same as in (b).

Figs. 11-13 show the variation of the normalized SIFs K_{III}^+ for the right interface crack tip with the inclination angle α. It is assumed that $\alpha_k = \alpha$. The interface crack size is $2a_0 = 2$ and the internal cracks have the same sizes $2a_k = 0.2$, i.e. $\lambda = 0.1$. The shear moduli of the used materials Al_2O_3 and SiC are 151 and 179 (MPa), and, as a result, the ratio $\mu_1/\mu_2 = 0.844$ or 1.19. One case for artificially selected material parameters $\mu_1/\mu_2 = 0.1$ and 10 is presented in Figs. 11-13, as well as one result for a homogeneous material where $\mu_1/\mu_2 = 1$.

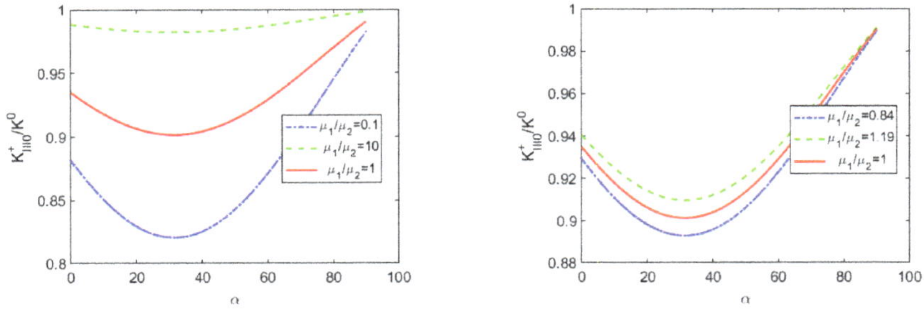

(a) (b)

Figure 12. Nondimensional SIFs for the right interface crack tip vs. the inclination angle α for the system of microcracks depicted in Fig. 10b: (a) for $\mu_1/\mu_2 = 0.1$, 10 and 1.0; (b) for $\mu_1/\mu_2 = 0.844$, 1.19 and 1.0.

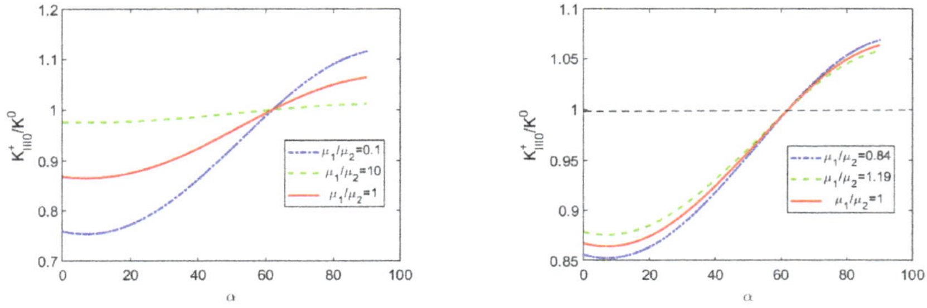

(a) (b)

Figure 13. Nondimensional SIFs for the right interface crack tip vs. the inclination angle α for the system of microcracks depicted in Fig. 10c: (a) for $\mu_1/\mu_2 = 0.1$, 10 and 1.0; (b) for $\mu_1/\mu_2 = 0.844$, 1.19 and 1.0.

Influence of geometry. It can be seen from Figs. 11-13 that there is a strong influence of the inclination angles and the microcrack patterns on the SIFs. The microcrack system ahead of the right crack tip (Fig. 10a) always amplifies the interface crack. The magnitude of SIFs in Fig. 11 is greater than 1 for all angles α and material parameters μ_1/μ_2. The graphs for SIFs for the system of cracks (Fig. 11b) and for one crack (Fig. 11d) show similar behavior, and the maximum difference between the values of SIFs is about 4%.

Fig. 11c presents the results for the case, when the angle α for one crack is varied (the crack is shown in a red circle in Fig. 10a) and other angles α are fixed with the value $\alpha = 45°$. Comparison of the SIFs with the results in Fig. 11b shows that the SIFs have the same maximal values (for $\alpha = 45°$), but the minimum values for the case in Fig. 11c are higher. In all four cases in Fig. 11 the maximum SIFs are at $\alpha = 45°$ for all material parameters.

From previous investigations [50, 54] and from a physical point of view it is known that nearby small cracks have a major effect on the stress intensity factors of a macrocrack. Fig. 11 demonstrates this phenomenon.

Another crack pattern, the system of microcracks located directly above the interface crack, Fig. 10b, causes shielding for all α and μ_1/μ_2, see Fig. 12. The SIFs are less than 1 in this case. The minimum value of the SIF is approximately at $\alpha = 30°$.

The results for the SIFs for the system of cracks above the interface, as shown in Fig. 10c, are presented in Fig. 13. In this case, the amplification/shielding depends on the inclination angle α, that is, for $0 < \alpha < 60°$ the shielding is observed, and for $60° < \alpha < 90°$ – amplification for all ratios of material parameters μ_1/μ_2. For $\alpha = 60°$, we have a neutral position, when this system of cracks does not affect the SIF near the right interface crack tip.

Influence of bimaterial parameters. As was shown above, the effect of material parameters on the SIFs is determined by the behavior of the function $F(\mu_1/\mu_2)$, Fig. 8. For the crack system in Fig. 10a, the larger the ratio μ_1/μ_2 – the smaller the SIF value. This effect is more pronounced for the artificially selected parameters as shown in Fig. 11a. The reverse behavior with respect to the previous case is observed for the crack system with shielding effect (Fig. 12 for the system depicted in Fig. 10b). In this case, the larger the ratio μ_1/μ_2 – the larger the values for the SIFs. The effect of μ_1/μ_2 on the SIFs for the crack system in Fig. 10c is a combination of the two previous cases and depends on the angle of α, as can be seen in Fig. 13.

Conclusions

Summarizing the presented overview the following can be stated. The solution to the classical Mode III interface problem (i.e. a crack is located at an interface between two perfectly bonded isotropic materials with different material properties and subjected to self-equilibrium antiplane shear loading) with respect to the stress intensity factors does not contain bimaterial parameters, as it was proved in earlier works and cited in SIFs handbooks.

When the interface bond outside of the location of the crack is not ideal (e.g. special continuity conditions at the interface or a thin interface layer), or the crack faces or the outer boundaries of bimaterials are subjected to mixed boundary conditions, or the materials are non-homogeneous (e.g. composites, in particular, functionally graded materials, or materials contain defects, cracks or dislocations) the SIF solution contains the bimaterial parameters. An example of such dependence was demonstrated for the solution for the interface crack interacting with internal small cracks. The resulting form of this SIF solution (36) is very simple and its structure is easily examined. This solution, when an interface crack is much larger in size than internal cracks, is presented in an explicit analytical form. The structure of the solution consists of two terms: one for a single interface crack and another for the "influence function" of internal cracks on the interface crack. The effect of the bimaterial parameter is only revealed in the second term, which is responsible for the influence of the internal cracks.

From a physical point of view, it is obvious that the ideal classical case of an interface crack is rarely realized. Consequently, for a physically reasonable description of the stress-strain state in the vicinity of a Mode III interface crack, other fracture characteristics should be used, for example, the energy release rate, which contains material parameters.

Acknowledgments

The authors V. Petrova and S. Schmauder would like to acknowledge the financial support of the German Research Foundation under Grant Schm 746/139-2 and 746/209-1.

References

[1] A.G. Metcalfe, Interfaces in Metal Matrix Composites, Acad. Pr., New York, 1974.

[2] Y. Murakami, Stress Intensity Factors Handbook, Pergamon Press, Oxford, 1987.

[3] R. Talreja, J. Varna, Modeling Damage, Fatigue and Failure of Composite Materials, Woodhead Publishing Limited, Cambridge, 2016.

[4] L. Zhuang, R. Talreja, J. Varna, Transverse crack formation in unidirectional composites by linking of fibre/matrix debond cracks, Compos. Part A: Appl. Sci. Manuf. 107 (2018) 294-303. https://doi.org/10.1016/j.compositesa.2018.01.013

[5] R. Krueger, K. N. Shivakumar, I.S. Raju, Fracture mechanics analyses for interface crack problems - A review, 54th AIAA/ASME/ASCE/AHS/ASC Structures, Structural Dynamics, and Materials Conference (2013). https://doi.org/10.2514/6.2013-1476

[6] L. Banks-Sills, Interface fracture mechanics: theory and experiment, Int. J. Fracture 191 (2015) 131-146. https://doi.org/10.1007/s10704-015-9997-1

[7] V. Govorukha, M. Kamlah, V. Loboda, Y. Lapusta, Interface cracks in piezoelectric materials, Smart Mater. Struct. 25 (2016) 023001. https://doi.org/10.1088/0964-1726/25/2/023001

[8] J. Dundurs, Effect of elastic constants on stress in a composite under plane deformation, J. Compos. Mater. 1 (1967) 310-322. https://doi.org/10.1177/002199836700100306

[9] J. Dundurs, Discussion: "Edge-Bonded dissimilar orthogonal elastic wedges under normal and shear loading" (Bogy, D. B., 1968, ASME J. Appl. Mech., 35, 460–466), J. Appl. Mech 36(3) (1969) 650-652. https://doi.org/10.1115/1.3564739

[10] T. Suga, G. Elssner, S. Schmauder, Composite parameters and mechanical compatibility of material joints, J. Compos. Mat. 22 (1988) 917-934. https://doi.org/10.1177/002199838802201002

[11] M. Paggi, A. Carpinteri, On the stress singularities at multimaterial interfaces and related analogies with fluid dynamics and diffusion, Appl. Mech. Rev. 61 (1-6) (2008) 02080110-02080122. https://doi.org/10.1115/1.2885134

[12] G.L. Golewski, T. Sadowski, A study of Mode III fracture toughness in young and mature concrete with fly ash additive, Solid State Phenomena 254 (2016) 120-125. https://doi.org/10.4028/www.scientific.net/SSP.254.120

[13] A.T. Zehnder, Fracture Mechanics, Lecture Notes in Applied and Computational Mechanics 62, Springer Science + Business Media B.V., 2012. https://doi.org/10.1007/978-94-007-2595-9

[14] V. Tvergaard, J.W. Hutchinson, Mode III effects on interface delamination, J. Mech. Phys. Solids 56 (2008) 215-229. https://doi.org/10.1016/j.jmps.2007.04.013

[15] K.Y. Lee, S.J. Park, Thermal stress intensity factors for partially insulated interface crack under uniform heat flow, Eng. Fract. Mech. 50 (4) (1995) 475-482. https://doi.org/10.1016/0013-7944(94)00243-B

[16] N.A. Weiss, L.M. Keer, Heat flow disturbed by periodic array of insulated interface cracks, Mech. Res. Commun. 54 (2013) 50-55. https://doi.org/10.1016/j.mechrescom.2013.09.009

[17] X.-F. Li, G.J. Tang, Antiplane interface crack between two bonded dissimilar piezoelectric layers, Eur. J. Mech. A/Solids 22 (2003) 231-242. https://doi.org/10.1016/S0997-7538(03)00028-7

[18] X.-F. Wu, S. Cohn, Y.A. Dzenis, Screw dislocation interacting with interfacial and interface cracks in piezoelectric bimaterials, Int. J. Eng. Sci. 41 (2003) 667-682. https://doi.org/10.1016/S0020-7225(02)00155-6

[19] Y. Lapusta, O. Onopriienko, V. Loboda, An interface crack with partially electrically conductive crack faces under antiplane mechanical and in-plane electric loadings, Mech. Res. Commun. 81 (2017) 38-43. https://doi.org/10.1016/j.mechrescom.2017.02.004

[20] R.K.L. Su, W.J. Feng, J. Liu, Transient response of interface cracks between dissimilar magneto-electro-elastic strips under out-of-plane mechanical and in-plane magneto-electrical impact loads, Compos. Struct. 78 (2007) 119-128. https://doi.org/10.1016/j.compstruct.2005.08.017

[21] A. Zhang, B. Wang, Temperature and electric potential fields of an interface crack in a layered thermoelectric or metal/thermoelectric material, Int. J. Therm. Sci. 104 (2016) 396-403. https://doi.org/10.1016/j.ijthermalsci.2016.01.023

[22] J.R. Barber, Elasticity, Springer, Dordrecht, 2010. https://doi.org/10.1007/978-90-481-3809-8

[23] F. Erdogan, G.D. Gupta, Bonded wedges with an interface crack under anti-plane shear loading, Int. J. Fracture 11(4) (1975) 583-593. https://doi.org/10.1007/BF00116366

[24] X.S. Zhang, A central crack at the interface between two different media in a rectangular sheet under anti-plane shear, Eng. Fract. Mech. 19 (4) (1984) 709-715. https://doi.org/10.1016/0013-7944(84)90103-6

[25] F. Erdogan, Elastic-plastic anti-plane problems for bonded dissimilar media containing cracks and cavities, Int. J. Solids Struct. 2 (3) (1966) 447-465. https://doi.org/10.1016/0020-7683(66)90032-1

[26] F. Delale, Mode-III fracture of bonded non-homogeneous materials, Eng. Fract. Mech. 22 (2) (1985) 213-226. https://doi.org/10.1016/S0013-7944(85)80023-0

[27] X.F. Li, Closed-form solution for a Mode-III interface crack between two bonded dissimilar elastic layers, Int. J. Fracture 109 (2) (2001) 3-8.

[28] X.-F. Wu, Y.A. Dzenis, Closed-form solution for a mode-III interfacial edge crack between two bonded dissimilar elastic strips, Mech, Res. Commun. 29 (2002) 407-412. https://doi.org/10.1016/S0093-6413(02)00317-8

[29] V.V. Silvestrov, I.O. Urakova, I.I. Ilina, Longitudinal shear of an elastic piecewise-

homogeneous space with a semi-infinite cut under mixed boundary conditions on the sides of the cut, Bulletin of the Yakovlev Chuvash state pedagogical university. Series: Mechanics of limit state 4(26) (2015)180-184 (in Russian).

[30] P. Yousefi, J. Fariborz, S.J. Fariborz, Half-layers with interface cracks under anti-plane impact, Theor. Appl. Fract. Mech. 85 (2016) 367-374. https://doi.org/10.1016/j.tafmec.2016.04.008

[31] E.M. Craciun, A. Carabineanu, N. Peride, Antiplane interface crack in a pre-stressed fiber-reinforced elastic composite, Comput. Mater. Sci. 43 (1) (2008) 184-189. https://doi.org/10.1016/j.commatsci.2007.07.028

[32] G. Mishuris, N. Movchan, A. Movchan, Steady-state motion of a mode-III crack on imperfect interfaces, Q. J. Mech. Appl. Math. 59 (4) (2006) 487-516. https://doi.org/10.1093/qjmam/hbl013

[33] Z. Suo, Singularities interacting with interfaces and cracks, Int. J. Solids Struct. 25 (10) (1989) 1133-1142. https://doi.org/10.1016/0020-7683(89)90072-3

[34] Z. Suo, Singularities, interfaces and cracks in dissimilar anisotropic media, Proc. R. Soc. Lond. A427 (1990) 331-358. https://doi.org/10.1098/rspa.1990.0016

[35] S.T. Choi, H. Shin, Y.Y. Earmme, On the unified approach to anisotropic and isotropic elasticity for singularity, interface and crack in dissimilar media, Int. J. Solids Struct. 40 (2003) 1411-1431. https://doi.org/10.1016/S0020-7683(02)00671-6

[36] X.D. Wang, S.A. Meguid, On the general treatment of interacting cracks near an interfacial crack, Int. J. Eng. Sci. 34(12) (1996) 1397-1408. https://doi.org/10.1016/0020-7225(96)00041-9

[37] X.D. Wang, S.A. Meguid, The interaction between an interfacial crack and a microcrack under antiplane loading, Int. J. Fracture 76 (1996) 263-278. https://doi.org/10.1007/BF00048290

[38] G. Mishuris, A. Movchan, N. Movchan, A. Piccolroaz, Interaction of an interfacial crack with linear small defects under out-of-plane shear loading, Comput. Mater. Sci. 52 (2012) 226-230. https://doi.org/10.1016/j.commatsci.2011.01.023

[39] A. Piccolroaz, G. Mishuris, A. Movchan, N. Movchan, Perturbation analysis of Mode III interfacial cracks advancing in a dilute heterogeneous material, Int. J. Solids Struct. 49 (2012) 244-255. https://doi.org/10.1016/j.ijsolstr.2011.10.006

[40] V.V. Panasyuk, M.P. Savruk, A.P. Datsyshin, Stress Distribution near Cracks in Plates and Shells, Naukova Dumka, Kiev, 1976 (in Russian).

[41] N.I. Muskhelishvili, Some Basic Problems of Mathematical Theory of Elasticity, Noordhoff, Groningen, The Netherlands, 1953.

[42] F.D. Gakhov, Boundary Value Problems, Pergamon Press, Oxford, 1966. https://doi.org/10.1016/B978-0-08-010067-8.50007-4

[43] M. Kuna, Finite Elements in Fracture Mechanics. Theory – Numerics – Applications, Springer, Dordrecht, 2013. https://doi.org/10.1007/978-94-007-6680-8

[44] M. Ordyan, V. Petrova, Thermal problem of interaction of partially insulated cracks in a

bimaterial subjected by a heat flux, Proc. Voronezh State University. Series: Physics (1) (2009) 141-149 (in Russian).

[45] V. Petrova, S. Schmauder, Thermal fracture of a functionally graded/homogeneous bimaterial with a system of cracks, Theor. Appl. Fract. Mech. 55 (2011) 148-157. https://doi.org/10.1016/j.tafmec.2011.04.005

[46] N.I. Muskhelishvili, Singular Integral Equations, Dover, New York, 1992.

[47] F. Erdogan, G. Gupta, On the numerical solution of singular integral equations, Quart. Appl. Math. 29 (1972) 525-534. https://doi.org/10.1090/qam/408277

[48] M. Ordyan, V. Petrova, Interaction of cracks in an elastic two-component material under anti-plane shear loading, Vestnik Volgogradskogo gosudarstvennogo universiteta. Seriya 1. Mathematica. Physica 3(34) (2016) 53–62 (in Russian). https://doi.org/10.15688/jvolsu1.2016.3.5

[49] V. Petrova, V. Tamuzs, N. Romalis, A survey of macro-microcrack interaction problems, ASME Appl. Mech. Rev. 53 (2000)117-146. https://doi.org/10.1115/1.3097344

[50] V. Tamuzs, N. Romalis, V. Petrova, Fracture of Solids with Microdefects, NOVA Science Publishers Inc., New York, 2000.

[51] V.E. Petrova, V.P. Tamuzs, Interaction of a main crack with microcracks under longitudinal shear, Three-Dimensional Problems for Structurally Inhomogeneous Media, Izd. VGU, Voronezh (1991) 135-140 (in Russian).

[52] X. Wang, P. Schiavone, Anti-plane eigenstrain problem of an inclusion of arbitrary shape in an anisotropic bimaterial with a semi-infinite interface crack, Continuum Mech. Thermodyn. 31 (2019) 71-77 (https://doi.org/10.1007/s00161-018-0630-1). https://doi.org/10.1007/s00161-018-0630-1

[53] V. Tamuzs, V. Petrova, Modified model of macro-microcrack interaction, Theor. Appl. Fract. Mech. 32 (1999) 111-117. https://doi.org/10.1016/S0167-8442(99)00031-2

[54] S. Loehnert, T. Belytschko, Crack shielding and amplification due to multiple microcracks interacting with a macrocrack, Int. J. Fracture 145 (1) (2007) 1-8. https://doi.org/10.1007/s10704-007-9094-1

Nomenclature

a, $2a$, $b-a$	lengths of cracks (mm)
a_0	half length of an interface crack (mm)
a_k	half length of the kth internal crack (mm)
$f_j(z)$	analytic function of complex variable $z = x + iy$
$F_j(z)$	complex potential for the antiplane shear problem
$F^0(z)$	complex potential for a crack in a homogeneous material

Materials Research Foundations **159** (2024) https://doi.org/10.21741/9781644902950

$F^1(z)$, $F^2(z)$	analytic functions in material 1 and 2, correspondingly
$G(z)$	analytic function in the whole bimaterial compound
G_{III}	energy release rate
H	width of a strip
$K_{IIIn}^{\pm}$	Mode III stress intensity factor at the right K_{IIIn}^{+} and at the left K_{IIIn}^{-} crack tips (MPa mm$^{0.5}$)
$K_{IIIn}^{*\pm}$	non-dimensional form of stress intensity factors Mode III
$K^0 = (\pi a_0)^{1/2}\tau$	Mode III stress intensity factor for a single crack (MPa mm$^{0.5}$)
$K(\cdot)$	complete elliptic integral of the first kind
$M = (\mu_2 - \mu_1)/(\mu_2 + \mu_1)$	bimaterial parameter
N	number of cracks
P	pair of uniform line forces
$P_{nk}(t,x)$	regular kernels that contain the geometry of the problem
V_1, V_2	domains with material 1 and material 2 above and below the x-axis respectively
w_j	displacement component in z-direction (mm)
$w^+ - w^- = h(x)$	displacement jump on an interface crack
w_j^0, w_{nj}^0	displacements on an interface crack and on internal n-th crack
x, y	Cartesian coordinates
x_n, y_n	local coordinate system connected with each crack
$z = x + iy$	complex number
$z_n^0 = x_n^0 + iy_n^0$	midpoint coordinates of cracks (mm)
α_k	inclination angles of a crack to the interface
α, β	Dundurs parameters for the plane interface crack problem
$\delta_n = z_n^0 / a_0 = (x_n^0 + iy_n^0)/a_0$	non-dimensional coordinates of crack centers
θ, $t = \theta a_n$	non-dimensional coordinates and coordinate transformation
θ_j (j=1,2)	angles of a wedge
$\lambda = a / a_0$	non-dimensional small parameter

μ_j	shear modulus, j = 1, 2 (MPa)
τ	antiplane shear loading (MPa)
τ_0	tractions on interface crack surface (MPa)
τ_n	tractions on internal crack surface (MPa)
τ_{xzj}, τ_{yzj}	shear stress components, j = 1, 2 (MPa)
ϕ_0, ϕ_{nj}	functions of displacement jumps on the interface and internal crack
ϕ_0', ϕ_{nj}'	derivative of functions of displacement jumps on the interface and internal crack
χ, $x = \chi a_n$	non-dimensional coordinates and coordinate transformation
$\overline{(...)}$	complex conjugate

Appendix A. An interface crack and internal cracks. Formulation of equations

Boundary conditions

The partial derivatives of the function $\mu_j w_j(x,y)$ with respect to the variables x and y can be expressed in terms of one analytic function $F_j(z)$ [42, 46]

$$\mu_j \frac{\partial w_j(x,y)}{\partial x} = \mathrm{Re}[F_j(z)] = \frac{1}{2}[F_j(z) + \overline{F_j(z)}] \tag{A.1}$$

$$\mu_j \frac{\partial w_j(x,y)}{\partial y} = -\,\mathrm{Im}[F_j(z)] = -\frac{1}{2i}[F_j(z) - \overline{F_j(z)}] \tag{A.2}$$

The physical transmission (continuity) and boundary conditions, Eqs. (11)-(18), in terms of the limiting values of the complex potential F_j are the following. The continuity of displacements, which are defined by the first equation in (14), after differentiation with respect to x yields

$$\mu_2[F_1^+(x) + \overline{F_1^+}(x)] = \mu_1[F_2^-(x) + \overline{F_2^-}(x)], \quad |x| \geq a_0,\ y = 0 \tag{A.3}$$

and the continuity of the tractions at the interface (the second equation in (14)) can be written as

$$-\frac{1}{2i}[F_1^+(x) - \overline{F_1^+}(x)] = -\frac{1}{2i}[F_2^-(x) - \overline{F_2^-}(x)], \quad |x| \geq a_0,\ y = 0\,. \tag{A.4}$$

The boundary conditions on the cracks, Eqs. (15) and (16), are given by

$$-\frac{1}{2i}[F_1^+(x) - \overline{F_1^+}(x)] = \tau_0(x), \quad -\frac{1}{2i}[F_2^-(x) - \overline{F_2^-}(x)] = \tau_0(x), \quad |x| \leq a_0 \tag{A.5}$$

and

$$-\frac{1}{2i}[F_j^+(x)-\overline{F_j^+}(x)]=\tau_n(x), \qquad |x|\le a_n. \tag{A.6}$$

In expressions (A.3)-(A.6) the differentiation rules (A.1) and (A.2) were used.

Applying (A.1) and (A.2), the functions ϕ_0' and ϕ_{nj}', Eqs. (11) and (12), are written as

$$4\mu_1\mu_2\phi_0'(x)=\mu_2F_1^+(x)-\mu_1\overline{F_2^-}(x)-(\mu_1F_2^-(x)-\mu_2\overline{F_1^+}(x)), \quad |x|\le a_0 \tag{A.7}$$

$$2\mu_j\phi_{nj}'(x)=[F_j^+(x)-\overline{F_j^-(x)}]-[F_j^-(x)-\overline{F_j^+(x)}], \quad |x|<a_n. \tag{A.8}$$

Interface crack

Denote by V_1 and V_2 the domains above and below the x-axis, respectively. The continuity condition (A.3) leads to a new complex function $G(z)$ analytic in the whole bimaterial compound ($V_1 \cup V_2$) in exception of the interface crack. That is, rewritten in the form

$$\mu_2F_1^+(x)-\mu_1\overline{F_2^-}(x)=\mu_1F_2^-(x)-\mu_2\overline{F_1^+}, \tag{A.9}$$

the left-hand side of this expression is the boundary value of an analytic function in the domain V_1 and the right-hand side is a boundary value of another analytic function in the domain V_2. Eq. (A.9) means that both functions can be analytically continued into the entire plane $V_1 \cup V_2$, i.e. we get the analytic function in the whole plane, i.e. G(z). Thus, the solution for the problem in full bimaterial region $V_1 \cup V_2$ is written via the new complex potential $G(z)$ as

$$G(z)=\begin{cases}\mu_2F_1(z)-\mu_1\overline{F_2}(z), & z\in(1)\\ \mu_1F_2(z)-\mu_2\overline{F_1}(z), & z\in(2)\end{cases} \tag{A.10}$$

and the derivative of the displacement jumps ϕ_0', Eqs. (23) and (A.7), is written as

$$4\mu_1\mu_2\phi_0'(x)=G^+(x)-G^-(x)$$

A formal solution to this equation is the sectionally holomorphic function defined by the Cauchy type integral as [46]

$$G(z)=\frac{\mu_1\mu_2}{\pi i}\int_{-a_0}^{a_0}\frac{2\phi_0'(t)}{t-z}dt \qquad (|x|<a_0). \tag{A.11}$$

Then, the boundary conditions (A.5) on the interface crack are used. Multiplying the first equality in (A.5) by μ_2, and the second by μ_1, and adding the result, we obtain

$$\mu_2F_1^+(x)-\mu_1\overline{F_2^-}(x)+(\mu_1F_2^-(x)-\mu_2\overline{F_1^+}(x))=-2i(\mu_1+\mu_2)\tau_0(x). \tag{A.12}$$

In view of the definition of G(z) (Eq. (A.10)), we get

$$G^+(x)+G^-(x)=-2i(\mu_1+\mu_2)\tau_0(x) \qquad (|x|<a_0). \tag{A.13}$$

On the other hand, from Eq. (A.11) and by virtue of the Plemelj formula [46], it is obtained

$$G^{+}(x)+G^{-}(x)=\frac{4\mu_1\mu_2}{\pi i}\int_{-a_0}^{a_0}\frac{\phi_0'(t)}{t-x}dt \qquad (|x|<a_0). \tag{A.14}$$

Combining Eqs. (A.13) and (A.14), the following singular integral equation for determination of the unknown function $\phi_0'(t)$ is obtained as Eq. (20).

Recalling the determination of function G(z), Eqs. (A.10) and (A.11), the complex potential for this problem is written as Eq. (22).

Internal cracks in a bimaterial

The solution of the problem is found in the form given in [33]

$$F(z)=\begin{cases}F^{1}(z)+F^{0}(z), & z\in(1),\\ F^{2}(z)+F^{0}(z), & z\in(2).\end{cases} \tag{A.15}$$

F^0 is the complex potential for the problem for a crack in a homogeneous material. $F^1(z)$ and $F^2(z)$ are additional complex functions which have to be determined in terms of F^0. It is assumed that the crack is located in material V_1, and in this case F^0 contains the shear modulus μ_1. If so, $F^1(z)$ is the analytic function in V_1, while the functions $F^2(z)$ and F^0 are analytic in V_2.

To determine $F(z)$, the continuity conditions (Eq. (14)) are used in the form of Eqs. (A.3) and (A.4). At first, Eqs. (A.4) and (A.10) yield

$$F^{1}(x)+F^{0+}(x)-\overline{F^{1}}(x)-\overline{F^{0+}}(x)=F^{2}(x)+F^{0-}(x)-\overline{F^{2}}(x)-\overline{F^{0-}}(x)$$

or (by virtue of $F^{0+}(x)=F^{0-}(x)$)

$$F^{1}(x)+\overline{F^{2}}(x)=F^{2}(x)+\overline{F^{1}}(x). \tag{A.16}$$

On the basis of the analytic continuation theorem and Liouville's theorem [42, 46] (and with condition (18) at infinity), expression (A.16) gives

$$\overline{F^{1}}(z)=-F^{2}(z),\quad \overline{F^{2}}(z)=-F^{1}(z) \tag{A.17}$$

Then, Eqs. (A.3) and (A.10) are used. They lead to equality

$$\mu_2[F^{1}(x)+F^{0}(x)+\overline{F^{1}}(x)+\overline{F^{0}}(x)]=\mu_1[F^{2}(x)+F^{0}(x)+\overline{F^{2}}(x)+\overline{F^{0}}(x)],$$

and accounting Eq. (A.17) this equality is written as

$$(\mu_1+\mu_2)F^2(x)-(\mu_2-\mu_1)F^0(x)=(\mu_1+\mu_2)F^1(x)+(\mu_2-\mu_1)\overline{F^0}(x) \tag{A.18}$$

In Eq. (A.18) the functions on the right and at the left are analytic in V_1(y>0) and V_2 (y<0), respectively. Again, using the analytic continuation and Liouville's theorem for the functions in Eq. (A.18), one has

$$F^1(x)=-\frac{\mu_2-\mu_1}{\mu_1+\mu_2}\overline{F^0}(x),\quad F^2(x)=\frac{\mu_2-\mu_1}{\mu_1+\mu_2}F^0(x). \tag{A.19}$$

Finally, after substitution of (A.19) in (A.15) the complex potential, Eq. (27), for an internal crack is obtained.

Internal cracks and an interface crack

Now consider an internal crack (located in material V_1 (y > 0)) interacting with an interfacial crack. Using Eqs. (27) and (A.2), an additional displacement arising on the interface due to the influence of the internal crack is determined as follows

$$\mu_1\frac{\partial w_1}{\partial y}=-\frac{1}{2i}\lim_{z\to x}[F^+(z)-\overline{F^+}(z)]=-\frac{\mu_2}{i(\mu_1+\mu_2)}[F^0(x)-\overline{F^0}(x)]=-\tau_s(x), \tag{A.20}$$

where the complex potential $F^0(z)$ for the same crack in a homogeneous material is used in the integral form [40]

$$F^0(z)=\frac{\mu_1}{\pi i}\int_{-a_k}^{a_k}e^{-i\alpha_k}\frac{\phi_k'(t)}{t-z_k}dt,\ z_k=e^{-i\alpha_k}(z-z_k^0). \tag{A.21}$$

Here, the unknown function ϕ_k' is the derivative of the discontinuity jump on the k-th crack.

$$\phi_k'=\frac{\partial\phi_k(x)}{\partial x},\ 2\phi_k(x)=w_{jk}^+-w_{jk}^-,\ |x|<a_k. \tag{A.22}$$

Substitution of Eq. (A.21) into (A.20) gives an additional load $\tau_s(x)$ acting on the interface, and which is written in the following form

$$\tau_s(x)=-\frac{2\mu_1\mu_2}{\pi(\mu_1+\mu_2)}\int_{-a_k}^{a_k}\phi_k'(t)\mathrm{Re}[\frac{1}{te^{i\alpha_k}+z_k^0-x}]dt,\qquad |x|<a_0. \tag{A.23}$$

$\tau_s(x)$ is the load caused by an internal crack. Adding $\tau_s(x)$ to the right side of the integral equation (20) for the interface crack, we obtain

$$\int_{-a_0}^{a_0} \frac{\phi_0'(t)}{t-x} dt = \frac{1}{2}\pi[\tau_0(x)+\tau_s(x)]\frac{\mu_1+\mu_2}{\mu_1\mu_2}, \qquad |x|<a_0. \tag{A.24}$$

Substituting (A.23) into (A.24), the following singular integral equation is derived

$$\int_{-a_0}^{a_0} \frac{\phi_0'(t)}{t-x} dt + \int_{-a_k}^{a_k} P_{0k}(t,x)\phi_k'(t)dt = \frac{1}{2}\pi\tau_0(x)\frac{\mu_1+\mu_2}{\mu_1\mu_2}, \qquad |x|<a_0, \tag{A.25}$$

where P_{0k} are regular kernels that contain the geometry

$$P_{0k}(t,x) = \mathrm{Re}[\frac{1}{te^{i\alpha_k}+z_k^0-x}]. \tag{A.26}$$

To find the unknown functions $\phi_0'(t)$ and $\phi_k'(t)$, the second equation is required. Using the complex potentials (22) for the interface crack problem, the load $\tau_a(x)$ on the internal crack line caused by the interface crack is determined using the second relation in Eq. (19) as follows

$$\tau_a(x) = -\mu_1\frac{\partial w_1^0}{\partial y} = \mathrm{Im}[F_1(x)] = -\frac{2\mu_1\mu_2}{\pi(\mu_1+\mu_2)}\int_{-a_0}^{a_0} \phi_0'(t)\mathrm{Re}[\frac{e^{i\alpha_n}}{t-xe^{i\alpha_n}-z_n^0}]dt$$

$$(|x|<a_n) \tag{A.27}$$

Adding the stress τ_a from (A.27) to the right side of the integral equation for the internal crack

$$\mu_1\int_{-a_n}^{a_n} \frac{\phi_n'(t)}{t-x} dt = \pi[\tau_n(x)+\tau_a(x)], \qquad (|x|<a_n) \tag{A.28}$$

and after some transformations, we obtain the following integral equation

$$\mu_1\int_{-a_n}^{a_n} \frac{\phi_n'(t)}{t-x} dt + \frac{2\mu_1\mu_2}{\mu_1+\mu_2}\int_{-a_0}^{a_0} P_{n0}(t,x)\phi_0'(t)dt = \pi\tau_n(x), \qquad (|x|<a_n) \tag{A.29}$$

with the regular kernel

$$P_{n0}(t,x) = \mathrm{Re}[\frac{e^{i\alpha_n}}{t-xe^{i\alpha_n}-z_n^0}]. \tag{A.30}$$

If there is a system of N internal cracks in the material V_1 ($y > 0$), then the system of equations (A.25) and (A.29) takes the form presented in Eqs. (28) and (29).

For simplicity, the variable x instead of x_n (for the n-th crack) is used in Eq. (A.27) and in other relevant formulas. This does not cause confusion due to the indication $|x| < a_n$.

Appendix B. Solution of singular integral equations, analytical approximate solution

Introducing dimensionless coordinates $x = \chi a_k$ and $t = \theta a_k$, and the dimensionless coordinates of crack centers $\delta_k = z_k^0 / a_0$ ($k = 0,1,2,...,N$), Eqs. (28) and (29) are rewritten as

$$\int_{-1}^{1} \frac{\tilde{\phi}_0'(\theta)}{\theta - \chi} d\theta + \sum_{k=1}^{N} \int_{-1}^{1} P_{0k}(a_k\theta, a_0\chi)\tilde{\phi}_k'(\theta)d\theta = \frac{1}{2}\pi\tilde{\tau}_0 \frac{\mu_1 + \mu_2}{\mu_1\mu_2}, \quad |\chi| < 1 \tag{B.1}$$

$$\frac{\mu_1}{a_k}\int_{-1}^{1} \frac{\tilde{\phi}_n'(\theta)}{\theta - \chi} d\theta + \frac{2\mu_1\mu_2}{\mu_1 + \mu_2}\int_{-1}^{1} P_{n0}(a_0\theta, a_k\chi)\tilde{\phi}_0'(a_0\theta)d\theta +$$

$$\mu_1 \sum_{\substack{k=1 \\ k\neq n}}^{N} \int_{-1}^{1} P_{nk}(a_k\theta, a_n\chi)\tilde{\phi}_k'(a_k\theta)d\theta = \frac{\pi}{a_0}\tilde{\tau}_n \quad |\chi| < 1,\ n = 1,2,...,N \tag{B.2}$$

$$\tilde{\tau}_n(\chi) = a_0\tau_n(\chi a_n) = \tilde{\tau}_n$$

The unknown functions $\tilde{\phi}_0'(\chi)$ and $\tilde{\phi}_n'(\chi)$ are presented as a power series with respect to the small parameter $\lambda = a / a_0 << 1$:

$$\tilde{\phi}_0'(\chi) = \sum_{p=0}^{\infty} \tilde{\phi}_{0p}'(\chi)\lambda^p, \quad \tilde{\phi}_n'(\chi) = \sum_{p=0}^{\infty} \tilde{\phi}_{np}'(\chi)\lambda^p, \tag{B.3}$$

and all known functions P_{nk} (31) are expanded in the power series in λ

$$P_{0k}(\theta,\chi) = \sum_{u=0}^{\infty} P_{0ku}(\theta,\chi)\lambda^u, \quad P_{0ku}(a\theta, a_0\chi) = -\frac{1}{a_0}\mathrm{Re}[\frac{\theta^u e^{iu\alpha_k}}{(\theta - \delta_k)^{u+1}}],$$

$$P_{no}(\theta,\chi) = \frac{1}{\lambda}\sum_{u=1}^{\infty} P_{nou}(\theta,\chi)\lambda^u, \quad P_{n0u}(a_0\theta, a\chi) = \frac{1}{a_0}\mathrm{Re}[\frac{\chi^{u-1}e^{iu\alpha_n}}{(\theta - \delta_n)^u}], \tag{B.4}$$

$$P_{nk}(\theta,\chi) = \frac{1}{\lambda}\sum_{u=1}^{\infty} P_{nku}(\theta,\chi)\lambda^u, \quad P_{nku}(a\theta, a\chi) = \frac{1}{a_0}\mathrm{Re}[\frac{e^{i\alpha_n}(\chi e^{i\alpha_n} - \theta e^{i\alpha_k})^{u-1}}{(\delta_k - \delta_n)^u}].$$

For convergence of series (B.4) the following inequality must be fulfilled:

$$\left|\frac{\lambda}{\theta - z_k^0 / a_0}\right| < 1,\ |\theta| < 1, \tag{B.5}$$

which is satisfied if the cracks do not intersect.

For the regularization of Eqs. (B.1), (B.2) Carleman-Vekua method is used [42, 46], which defined by the formula

$$\tilde{\varphi}_n'(\chi) = \frac{1}{\pi\sqrt{1-\chi^2}}\left[-\int_{-1}^{1}\frac{f_n(\xi)\sqrt{1-\xi^2}}{\xi-\chi}d\xi + C\right],\ |\chi|<1,\ n=0,1,2,\dots,N \quad \text{(B.6)}$$

where constant C is defined from the condition at infinity (18) and equal to zero, C = 0, in the considered case. Thus, after regularization of Eqs. (B.1) and (B.2), the following system of Fredholm equations of the second kind is written

$$\tilde{\varphi}_0'(\chi) = \frac{1}{\pi\sqrt{1-\chi^2}}\left[-\frac{1}{2}\pi\tilde{\tau}_0\frac{\mu_1+\mu_2}{\mu_1\mu_2}\int_{-1}^{1}\frac{\sqrt{1-\xi^2}}{\xi-\chi}d\xi + \sum_{k=1}^{N}\int_{-1}^{1}\left[\int_{-1}^{1}\frac{\sqrt{1-\xi^2}P_{0k}(a_k\theta,a_0\xi)}{\xi-\chi}d\xi\right]\tilde{\varphi}_k'(\theta)d\theta\right],\quad |\chi|<1 \quad \text{(B.7)}$$

$$\tilde{\varphi}_n'(\chi) = \frac{1}{\pi\sqrt{1-\chi^2}}\left[-\frac{a_k}{k_1}\frac{\pi}{a_0}\tilde{\tau}_n\int_{-1}^{1}\frac{\sqrt{1-\xi^2}}{\xi-\chi}d\xi + \frac{a_k}{\mu_1}\frac{2\mu_1\mu_2}{\mu_1+\mu_2}\int_{-1}^{1}\left[\int_{-1}^{1}\frac{\sqrt{1-\xi^2}P_{n0}(a_0\theta,a_k\xi)}{\xi-\chi}d\xi\right]\tilde{\varphi}_0'(a_0\theta)d\theta + \right.$$

$$\left. + a_k\sum_{\substack{k=1\\k\neq n}}^{N}\int_{-1}^{1}\left[\int_{-1}^{1}\frac{\sqrt{1-\xi^2}P_{nk}(a_k\theta,a_k\xi)}{\xi-\chi}d\xi\right]\tilde{\varphi}_k'(a_k\theta)d\theta\right],\quad |\chi|<1\ \ n=1,2,\dots,N \quad \text{(B.8)}$$

By substituting the series (B.3) and (B.4) into Eqs. (B.7) and (B.8), we obtain

$$\sum_{p=0}^{\infty}\tilde{\varphi}_{0p}'(\chi)\lambda^p = \frac{1}{\pi\sqrt{1-\chi^2}}\{-\frac{\mu_1+\mu_2}{2\mu_1\mu_2}\int_{-1}^{1}\frac{\sqrt{1-\theta^2}}{\theta-\chi}\tilde{\tau}_0 d\theta + \sum_{k=1}^{N}\int_{-1}^{1}\sum_{u=0}^{\infty}\tilde{P}_{0ku}(a\theta,a_0\chi)\lambda^u\sum_{p=0}^{\infty}\tilde{\varphi}_{kp}'(\tau)\lambda^p d\theta\},\quad |\chi|<1 \quad \text{(B.9)}$$

$$\sum_{p=0}^{\infty}\tilde{\varphi}_{np}'(\chi)\lambda^p = \frac{1}{\pi\sqrt{1-\chi^2}}\frac{1}{\mu_1}\{-\lambda\int_{-1}^{1}\frac{\sqrt{1-\theta^2}}{\theta-\chi}\tilde{\tau}_n d\theta + \frac{2\mu_1\mu_2}{\mu_1+\mu_2}\int_{-1}^{1}\sum_{u=1}^{\infty}\tilde{P}_{n0u}(a_0\theta,a\chi)\lambda^u \sum_{p=0}^{\infty}\tilde{\varphi}_{0p}'(\theta)\lambda^p d\theta + \mu_1\sum_{\substack{k=1\\k\neq n}}^{N}\int_{-1}^{1}\sum_{u=1}^{\infty}\tilde{P}_{nku}(a\theta,a\chi)\lambda^u\sum_{p=0}^{\infty}\tilde{\varphi}_{kp}'(\theta)\lambda^p d\theta\}$$

$$|\chi|<1,\ \ n=1,2,\dots,N \quad \text{(B.10)}$$

where

$$\tilde{P}_{nku}(a_k\theta, a_n\chi) = \frac{a_0}{\pi}\int_{-1}^{1}\frac{\sqrt{1-\xi^2}}{\xi-\chi}P_{nku}(a_k\theta, a_n\xi)d\xi, \qquad k,n = 0,1,2,\ldots,N.$$

Equating the expressions for the same powers of λ, the recurrent system of equations for the determination of unknown coefficients $\tilde{\phi}'_{0k}$ and $\tilde{\phi}'_{nk}$ in series (B.3) is obtained as

$$\tilde{\phi}'_{0p}(\chi) = \frac{1}{\pi\sqrt{1-\chi^2}}\left\{-\varepsilon_p \frac{\mu_1+\mu_2}{2\mu_1\mu_2}\int_{-1}^{1}\frac{\sqrt{1-\theta^2}}{\theta-\chi}\tilde{\tau}_0 d\theta + \sum_{k=1}^{N}\int_{-1}^{1}\sum_{q=0}^{p}\tilde{\phi}'_{kq}(\theta)\tilde{P}_{0k(p-q)}(a\theta, a_0\chi)d\theta\right\} \tag{B.11}$$

$$\begin{aligned}\tilde{\phi}'_{np}(\chi) = \frac{1}{\pi\sqrt{1-\chi^2}}\frac{(\varepsilon'_p)^2}{\mu_1}\Bigg\{&-\frac{(1-\varepsilon'_p)}{2}\int_{-1}^{1}\frac{\sqrt{1-\theta^2}}{\theta-\chi}\tilde{\tau}_n d\theta \\ &+\frac{2\mu_1\mu_2}{\mu_1+\mu_2}\int_{-1}^{1}\sum_{q=0}^{p-1}\tilde{\phi}'_{0q}(\theta)\tilde{P}_{n0(p-q)}(a_0\theta, a\chi)d\theta \qquad , n = 1,2,\ldots,N \\ &+\mu_1\sum_{k=1,k\neq n}^{N}\int_{-1}^{1}\sum_{q=0}^{p-1}\tilde{\phi}'_{kq}(\tau)\tilde{P}_{nk(p-q)}(a\theta, a\chi)d\theta\Bigg\}\end{aligned} \tag{B.12}$$

here

$$\varepsilon_p = \begin{cases} 1, & p = 0, \\ 0, & p = 1,2,\ldots \end{cases} \qquad \varepsilon'_p = \begin{cases} 0, & p = 0, \\ -1, & p = 1, \\ 1, & p = 2,3,\ldots \end{cases}$$

The problem was solved with the accuracy up to second order of λ^2 and the necessary equations from the recurrent system (B.11) and (B.12) are the following

$$\begin{aligned}\tilde{\phi}'_{00}(\chi) &= \frac{1}{\pi\sqrt{1-\chi^2}}\left\{-\frac{\mu_1+\mu_2}{2\mu_1\mu_2}\int_{-1}^{1}\frac{\sqrt{1-\theta^2}}{\theta-\chi}\tilde{\tau}_0 d\theta + \sum_{k=0}^{N}\int_{-1}^{1}\tilde{P}_{0k0}(a_0\chi)\tilde{\phi}'_{k0}(\theta)d\theta\right\}, \\ \tilde{\phi}'_{01}(\chi) &= \frac{1}{\pi\sqrt{1-\chi^2}}\sum_{k=1}^{N}\int_{-1}^{1}\left[\tilde{\phi}'_{k1}(\theta)\tilde{P}_{0k0}(a_0\chi) + \tilde{\phi}'_{k0}(\theta)\tilde{P}_{0k1}(a\theta, a_0\chi)\right]d\theta, \\ \tilde{\phi}'_{02}(\chi) &= \frac{1}{\pi\sqrt{1-\chi^2}}\sum_{k=1}^{N}\int_{-1}^{1}[\tilde{\phi}'_{k1}(\theta)\tilde{P}_{0k1}(a\theta, a_0\chi) + \tilde{\phi}'_{k2}(\theta)\tilde{P}_{0k0}(a\theta, a_0\chi) \\ &\quad + \tilde{\phi}'_{k0}(\theta)\tilde{P}_{0k2}(a\theta, a_0\chi)]d\theta,\end{aligned} \tag{B.13}$$

and

$$\tilde{\phi}'_{n0}(\chi)=0\,,$$

$$\tilde{\phi}'_{n1}(\chi)=\frac{1}{\pi\sqrt{1-\chi^2}}\frac{1}{\mu_1}\left\{-\int_{-1}^{1}\frac{\sqrt{1-\theta^2}}{\theta-\chi}\tilde{\tau}_n\,d\theta+\frac{2\mu_1\mu_2}{\mu_1+\mu_2}\int_{-1}^{1}\tilde{\phi}'_{00}(\theta)\tilde{P}_{n01}(a_0\theta,a\chi)d\theta\right.$$

$$\left.+\mu_1\sum_{k=1,k\neq n}^{N}\int_{-1}^{1}\tilde{\phi}'_{k0}(\theta)\tilde{P}_{nk1}(a_0\theta,a\chi)d\theta\right\}, \qquad \text{(B.14)}$$

$$\tilde{\phi}'_{n2}(\chi)=\frac{1}{\pi\sqrt{1-\chi^2}}\frac{1}{\mu_1}\left\{\frac{2\mu_1\mu_2}{\mu_1+\mu_2}\int_{-1}^{1}[\tilde{\phi}'_{00}(\theta)\tilde{P}_{n02}(a_0\theta,a\chi)+\tilde{\phi}'_{01}(\theta)\tilde{P}_{n01}(a_0\theta,a\chi)]d\theta\right.$$

$$\left.+\mu_1\sum_{k=1,k\neq n}^{N}\int_{-1}^{1}[\tilde{\phi}'_{k1}(\theta)\tilde{P}_{nk1}(a\chi)+\tilde{\phi}'_{k0}(\theta)\tilde{P}_{nk2}(a\theta,a\chi)]d\theta\right\},\quad n=1,2,...,N$$

Calculating the integrals in expressions (B.13) and (B.14), one obtains the solution for the system (28) and (29) for the interface and internal cracks up to λ^2, Eqs. (33) and (34)

CHAPTER 7

Theoretical Modelling and Analysis of Thermal Fracture of Semi-Infinite Functionally Graded Materials with Edge Cracks

Vera Petrova [1,2]*, Tomasz Sadowski [2]

[1]Voronezh State University, University Sq.1, Voronezh 394006, Russia

[2]Lublin University of Technology, Nadbystrzycka 40, 20-618 Lublin, Poland

veraep@gmail.com *, t.sadowski@pollub.pl

Abstract

The present investigation is devoted to a problem of the interaction of two edge cracks inclined arbitrary to the boundary of a non-homogeneous half-plane, which is a functionally graded layer on a homogeneous substrate. The functionally graded properties vary exponentially in thickness direction. One cycle of cooling from sintering temperature is considered. An approach based on integral equations is used and a solution is obtained, then stress intensity factors are calculated and direction of the initial crack propagation is evaluated by using the maximum circumferential stress criterion. Influence of geometrical and material (inhomogeneity) parameters on the fracture characteristics is investigated. This study can serve as a part of the modeling of the fracture process in FGM coatings under cyclic heating-cooling thermal loading.

Keywords

Edge Cracks, Singular Integral Equations, Stress Intensity Factors, Thermal Fracture

1. Introduction

In different engineering applications, e.g., nuclear energy, aerospace, energy conversions, thermal barrier coating are used to protect metallic or composite components from extremely high temperatures [1,2,3]. Last years for these purposes the so-called functionally graded materials (FGMs) are used. FGMs are composite materials with continuously varying properties in one direction. The application of FGM coatings can reduce bimaterial mismatch at interfaces between the coating and the substrate and prevent delamination and debonding along interfaces. However, cracks can initiate from initial defects or microcracks appear during manufacturing or service. Experimental results [4] showed that when functionally graded plates are subjected to thermal shock, multiple cracks often occur on the ceramic surface during cooling-heating cycles. This fracture process begins from formation of a single crack from initial defects and then a system of edge cracks is formed. Therefore, the study of fracture of FGM coatings is important for a better understanding of the fracture processes in FGM structures and to improve their fracture resistance.

Numerous papers are devoted to different problems of modelling and analysis of fracture processes in FGMs, references can be found in the review papers [2, 5]. Different methods are widely used for modeling of FGMs and structures under thermal and mechanical loadings, among them FE methods [6-8] the boundary integral methods [9-14] and their modifications [15, 16]. In spite of many available solutions the problems of interaction of arbitrary located cracks in FGMs are still important.

An approximation method for determining stress intensity factors for a periodic system of edge cracks in an FGM coating in a semi-infinite medium was introduced in [9]. The method is based on singular integral equations. The validity of this approach is discussed in [17] and good accuracy is demonstrated for some gradient (inhomogeneity) parameters of FGMs and crack lengths.

The present work is devoted to the theoretical modeling of fracture of a FGM coating on a semi-infinite homogeneous substrate under thermal and mechanical loading. One cycle of cooling from sintering temperature is considered. It is supposed that two edge cracks arbitrary inclined to the boundary are located in the FGM. The FGM properties are presented by exponential functions. The method of singular integral equations is used and approach similar to the presented in [9] is applied. It is supposed that the inhomogeneity of material is revealed in non-homogeneous residual stresses on crack surfaces. An example of accounting such residual forces can be found in [9] where a semi-infinite functionally graded material (FGM) with edge cracks is considered. In [18] the influence of an additional loading, which vary with a coordinate along the crack lines (and can be considered as a residual stresses), on both stress intensity factors Mode I and Mode II was considered in the problem for two parallel cracks under shear loading (pure Mode II) corresponding to the loading in a Compact Shear specimen. In the present investigation the interaction of two edge cracks is considered and this study can serve as a part of the modeling of the fracture process in FGM coatings and further formation of a system of cracks under cyclic thermal loading.

2. Formulation of the problem

The present investigation is devoted to a problem of the interaction of two edge cracks (with length $2a_n$, $n = 1, 2$) inclined arbitrary to the boundary of a non-homogeneous half-plane (Fig. 1). Cartesian coordinates (x,y) have x-axis along the boundary of the half-plane, local coordinate systems (x_n, y_n) are attached to each crack. The crack position is determined by the crack midpoint coordinate $z_n^0 = x_n^0 + iy_n^0$ ($i = \sqrt{-1}$ is the imaginary unity) and an inclination angle β_n to the boundary, i.e. to the x-axis (Fig. 1).

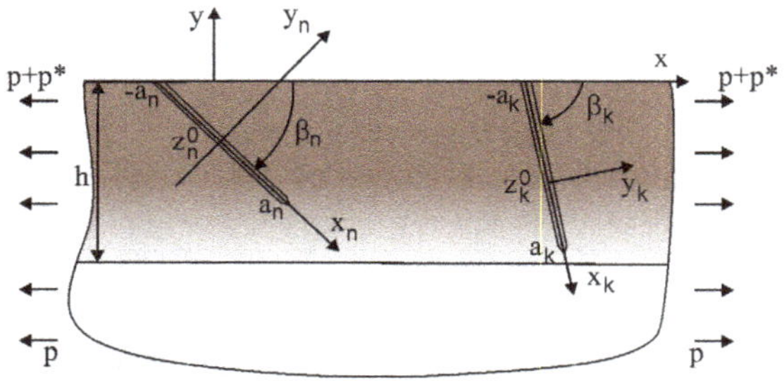

Figure 1. Two edge cracks in a non-homogeneous half-plane.

A functionally graded material (FGM) is located in the region $0 \le y \le -h$ with width h. The Poisson's ratio v is assumed to be a constant because the effect of its variation on the crack-tip stress intensity factors is negligible [19,20]. The remaining thermo-mechanical properties, i.e. the Young's modulus $E(y)$ and the coefficient of thermal expansion $\alpha_t(y)$, depend on the y-coordinate only and are modeled by the exponential function [12-14, 19,20].

For arbitrary located cracks in a half-plane the system of singular integral equations is written as [21, 22]

$$\int_{-a_n}^{a_n} \frac{g_n'(t)dt}{t-x} + \sum_{\substack{k=1 \\ k\neq n}}^{N} \int_{-a_k}^{a_k} [g_k'(t)R_{nk}(t,x) + \overline{g_k'(t)}S_{nk}(t,x)]dt = \pi p_n(x), \quad |x| < a_n,$$

$$n = 1,2 \tag{1}$$

and for two cracks we have two equations. The unknown functions

$$g_n'(x) = \frac{2\mu}{i(\kappa+1)}\frac{\partial}{\partial x}([u_n] + i[v_n]) \tag{2}$$

are the derivative of displacement jumps on the crack faces; $[u_n]$ and $[v_n]$ are shear and vertical displacement jumps, respectively, on the n-th crack line, $\mu = E/2(1+\nu)$ is the shear modulus, E - Young's modulus, ν - Poisson's ratio, $\kappa = 3 - 4\nu$ for the plane strain state and $\kappa = (3-\nu)/(1+\nu)$ for the plane stress state.

The functions p_n are determined by the applied load. The regular kernels R_{nk}(t,x) and $S_{nk}(t,x)$ contain geometry of the problem and are cited in the Appendix in Eqs. (A.1) – (A.5).

The Eqs. (1) and (2) are for common case of two arbitrary cracks in a half plane. For edge cracks we should put $z_n^0 = x_n^0 - ia_n \sin\beta_n$.

The FGM is cooled from sintering temperature. The FGMs inhomogeneity is accounted via continuously varying residual stresses arising due to mismatch in the coefficients of thermal expansion. The additional stresses p^* are the following [9]:

Materials Research Foundations **159** (2024) https://doi.org/10.21741/9781644902950

$$p^*: \ \sigma^T_{xx}(y) = [\alpha_t(y) - \alpha_{t0}]\Delta T E(y), \ \sigma^e_{xx}(y) = [E(y)/E_0 - 1]\sigma^0_{xx}. \tag{3}$$

These functions will be added to the right sides of the Eqs. (1) and (2). α_{t0} is the thermal expansion coefficient of a homogeneous material (in the region $y < -h$) and E_0 is the Young's modulus of this material.

The thermal expansion coefficient is presented in exponential form

$$\alpha_{t1} = \alpha_{t0}\exp(\varepsilon(y+h)), \qquad -h \le y \le 0, \tag{4}$$

where ε is inhomogeneity parameter of this coefficient. The Young's modulus is

$$E = E_0\exp(\delta(y+h)), \qquad -h \le y \le 0 \tag{5}$$

with inhomogeneity parameter δ.

The relation between the global coordinates (x,y) and the local coordinate systems (x_k,y_k) can be written in the complex form as follows $z = z^0_k + z_k e^{-i\beta_k}$, where $z_k = x_k + iy_k$ and $i = \sqrt{-1}$. $z^0_k = x^0_k + iy^0_k$ is the origin coordinate of the system (x_k,y_k) in the global system. In the local coordinate system (x_k,y_k) connected with each crack the coefficient α_t Eq. (5) possess the form

$$\alpha_{t1}(x_k, y_k) = \alpha_{t0}e^{\varepsilon(h+y^0_k)}e^{\varepsilon_1 x_k + \varepsilon_2 y_k}, \ \varepsilon_1 = \varepsilon\sin(-\beta_k), \ \varepsilon_2 = \varepsilon\cos\beta_k$$

and on crack lines, where $y_k = 0$, we will have

$$\alpha_t = \alpha_{t0}\exp(\varepsilon(h+y^0_k))\exp(-\varepsilon x_k\sin\beta_k) \tag{6}$$

Similar expressions can be written for the Young's modulus.

If we will suppose that the Young's moduli of materials in the FGM have approximately same values, then in this case – $\sigma^e_{xx}(y) = 0$ in Eq. (4). It means that the material is elastically homogeneous. The examples of such materials are the following: ceramic/ceramic TiC/SiC, $MoSi_2/Al_2O_3$ and $MoSi_2$/SiC, and also ceramic/metal FGMs, e.g., zirconia/nickel and zirconia/steel. For this special case we will investigate the influence of the inhomogeneity parameter ε on the fracture characteristics of the material. For fully non-homogeneous material we will have two inhomogeneity parameters ε and δ.

The equations (1), (2) are rewritten in dimensionless form with the non-dimension coordinates $\xi = t/a$ and $\eta = x/a$, where $2a$ is a length of the crack (here we suppose that $a_1 = a_2 = a$). In the considered case of the edge cracks functions $g'_n(\eta)$ are bounded in the edge point, i.e. at the points η= – 1. At the other tip of the crack, for η= 1, the functions $g'_n(\eta)$ as well as the stresses have square root singularity. The stress intensity factors (SIFs) at the internal tips of the edge cracks are obtained as

$$k_{nI} - ik_{nII} = -\lim_{\eta \to +1} \sqrt{a}\sqrt{1-\eta^2}\, g_n'(\eta) \quad (n = 1,2). \tag{7}$$

3. Solution

3.1 Numerical solution

The system of Eq. (1) is solved by method of mechanical quadrature [21-23] which is based on the Chebyshev polynomials. The solution for edge cracks is presented in the following form

$$g_n'(\eta) = u_n(\eta) / \sqrt{1-\eta^2} \tag{8}$$

Here $u_n(\eta)$ are regular functions on the segment [-1,1] and $1/\sqrt{1-\eta^2}$ is the weight function. The condition that the functions $g_n'(\eta)$ are bounded at the edge point $\eta = -1$ (or have singularity less than $1/\sqrt{1+\eta}$) is the following [22]

$$u_n(-1) = 0. \tag{9}$$

That is the exact singularity at the edge points is not taking into account, but the result obtained with this assumption have shown good accuracy [22].

Using Gauss's quadrature formulae for the regular and singular integrals the integral equations are reduced to the system of *NxM* (*N*=2 – number of cracks, *M* – number of nodes) algebraic equations

$$\frac{1}{M}\sum_{m=1}^{M}\sum_{k=1}^{N}\left[u_k(\xi_m)R_{nk}(\xi_m,\eta_r) + \overline{u_k(\xi_m)}S_{nk}(\xi_m,\eta_r)\right] = \pi p_n(\eta_r), \tag{10}$$

$$\sum_{m=1}^{M}(-1)^m u_n(\xi_m)\tan\frac{2m-1}{4M}\pi = 0 \qquad (n=1,2;\ r=1,2,\ldots,M-1) \tag{11}$$

where

$$\xi_m = \cos\frac{2m-1}{2M}\pi \ \ (\mathrm{m}=1,2,\ldots,\mathrm{M}); \ \ \eta_r = \cos\frac{\pi r}{M} \qquad (r=1,2,\ldots,\mathrm{M}-1) \tag{12}$$

M is the total number of the discrete points of the unknown functions $u_n(\eta)$ on the interval (-1,1). After solution of the algebraic system (10) and (11) the functions $u_n(\eta)$ are calculated by the interpolation formula:

$$u_n(\eta) = \frac{2}{M}\sum_{m=1}^{M}u_n(\xi_m)\sum_{r=0}^{M-1}T_r(\xi_m)T_r(\eta) - \frac{1}{M}\sum_{m=0}^{M}u_n(\xi_m). \tag{13}$$

Here T_r are Chebyshev polynomials of the first kind. Setting $\eta = 1$ in Eq. (13), it is obtained

$$u_n(+1) = \frac{1}{M}\sum_{m=1}^{M} (-1)^{m+1} u_n(\xi_m)\cot\frac{2m-1}{4M}\pi \tag{14}$$

and for $\eta = -1$ we have

$$u_n(-1) = \frac{1}{M}\sum_{m=1}^{M} (-1)^{M+m} u_n(\xi_m)\tan\frac{2m-1}{4M}\pi$$

This equation and the condition (9) yield the Eq. (11).

Applying the conjugate operation to the system (9) additional NxM equations are obtained, i.e. $2NxM$ equations should be solved, for two cracks we have 4xM equations.

Inserting Eq. (14) into the formula (8) and then into (7) the SIFs are obtained

$$k_{In} - ik_{IIn} = -\sqrt{a_n}\, u_n(+1) \quad (n = 1,2). \tag{15}$$

3.2 Validation

To validate this model and verify computational results, we consider some numerical examples and compare our results for SIFs with other available in the literature. The validation of the approximation method similar to the present method with respect to stress intensity factors was discussed in [17]. It was shown that the error depends on the gradient of the profile of Young's modulus of FGMs and crack lengths. For a small crack length the error remains within acceptable limits even for a large gradient (a large inhomogeneity parameter) at the crack tip. However, for a large crack length the gradient (inhomogeneity parameter) should be small at the crack tip for the error to be small. In our modeling the inhomogeneity parameters are used in the range from -1 to 1.

For verification of the numerical results and validation of the computer program a particular case of two inclined edge cracks in a homogeneous half-plane under uniform tension p (Fig. 1, p^*=0) is considered. The cracks are assumed to have same length $2a_i = 2a$ and same slope angle β. The numerical results with respect to stress intensity factors are obtained for different angles β in the range of 15^0 to 90^0 and for different non-dimensional distances d/a between the cracks. The SIFs Mode I (k_I) and Mode II (k_{II}) are presented in the non-dimensional form, they are normalized by $k^0 = p\sqrt{2a}$.

The convergence of the numerical results is checked by comparing the values for SIFs for different number of collocation points M. In the case of the angle $\beta = 90^0$ M=40 is enough for good accuracy, the results obtained for M=40 and M=80 are very close, they differ only in forth sign in the decimal digit, the relative error is about 10^{-2}. At the same time for 15^0 similar good accuracy can be achieved for M = 80. For the problem of a single oblique edge crack under constant normal tractions applied to the crack surfaces the results showed good accuracy for M=30 and good agreement with the results cited in [21, 22] is demonstrated.

Comparison with published in the literature solutions for SIFs can be done for the homogeneous half-plane. The values for SIF for oblique edge cracks can be found in [10, 21-25]. Detailed analysis of these problems was done for one crack in [10] and for periodic edge cracks of unequal cracks in a semi-infinite tensile sheet in [25] for β = 45^0. Unfortunately, the results for SIFs for two oblique edge cracks by Nisitani, 1977 cited in the handbook of Murakami [24] are not clear for comparison with our results (besides, the original paper by Nisitani, 1977 is not available). The comparison have been done with results for SIFs for single inclined edge crack cited in [10] and with SIFs for periodic edge cracks cited in [25]. Table 1 shows that with increasing the distance d/a between the cracks the SIFs tend to the value for a single edge crack cited in [10]. It should be noted that for small slope angles β = 15^0 and 30^0 and for distances for $d/a = 10$ the values of SIFs are close to the values of SIFs for a single edge crack [10]. With increasing the angle β larger distances d/a should be taken to achieve the same result, e.g., for $\beta = 90^0$ good result is for the distance $d/a = 100$. It means that interaction effect stronger for the cracks perpendicular to the surface and nearly perpendicular and for small inclination angles the influence of the boundary of the half-plane is prevailed.

Comparison of the values for SIFs for two interacting edge cracks with the SIFs [25] for a crack in a periodic system of edge cracks inclined on the angle $\beta = 45^0$ shows that the values of SIFs are similar for $d/a = 10$ and differ considerably for close located cracks with $d/a = 1$, see Table 1.

Table 1. Non-dimension SIFs (k_I, k_{II}) of the edge cracks in a semi-infinite homogeneous plane

β / d/a	15^0	30^0	45^0	60^0	75^0	90^0
1 [25]	0.134; 0.157	0.263; 0.258	0.418; 0.323 0.28; 0.18	0.582; 0.332	0.725; 0.275	0.817; 0.159
10 [25]	0.229; 0.224	0.448; 0.328	0.667; 0.352 0.67; 0.35	0.855; 0.298	0.986; 0.179	1.037; 0.019
60	0.232; 0.227	0.462; 0.336	0.704; 0.364	0.918; 0.305	1.066; 0.173	1.118; 0.000
100						1.121; 0.000
[10]	0.232; 0.226	0.462; 0.336	0.705; 0.365	0.920; 0.306	1.069; 0.174	1.121; 0.000

For all angles β the SIF k_I increases with increasing the distance d/a between the cracks and tends to the value for a single crack, while with decreasing d/a SIF k_I decreases, i.e. the so-called shielding effect is observed which is known for the parallel cracks under tension normal to the crack lines. The behavior of k_{II} is more complicated than k_I. For angles $\beta = 15^0$, 30^0, 45^0 and 60^0 k_{II} increases with increasing the distance between cracks, and for $\beta = 75^0$ and 90^0 k_{II} decreases.

4. Results

4.1 Material parameters

For analysis of the fracture development in the FGM/homogeneous half plane with pre-existing two edge cracks the parameters of the materials should be chosen. Besides, the model for the functional gradation has to be selected. In this study the exponential form for FGMs is used, Eqs (4) - (6). Then, on the basis of this model and the real material combinations of the structure the special inhomogeneity parameters should be estimated.

Functionally graded materials are used in thermal barrier coating to protect details from high temperatures as well as from wear and corrosion. The materials for protecting from high temperatures should have a low thermal conductivity and at the same time they are desired to have a thermal expansion coefficient close to that of the material for the protected substrate. Consider some actual material combinations ceramic/ceramic and ceramic/metal which can be used in the model. The parameters of these materials, which are available in the literature, are presented in the Table 2 [25]. The Young's modules of these materials are similar; it means that these FGMs are elastically homogeneous, in this case – $\sigma_{xx}^{e}(y) = 0$ in Eq. (3). For this special case we can investigate the influence of the inhomogeneity parameter ε on the fracture characteristics of the material, such as stress intensity factors at crack tips and fracture angles which determines the direction of further propagation of the cracks.

Table 2. Thermal properties of some FGMs and the inhomogeneity coefficient ε

Thermal expansion coeff. ($*10^{-6}$ K^{-1})			Thermal conductivity ($Wm^{-1}K^{-1}$)		
FGM/H (Al_2O_3/ $MoSi_2$)/ $MoSi_2$ (Ceramic/Ceramic)					
Al_2O_3	α_{t1}	5	$\alpha_{t1}/\alpha_{t2}=1$	k_1	25
$MoSi_2$	α_{t2}	5	$\varepsilon = 0$	k_2	52
FGM/H ($MoSi_2$/ Al_2O_3)/ Al_2O_3			$\varepsilon = 0$		
FGM/H ($MoSi_2$/ SiC)/ SiC (Ceramic/Ceramic)					
$MoSi_2$	α_{t1}	5	$\alpha_{t1}/\alpha_{t2}>1$	k_1	52
SiC	α_{t2}	4	$\varepsilon > 0$	k_2	60
FGM/H (SiC/ $MoSi_2$)/ $MoSi_2$			$\varepsilon < 0$		
FGM/H (TiC/ SiC)/ SiC (Ceramic/Ceramic)					
TiC	α_{t1}	7	$\alpha_{t1}/\alpha_{t2}>1$	k_1	20
SiC	α_{t2}	4	$\varepsilon > 0$	k_2	60
FGM/H (SiC/ TiC)/ TiC			$\varepsilon < 0$		
FGM/H (ZrO_2/ Ni)/ Ni (Ceramic/Metal)					
ZrO_2	α_{t1}	10	$\alpha_{t1}/\alpha_{t2}<1$	k_1	2
Ni	α_{t2}	18	$\varepsilon < 0$	k_2	90
FGM/H (Ni/ ZrO_2)/ ZrO_2			$\varepsilon > 0$		
FGM/H (ZrO_2/ Steel)/ Steel (Ceramic/Metal)					
ZrO_2	α_{t1}	10	$\alpha_{t1}/\alpha_{t2}<1$	k_1	2
Steel	α_{t2}	12	$\varepsilon < 0$	k_2	20
FGM/H (Steel/ ZrO_2)/ ZrO_2			$\varepsilon > 0$		

From Eq. (4) $\alpha_{t1} / \alpha_{t0} = \exp(\varepsilon(y+h))$ and the inhomogeneity parameter ε is written as

$$\varepsilon = \ln(\alpha_{t1} / \alpha_{t0}) / (y + h) \tag{16}$$

The non-dimensional parameter is εa_1, (here a_1 is half size of one crack), but the designation ε remains in the figures.

Eq. (16) in combination with data in Table 2 shows that the values of ε for a thick FGM layer are not large and, accordingly, the actual variation of the residual stresses with coordinate y is not very strong. Fig. 2 illustrates the variations of exponential functions $\exp(\varepsilon(y / h + 1))$ with non-dimensional coordinate y/h for different parameters ε, positive at the Fig. 2a and negative at the Fig. 2b. The value of the exponential function (and, hence, the other values, containing this function, e.g., thermal expansion coefficient and residual stresses) increases by 35% for $\varepsilon = 0.3$ and decreases by the same value for the negative ε equals to $\varepsilon = -1$.

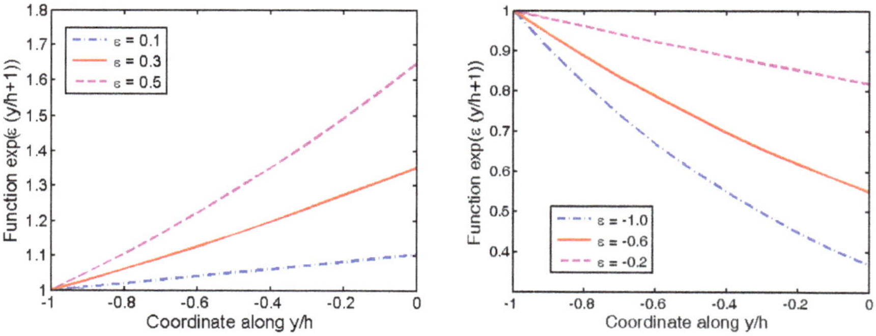

(a) (b)

Figure 2. Variation of the exponential function in Eq. (1) with non-dimensional coordinate y/h: (a) for positive ε, (b) for negative ε.

4.2 Stress intensity factors and fracture angles

To study the effects of inhomogeneity parameters of FGMs on the SIFs at the tips of the edge cracks, as well as on the other fracture characteristic, such as fracture angles, some examples for the problem with respect to loading, material parameters and geometry are considered.

The direction of the initial crack propagation (fracture angle) is evaluated by using the maximum circumferential stress criterion (Cherepanov, 1963; Erdogan and Sih, 1963; Panasyuk and Berezhnitskij, 1964, see for the references [26]) and is written as

$$\varphi_0 = 2\,\mathrm{arctg}\left[\left(k_I - \sqrt{k_I^2 + 8k_{II}^2}\right) / 4k_{II}\right] \tag{17}$$

For cracks in pure Mode II loading ($k_I = 0$) the fracture angle is calculated as $|\varphi_0| \approx 70.5^\circ$. For elastically homogeneous materials we can use this criterion without any assumptions. In the case of elastically non-homogeneous material it is supposed that the material is elastically homogeneous in the vicinity of the crack tips.

We will suppose that the cracks have equal lengths $a_1=a_2=a$ and have same inclination angles $\beta_1=\beta_2=\beta$ to the boundary. These assumptions are made for simplicity of the parametric analysis. Other crack geometries will be studied in the future in other work.

4.2.1 Two edge cracks in a homogeneous half-plane

For this case we have only tensile loading p parallel to the boundary of the half-plane, which corresponds to the function

$$p_n = \sigma_n - i\tau_n = p(1-\exp(-2i\alpha_n))/2, \quad (n=1,2) \tag{18}$$

in the integral equations (1) and (9), here $\alpha_n = -\beta_n$. Non-dimensional SIFs ($k_{I,II}/\sigma\sqrt{2a}$) Mode I and Mode II and fracture angles Eq. (18) are presented in the Figs. 3 – 6. Variation of these values with inclination angle $60^0 \le \beta \le 120^0$ of the cracks to the half-plane boundary is shown for different distances between the cracks: Figs. 3 and 4 for d/a = 1, 1.5, 2.5, and Figs. 5 and 6 for d/a = 1, 5, 10. The designation d for non-dimensional distance remains in the figures.

Small variation of the magnitude of k_I with changing β is observed in the Fig. 3 and $k_I < 1$ for all β and d, hence k_I is smaller than k_I for a solitary crack and shielding effect is observed. The SIF k_{II} decreases with increasing β from 60^0 to 120^0 and changes sign at $\beta \approx 105^0$ for the crack 1 and at $\beta \approx 75^0$ for the crack 2. It means also that at these angles we have pure mode conditions. Fig. 4 shows that the fracture angles increases with increasing β and change orientation (sign) at $\beta \approx 105^0$ for the crack 1 (Fig. 4a) and at $\beta \approx 75^0$ for the crack 2 (Fig. 4b) and it corresponds to the change of the sign of k_{II}, which characterizes the fracture angles in the mixed-mode conditions.

For close located cracks (d/a = 1, 1.5, 2) the influence of d on k_I and k_{II} is small (Fig. 3) but for d/a = 1, 5, 10 this influence is stronger (Fig. 5). For d/a = 10 and $\beta \approx 90^0$ $k_I > 1$, i.e. greater than k_I for a single crack in an infinite plane and close to the value k_I for a single edge crack.

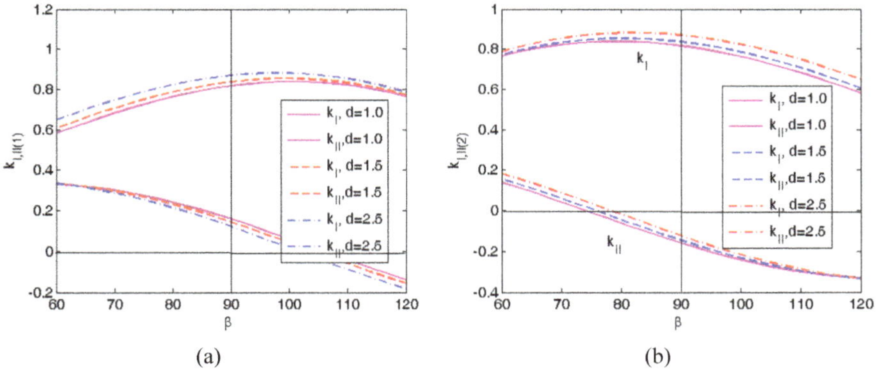

Figure 3. SIFs k_I and k_{II} as function of angle β ($\beta_n = \beta$, n=1,2) for two edge cracks: (a) for crack 1, (b) for crack 2. Homogeneous material.

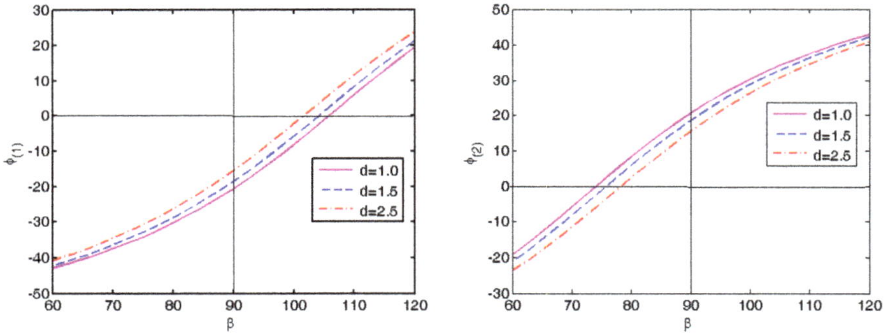

(a) (b)

Figure 4. Fracture angles as function of the angle β ($\beta_n = \beta$, n=1,2) for two edge cracks: (a) for crack 1, (b) for crack 2. Homogeneous material.

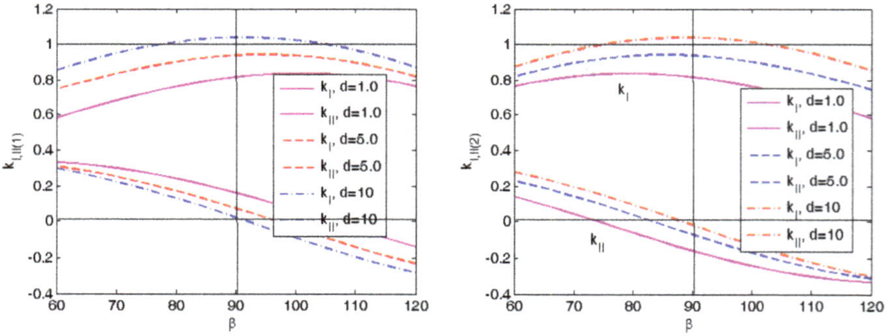

(a) (b)

Figure 5. SIFs k_I and k_{II} as function of angle β ($\beta_n = \beta$, n=1,2) for two edge cracks: (a) for crack 1, (b) for crack 2. Homogeneous material.

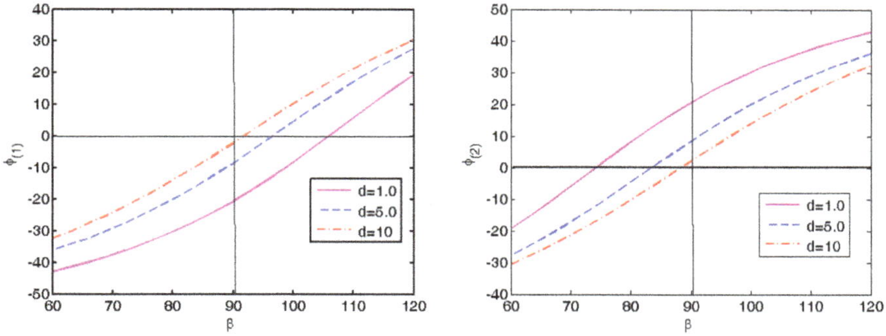

(a) (b)

Figure 6. Fracture angles as function of the angle β ($\beta_n = \beta$, n=1,2) for two edge cracks: (a) for crack 1, (b) for crack 2. Homogeneous material.

4.2.2 Elastically homogeneous material

It was mentioned above $E = E_0$ and $\sigma_{xx}^e(y) = 0$ for this case and, hence, the Eq. (3) is written as

$$p^* = \sigma_{xx}^T(y) = [\alpha_t(y) - \alpha_{t0}]\Delta T E(y)$$

$$p + p^* = Q[\, p / Q + [\exp(\varepsilon(h + y_k^0))\exp(-\varepsilon x_k \sin\beta_k) - 1]\,] \qquad (19)$$

$$Q = \alpha_{t0}\Delta T E_0$$

If we suppose that $p = Q$, i.e. mechanical and thermal loadings are equal, then

$$p + p^* = Q\exp(\varepsilon(h + y_k^0))\exp(-\varepsilon x_k \sin\beta_k) \qquad (20)$$

Otherwise the new parameter p / Q should be considered.

In the right side of the Eqs. (1) and (10) the function (18) with p determined by (20) should be used.

For elastically inhomogeneous materials under tension (without thermal loading, i.e. for $\sigma_{xx}^T(y) = 0$) the load Eq.(3) will be determined by the same expressions (19) and (20), where instead of thermal load Q will be the mechanical load p.

Figs. 7 - 10 show the results of calculation SIFs and fracture angles (maximum circumferential stress criterion was used) for two edge cracks. For this case of thermally non-homogeneous materials we have only one inhomogeneity parameter ε of the thermal expansion coefficient. The SIFs are presented in the non-dimensional form $k_{I,II} / Q\sqrt{2a}$. The calculation were performed for the non-dimensional distances d/a = 1, 1.5, 2.5 and the non-dimensional h/a = 4. (The designation d was used in the figures for the non-dimensional values.) Two values of the inhomogeneity parameter are used, $\varepsilon = -1$ and 0.5. The result is obtained on the basis of the solution of the system (10) – (11) for a special case where at right side of Eq. (1) and accordingly in Eq. (9) is the function (18) with (20).

Figs. 7 and 8 show k_I, k_{II} and fracture angles ϕ as functions of β for $\varepsilon = -1$, which corresponds to smaller value of the thermal expansion coefficient in the upper part of FGM layer. Comparing this case and the previous homogeneous case we see similar trends in the results, but with some non-linearity and with different magnitudes. The values of k_I, k_{II} and ϕ are small for this case.

Results for $\varepsilon = 0.5$ for the case, where the thermal expansion coefficient is larger in the upper part of the FGM layer, are presented in Fig. 9. The values of k_I are much larger than in the previous case for $\varepsilon = -1$ and for the homogeneous case. The variation of k_I with β is small, the changes of k_{II} with β are larger and for both k_I and k_{II} are almost linear. The variation of ϕ (Fig. 9) is similar to the previous cases.

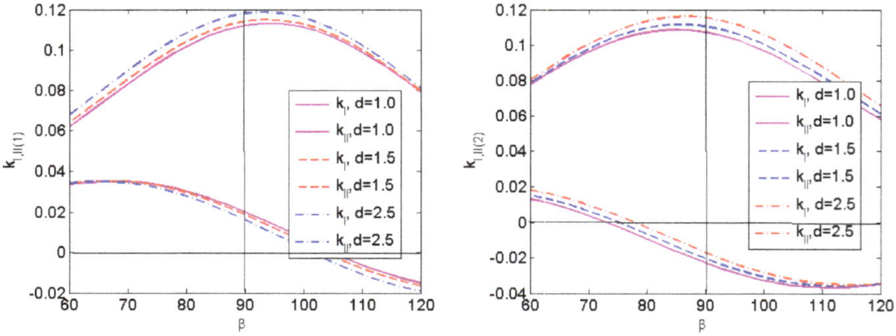

(a) (b)

Figure 7. SIFs k_I and k_{II} as function of angle β ($\beta n = \beta$, $n=1,2$) for two edge cracks: (a) for crack 1, (b) for crack 2; for $\varepsilon = -1$. Thermo-mechanical loading, $E = E_0$.

Materials Research Foundations **159** (2024) https://doi.org/10.21741/9781644902950

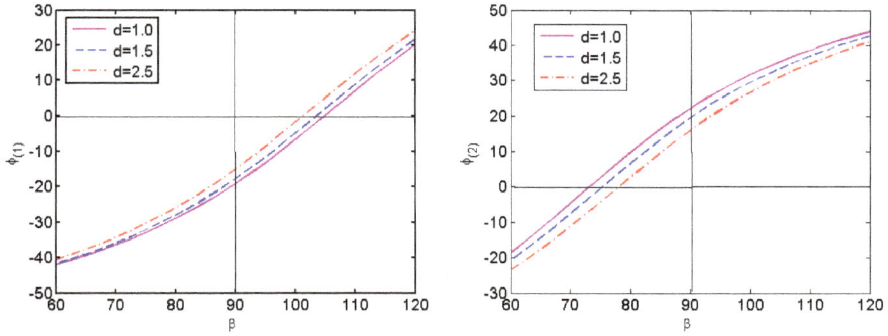

(a) (b)

Figure 8. Fracture angles as function of the angle β (βn = β, n=1,2) for two edge cracks: (a) for crack 1, (b) for crack 2; for ε= – 1. Thermo-mechanical loading, E = E*0*.

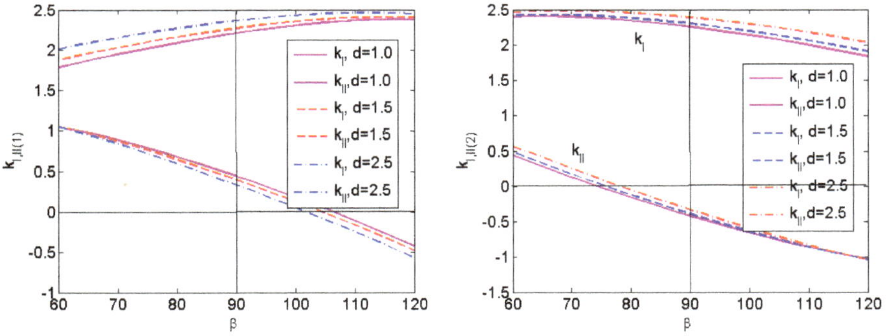

(a) (b)

Figure 9. SIFs k_I and k_{II} as function of angle β (βn = β, n=1,2) for two edge cracks: (a) for crack 1, (b) for crack 2; for ε= 0.5. Thermo-mechanical loading, E = E*0*.

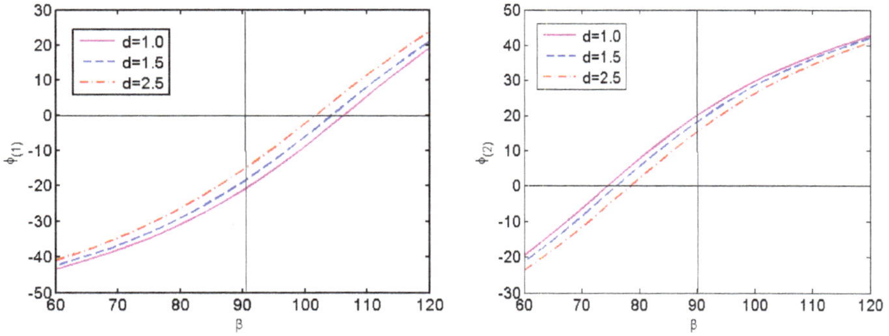

(a) (b)

Figure 10. Fracture angles as function of the angle β ($\beta_n = \beta$, n=1,2) for two edge cracks: (a) for crack 1, (b) for crack 2; for ε= 0.5. Thermo-mechanical loading, E = E_0.

4.2.3 Inhomogeneous material, thermal and mechanical loadings

For this case

$$p+p^{*}=p+\sigma_{xx}^{e}+\sigma_{xx}^{T}=p\exp(\delta(h+y_{k}^{0}))\exp(-\delta x_{k}\sin\beta_{k}) \\ +Q[\exp(\varepsilon(h+y_{k}^{0}))\exp(-\varepsilon x_{k}\sin\beta_{k})-1]\,] \\ =Q[p/Q\,\exp(\delta(h+y_{k}^{0}))\exp(-\delta x_{k}\sin\beta_{k})+\exp(\varepsilon(h+y_{k}^{0}))\exp(-\varepsilon x_{k}\sin\beta_{k})-1] \tag{21}$$

For simplicity of the parametric analysis we assume that the inhomogeneity parameters ε and δ of thermal expansion coefficient and Young's modulus are equal and, besides, $p = Q$ as it was in the previous case for thermally inhomogeneous materials.

Figs. 11 and 12 show the results of calculation of SIFs for this case of loading. The SIFs are presented in the non-dimensional form $k_{I,II}/Q\sqrt{2a}$. Other parameters are the same as in the previous cases, i.e. d/a = 1, 1.5, 2.5 and h/a = 4. The results for ε = –0.3, where the thermal expansion coefficient is larger in the upper FGM, are presented in Fig. 11 and they have similar trends as in the thermally inhomogeneous case (Fig. 7, ε = –1) with very small values of k_I and k_{II}. Fig. 12 show the values of k_I and k_{II} for ε = 0.3 and these results are similar to the previous case shown in Fig. 9 (for thermally inhomogeneous materials for ε = 0.5) with slightly different magnitudes of k_I and k_{II}.

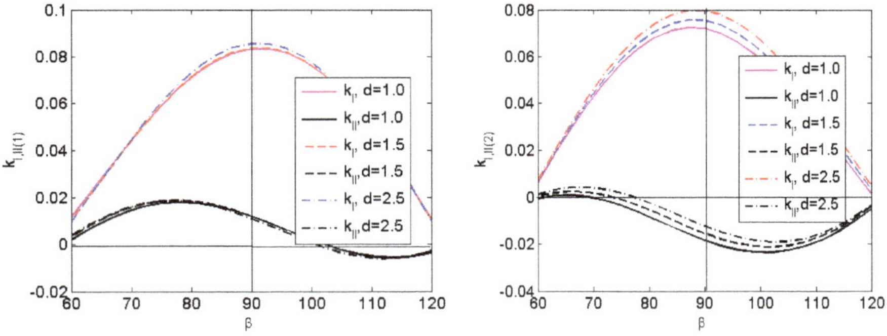

(a) (b)

Figure 11. SIFs kI and kII as function of angle β (βn = β, n=1,2) for two edge cracks: (a) for crack 1, (b) for crack 2; for ε= – 0.3. Thermo-mechanical loading.

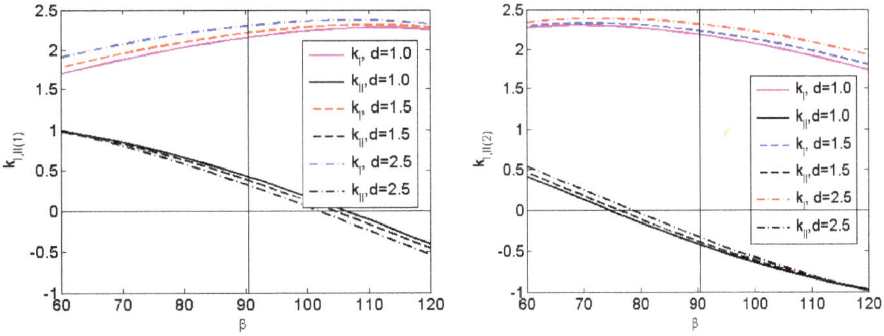

(a) (b)

Figure 12. SIFs kI and kII as function of angle β (βn = β, n=1,2) for two edge cracks: (a) for crack 1, (b) for crack 2; for ε= 0.3. Thermo-mechanical loading.

Conclusions

Theoretical modeling of thermal fracture of a semi-infinite FGM is presented for the case of one cycle of thermo-mechanical loading. This study can serve as a part of the modeling of the fracture process in FGM coatings under cyclic heating-cooling thermal loading.

Influence of geometrical and material (inhomogeneity) parameters on the fracture characteristics is investigated. Strong influence of the inhomogeneity parameter ε of the material on SIFs is observed. If ε is negative, which corresponds to smaller values of the thermal expansion coefficient in the upper part of the FGM layer, the SIFs k_I and k_{II} at the edge crack tips are small and have much smaller values in comparison to the homogeneous material. For positive ε, the case where magnitudes of the thermal expansion coefficient have larger values in the upper FGM layer, the SIF k_I is much larger than the values for the negative ε and for the homogeneous material. The magnitude of SIF k_{II} is also larger, but not so much as for k_I. The fracture angles have similar tendencies for all considered cases and depend mainly on the inclination angle of the cracks to the boundary. The maximum circumferential stress criterion was applied and probably other criterion should be used in order to take into account the inhomogeneity material properties.

Acknowledgement

The research leading to these results has received funding from the European Union Seventh Framework Programme (FP7/2007 – 2013), FP7 - REGPOT – 2009 – 1, under grant agreement No: 245479; CEMCAST. The support by Polish Ministry of Science and Higher Education - Grant No 1471-1/7.PR UE/2010/7 - is also acknowledged.

References

[1] Y. Miyamoto, W.A. Kaysser, B.H. Rabin, Functionally graded materials: design, processing and applications. Kluwer, Netherlands, Dordrecht, 1999. https://doi.org/10.1007/978-1-4615-5301-4

[2] V. Birman, L.W. Byrd, Modeling and analysis of functionally graded materials and structures, ASME Appl. Mech. Rev. 60 (2007) 195-216. https://doi.org/10.1115/1.2777164

[3] T. Sadowski, Non-symmetric thermal shock in ceramic matrix composite (CMC) materials. In: de Bost R, Sadowski T (eds) Lecture Notes on Composite Materials – Current Topics and Achievements. Springer, Berlin, 2008, pp. 99-148. https://doi.org/10.1007/978-1-4020-8772-1_4

[4] A. Kawasaki, R. Watanabe, Thermal fracture behavior of metal/ceramic functionally graded materials, Eng. Fract. Mech. 69 (2002) 1713–1728. https://doi.org/10.1016/S0013-7944(02)00054-1

[5] A. Shukla, N. Jain, R. Chona, A review of dynamic fracture studies in functionally graded materials, Strain 43 (2007) 76-95. https://doi.org/10.1111/j.1475-1305.2007.00323.x

[6] G. Qian, T. Nakamura, C.C. Berndt, Effects of thermal gradient and residual stresses on thermal barrier coating fracture, Mech. Materials 27 (1998) 91-110. https://doi.org/10.1016/S0167-6636(97)00042-2

[7] G. Anlas, M.H. Santare, J. Lambros, Numerical calculation of stress intensity factors in functionally graded materials, Int. J. Fract. 104 (2000) 131-143. https://doi.org/10.1023/A:1007652711735

[8] G. Anlas, J. Lambros, M.H. Santare, Dominance of asymptotic crack tip fields in elastic functionally graded materials, Int. J. Fract. 115 (2002) 193-204. https://doi.org/10.1023/A:1016372120480

[9] A.M. Afsar, H. Sekine, Crack spacing effect on the brittle fracture characteristics of semi-infinite functionally graded materials with periodic edge cracks, Int. J. Fract. 102 (2000) L61-L66.

[10] N.-A. Noda, K. Oda, Numerical solution of the singular integral equations in the crack analysis using the body force method, Int. J. Fract. 58 (1992) 285-304. https://doi.org/10.1007/BF00048950

[11] B. Yıldırım, Özge Kutlu, S. Kadıoglu, Periodic crack problem for a functionally graded half-plane an analytic solution, Int. J. Solids Struct. 48 (2011) 3020-3031. https://doi.org/10.1016/j.ijsolstr.2011.06.019

[12] S. El-Borgi, M.F. Djemel, R. Abdelmoula, A surface crack in a graded coating bonded to a homogeneous substrate under thermal loading, J. Therm. Stresses 31 (2008) 176-194. https://doi.org/10.1080/01495730701737886

[13] V. Petrova, S. Schmauder, Interaction of a system of cracks with an interface crack in functionally graded/homogeneous bimaterials under thermo-mechanical loading, Comp. Mater. Sci. 64 (2012) 229-233. https://doi.org/10.1016/j.commatsci.2012.04.032

[14] V. Petrova, S. Schmauder, Modelling of thermal fracture of functionally graded/ homogeneous bimaterial structures under thermo-mechanical loading. Key Eng. Mater. 592-593 (2014) 145-148. https://doi.org/10.4028/www.scientific.net/KEM.592-593.145

[15] T. Sadowski, A. Neubrand, Estimation of the crack length after thermal shock in FGM strip, Int. J. Fract. 127 (2004) L135-L140. https://doi.org/10.1023/B:FRAC.0000035087.34082.88

[16] J. Sladek, V. Sladek, Ch. Zhang, An advanced numerical method for computing elastodynamic fracture parameters in functionally graded materials, Comp. Mater. Sci. 32 (2005) 532-543. https://doi.org/10.1016/j.commatsci.2004.09.011

[17] A.M. Afsar, H. Sekine, Inverse problems of material distributions for prescribed apparent fracture toughness in FGM coating around a circular hole in infinite elastic media, Compos. Sci. Technol. 62 (2002) 1063-1077. https://doi.org/10.1016/S0266-3538(02)00049-0

[18] V. Petrova, T. Sadowski, Theoretical modeling and analysis of Mode II cracks in Compact Shear specimens. In: ECCOMAS 2012 - European Congress on Computational Methods in Applied Sciences and Engineering, e-Book Full Papers, (2012) 2601-2611.

[19] F. Erdogan, B.H. Wu, The surface crack problem for a plate with functionally graded properties, ASME J. Appl. Mech. 64 (1997) 449-456. https://doi.org/10.1115/1.2788914

[20] F. Erdogan, B.H. Wu, Crack problems in FGM layers under thermal stresses, J. Therm. Stress 19 (1996) 237-265. https://doi.org/10.1080/01495739608946172

[21] V.V. Panasyuk, M.P. Savruk, A.P. Datsyshin, Stress Distribution near Cracks in Plates and Shells (in Russian), Naukova Dumka, Kiev, 1976.

[22] M.P. Savruk, Two- Dimensional Problems of Elasticity for Body with Cracks (in Russian), Naukova Dumka, Kiev, 1981.

[23] F. Erdogan, G. Gupta, On the numerical solution of singular integral equations, Quart. Appl. Math. 29 (1972) 525-534. https://doi.org/10.1090/qam/408277

[24] Y. Murakami (ed.), Stress Intensity Factors Handbook. In 2 Volumes. Pergamon press, Oxford etc. (1987)

[25] C.E. Freese, Periodic edge cracks of unequal length in a semi-infinite tensile sheet, Int. J. Fract. 12 (1976) 125-134. https://doi.org/10.1007/BF00036015

[26] J.F. Shackelford, W. Alexander, CRC Materials Science and Engineering Handbook.

CRC Press, Boca Raton, 2001. https://doi.org/10.1201/9781420038408
[27] F. Erdogan, G.C. Sih, On the crack etension in plates under plane loading and transverse shear, J. Basic Eng. 85 (1963) 519-527. https://doi.org/10.1115/1.3656897
[28] G.C. Sih, Strain-energy-density factor applied to mixed-mode crack problems. Int. J. Fract. 10 (1974) 305-321. https://doi.org/10.1007/BF00035493

Appendix

The regular kernels R_{nk}(t,x) and $S_{nk}(t,x)$ contain geometry of the problem and are written as

$$R_{nk}(t,x)=(1-\delta_{nk})K_{nk}(t,x)+\frac{e^{i\alpha_k}}{2}\left\{\frac{1}{X_n-\overline{T}_k}+\frac{e^{-2i\alpha_n}}{\overline{X}_n-T_k}+\right.$$
$$\left.+(\overline{T}_k-T_k)\left[\frac{1+e^{-2i\alpha_n}}{(\overline{X}_n-T_k)^2}-\frac{2e^{-2i\alpha_n}(X_n-T_k)}{(\overline{X}_n-T_k)^3}\right]\right\}, \quad \text{(A.1)}$$

$$S_{nk}(t,x)=(1-\delta_{nk})L_{nk}(t,x)+\frac{e^{-i\alpha_k}}{2}\left[\frac{T_k-\overline{T}_k}{(X_n-\overline{T}_k)^2}+\frac{1}{\overline{X}_n-T_k}-e^{-2i\alpha_n}\frac{X_n-\overline{T}_k}{(X_n-\overline{T}_k)^2}\right] \quad \text{(A.2)}$$

$$T_k=te^{i\alpha_k}+z_k^0,\quad X_n=xe^{i\alpha_n}+z_n^0,\quad n,k=1,2. \quad \text{(A.3)}$$

and

$$\delta_{nk}=\begin{cases}0 & for\ \ n\neq k;\\ 1 & for\ \ n=k.\end{cases}$$

The kernels K_{nk}(t,x) and L_{nk}(t,x) are

$$K_{nk}(t,x)=\frac{e^{i\alpha_k}}{2}\left(\frac{1}{T_k-X_n}+\frac{e^{-2i\alpha_n}}{\overline{T}_k-\overline{X}_n}\right); \quad \text{(A.4)}$$

$$L_{nk}(t,x)=\frac{e^{-i\alpha_k}}{2}\left(\frac{1}{\overline{T}_k-\overline{X}_n}+\frac{T_k-X_n}{(\overline{T}_k-\overline{X}_n)^2}e^{-2i\alpha_n}\right), \quad \text{(A.5)}$$

are the same as for the system of cracks in an infinite plane. The additional terms in Eqs. (A.1) and (A.2) are taking into account the influence of the edge of the half plane.

α_n is the inclination angle of *n*-th crack to the *x*-axis coincide with the edge of the half plane and $\alpha_n=-\beta_n$, Fig. 1; z^0_n is the coordinate of the center of crack in global coordinate system (*x,y*).

Materials Research Foundations **159** (2024) https://doi.org/10.21741/9781644902950

CHAPTER 8

Modeling of Edge Cracks Interaction

Vera Petrova [1,2]*, Siegfried Schmauder [1], M. A. Shashkin [2]

[1] IMWF, University of Stuttgart, Pfaffenwaldring 32, D-70569 Stuttgart, Germany

[2] Voronezh State University, University Sq.1, Voronezh 394006, Russia

veraep@gmail.com *, Siegfried.Schmauder@imwf.uni-stuttgart.de, shashkin@amm.vsu.ru

Abstract

From experimental and theoretical investigations it is known that cracks are sensitive to geometry, e.g. to the inclination angle to the load. A small deviation of a crack from the normal direction to a tensile load causes mixed mode conditions near the crack tip which lead to deviation of the crack from its initial propagation direction. Besides, the presence of other cracks, inhomogeneities, surfaces and their interaction causes additional deformations and stresses which also have influence on the initiation of the crack propagation and on the direction of this propagation. The aim of this paper is to show the effects of the interaction of edge cracks on further crack formation. The main fracture characteristics, such as, stress intensity factors, fracture angles and critical loads are provided in this study. A series of illustrative examples is presented for different geometries of arbitrarily inclined edge cracks.

Keywords

Edge Cracking; Stress Intensity Factors; Fracture Criteria; Direction Of Crack Propagation; Shielding-Amplification Effects

1. Introduction

Surface cracking is observed in many engineering structures, e.g. aircraft structures, turbine blades, engine components and many others, see [1, 2] for some examples and references. In everyday life we can see asphalt pavement cracking, called also as crocodile cracking [3]. The structures are subjected to different mechanical and thermal loading as well have to resist high temperature, wear and aggressive environments. Cracks can initiate from initial defects or microcracks appear during manufacturing or service. An example of multiple surface cracking is the fracture of thermal barrier coatings (TBCs), where the upper layer is usually a ceramic - the brittle material. Investigations of thermal barrier coatings show that heating and then subsequent cooling of the coating causes the surface to experience a tensile stress leading to surface cracking [4].

There is abundant experimental results (e.g. [4-6]) showing that when TBCs are subjected to thermal shock, multiple cracks occur at the ceramic surface. Besides, the crack patterns strongly depend on the microstructure of the materials and on the type of loading.

Numerous investigations are devoted to different types of fracture including surface fracture. In previous papers of the authors [7-10] the fracture of FGM/homogeneous bimaterials (an infinite medium) under thermal and mechanical loadings were investigated, besides, in [11] some results for edge cracks in FGM/homogeneous structures (a semi-infinite medium) were obtained in the frame of the approach used in [7-10]. The results show that the fracture of materials (both composites and homogeneous) is significantly affected by a complex crack interaction mechanism, e.g. interacting cracks can enhance or suppress the propagation of each other.

During further studying of the fracture of functionally graded coatings on a homogeneous substrate and the preparation of the results for the influence of material non-homogeneity on surface fracture it became clear that a classical problem for edge cracks interaction is still not well examined. Before presenting the results for more complicated cases of non-homogeneous materials (FGMs, bimaterials, and others), modeling of the interaction of edge cracks should be done for a homogeneous medium.

From experimental and theoretical investigations it is known that cracks are sensitive to geometry, e.g. to the inclination angle to the load. A small deviation of a crack from the normal direction to a tensile load causes mixed mode conditions near the crack which lead to deviation of the crack from its initial propagation direction. Besides, the presence of other cracks, inhomogeneities, surfaces and their interaction causes additional deformations and stresses which are also influenced on the initiation of the crack propagation and on the direction of this propagation. That is, the picture of the fracture with respect to the crack pattern for a system of arbitrary inclined edge cracks will be different from the picture of regularly distributed cracks, e.g, for periodically distributed equal (and non-equal) cracks, this case was often studied, see [12-14].

The goal of this paper is to show the effects of the interaction of edge cracks on further fracture formation. The main fracture characteristics, such as, stress intensity factors, fracture angles and critical loads are provided for this study. A series of illustrative examples is presented for different geometries of arbitrarily inclined edge cracks.

2. Problem formulation and assumptions

The geometry of the problem is presented in Fig. 1 a. A homogeneous half-medium contains pre-existing edge cracks inclined arbitrarily on angles β_n to the surface. A Cartesian coordinate system (x, y) has x-axis along the boundary of the half-plane, and local coordinate systems (x_k, y_k) are attached to each crack. The lengths of the cracks are $2a_n$, and the midpoint coordinates are $z_n^0 = x_n^0 + iy_n^0$ ($i = \sqrt{-1}$ is imaginary unity). The homogeneous medium is subjected to tension p applied parallel to the free surface.

The problem is solved by using the method of singular integral equations. The cracks are modeled by displacement jumps on the crack faces and unknown functions in this formulation are the derivatives of displacement jumps

$$g_n'(x) = \frac{2\mu}{i(\kappa+1)} \frac{\partial}{\partial x}\left([u_n] + i[v_n]\right) \tag{1}$$

Here $[u_n]$ and $[v_n]$ are shear and vertical displacement jumps, respectively, on the n-th crack line, $\mu = E/2(1+\nu)$ is the shear modulus, E - Young's modulus, ν - Poisson's ratio, $\kappa = 3-4\nu$ for the plane strain state and $\kappa = (3-\nu)/(1+\nu)$ for the plane stress state.

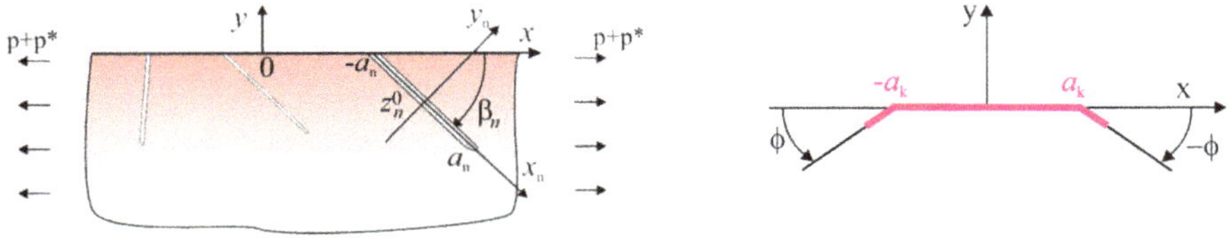

(a) (b)

Figure 1. (a) Edge cracks inclined arbitrarily with an angle βn to the surface of the medium. a_n – a half-length of n-th crack, zn0 = xn0 +i yn0 – the crack midpoint coordinate. (b) The angle ϕ of crack deflection (the fracture angle).

For arbitrary located cracks in a half-plane the system of singular integral equations is written as [15, 16]

$$\int_{-a_n}^{a_n} \frac{g_n'(t)dt}{t-x} + \sum_{\substack{k=1\\k\neq n}}^{N} \int_{-a_k}^{a_k} [g_k'(t)R_{nk}(t,x) + \overline{g_k'(t)}S_{nk}(t,x)]dt = \pi p_n(x), \quad |x| < a_n,$$

$$n = 1, 2, \ldots, N. \tag{2}$$

An overbar $\left(\overline{\ldots}\right)$ is the complex conjugate. N is number of cracks.

The method of superposition was used in deriving of Eq. (2) where the loads at infinity are reduced to the corresponding loads on the crack faces. The functions p_n in the right side of Eq. (2) are these loadings, and in the case of a homogeneous half-plane under tension p they are written as

$$p_n = \sigma_n - i\tau_n = p(1-\exp(-2i\alpha_n))/2, \quad (\mathrm{n} = 1, 2, .., N) \tag{3}$$

with $\alpha_n = -\beta_n$ (see Fig. 1 a).

If a non-homogeneous medium is considered, e.g. a functionally graded structure with continuous gradation of the thermo-mechanical properties with the coordinate y, and this structure is cooled, then tensile residual stresses are arising due to mismatch in the coefficients of thermal expansion [4, 14]. The influence of this inhomogeneity can be accounted via continuously varying residual stresses p^* which are written as follows [14]:

$$p^* = \sigma_{xx}^T(y) = [\alpha_t(y) - \alpha_{t0}]\Delta TE.$$

This function is added to the right side of Eq. (2). It should be noted, that in this case we also have the problem for a half-plane under tension.

3. Numerical solutions, Stress intensity factors

The solution of singular integral equations (Eq. 2) is obtained by a numerical method which is based on Gauss-Chebyshev quadrature. The method is similar to the method presented by Erdogan and Gupta [17], but we will follow the version formulated in [15, 16].

The equations (2) are rewritten in dimensionless form with the non-dimensionless coordinates $\xi = t / a_k$ and $\eta = x / a_n$, where $2a_k$ is a length of the k-th crack. The unknown function $g_n'(\eta)$ consists of a function $u_n(\eta)$ (a bounded continuous function in the segment [-1, 1]) and the weight function $1/\sqrt{1-\eta^2}$, that is,

$$g_n'(\eta) = u_n(\eta) / \sqrt{1-\eta^2}\ . \tag{4}$$

For edge cracks the function $g_n'(\eta)$ possess a singularity less than $1/\sqrt{1+\eta}$ at the edge point $\eta = -1$ and this condition is accounted as [15, 16]

$$u_n(-1) = 0\,. \tag{5}$$

In spite of the exact singularity at the edge points is not taking into account, the numerical results have shown good accuracy [15, 16].

Using Gauss's quadrature formulae for the regular and singular integrals the integral equations are reduced to the following system of NxM (N – number of cracks, M – number of nodes) algebraic equations

$$\frac{1}{M}\sum_{m=1}^{M}\sum_{k=1}^{N}\left[u_k(\xi_m)R_{nk}(\xi_m,\eta_r)+\overline{u_k(\xi_m)}S_{nk}(\xi_m,\eta_r)\right]=\pi p_n(\eta_r), \tag{6}$$

$$\sum_{m=1}^{M}(-1)^m u_n(\xi_m)\tan\frac{2m-1}{4M}\pi = 0 \qquad (n=1, 2, \ldots, N;\ r=1,2,\ldots, M-1) \tag{7}$$

with

$$\xi_m = \cos\frac{2m-1}{2M}\pi \ \ (m=1, 2,\ldots, M);\ \ \eta_r = \cos\frac{\pi r}{M} \qquad (r = 1, 2, \ldots, M-1)$$

M is the total number of discrete points of the unknown functions $u_n(\eta)$ within the interval (-1, 1). Applying the conjugate operation to the system (6) additional NxM equations are obtained, i.e. $2NxM$ equations should be solved, where N is the number of cracks.

Eq. (7) is obtained from the condition (5) and the interpolation formula for the functions $u_n(\eta)$:

$$u_n(\eta) = \frac{2}{M}\sum_{m=1}^{M} u_n(\xi_m)\sum_{r=0}^{M-1} T_r(\xi_m)T_r(\eta) - \frac{1}{M}\sum_{m=0}^{M} u_n(\xi_m). \tag{8}$$

Here T_r are Chebyshev polynomials of the first kind. Inserting (8) into Eq. (4) the derivative of displacement jumps on the crack lines are obtained and then the displacement jumps can be derived by integrating the function (4) with (8).

The stress intensity factors (SIFs) are calculated according to the following formula

$$K_{In} - iK_{IIn} = -\sqrt{a_n}u_n(+1)$$

$$= p_n\sqrt{a_n}\frac{1}{M}\sum_{m=1}^{M}(-1)^m u_n(\xi_m)\cot\frac{2m-1}{4M}\pi \quad (n = 1, 2,\dots, N). \tag{9}$$

4. Critical loads, Fracture angles

For general crack problems the stress intensity factors are both nonzero, i.e. mixed-mode conditions are in the vicinity of cracks. For this mixed-mode case the cracks deviate from their initial propagation direction. For the prediction of the crack growth and direction of this growth a fracture criterion should be applied. Using the maximum circumferential stress criterion (see [18] and for references [15, 16]) the direction of the initial crack propagation (Fig. 1 b) is evaluated as

$$\varphi = 2\arctan\left[\left(K_I - \sqrt{K_I^2 + 8K_{II}^2}\right)/4K_{II}\right] \tag{10}$$

and the critical stresses can be calculated from the expression

$$\cos^3(\varphi/2)\left(K_I - 3K_{II}\tan(\varphi/2)\right) = K_{Ic}/\sqrt{\pi}. \tag{11}$$

Here K_{Ic} is the fracture toughness of the material. The critical stresses are given as

$$p_{cr} = P_{cr}/p_0 = P_{cr}/(K_{Ic}/\sqrt{2\pi a}) = 1/[\cos^3(\varphi/2)\left(k_I - 3k_{II}\tan(\varphi/2)\right)]. \tag{12}$$

Here $k_{I,II}$ are non-dimensional SIFs

$$k_{I,II} = K_{I,II}/K^0, \quad K^0 = p\sqrt{2a} \tag{13}$$

and $p_0 = K_{Ic}/(2\pi a)^{1/2}$ is critical load for a single crack in a material with the fracture toughness K_{Ic}.

For the system of cracks the fracture starts from the crack tip where P_{cr} is minimal, i.e. $\min_k[P_{cr(k)}/p_0]$.

Materials Research Foundations **159** (2024) https://doi.org/10.21741/9781644902950

5. Results and discussion

Some examples for edge crack interaction is investigated and presented here for homogeneous materials. The verification of the method and the numerical outcomes has been done in [11], where the results for some special cases were compared with the results for SIFs for a single inclined edge crack cited in [19] and with SIFs for periodic edge cracks cited in [20].

The tensile loading p is applied parallel to the boundary and on the crack lines we have the loading Eq. (3). The non-dimensional stress intensity factors Mode I and Mode II ($k_{I,II}$) are defined by Eqs. (9) and (13). Non-dimensional k_I for a single edge crack normal to the surface is equal to $k_I = 1.12$ and SIF k_{II} is $k_{II} = 0$.

The non-dimensional distances $d = \hat{d} / a$ between the cracks are $d = 1, 2, 4, 6$, $a = \max_k a_k$ and we remind that $2a_k$ is the size of the *k-th* crack. After obtaining SIFs the fracture angles ϕ are calculated by Eq. (10) and critical loads p_{cr} by Eq. (12).

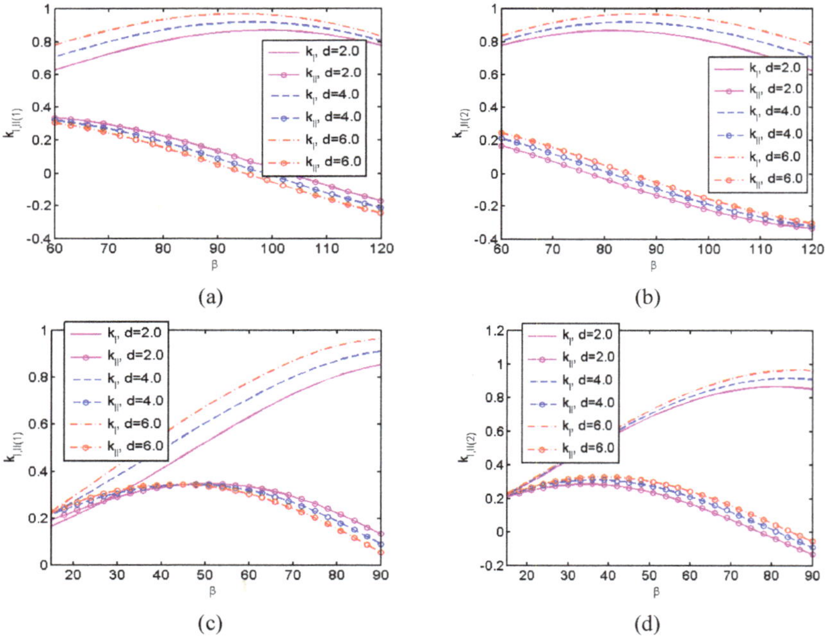

Figure 2. Stress intensity factors kI and kII as functions of the inclination angle β=βn of the edge cracks to the surface for different distances d between the cracks: (a) for crack 1 (60°≤ β ≤120°), (b) for crack 2 (60°≤ β ≤120°), (c) for crack 1 (15°≤ β ≤90°), (d) for crack 2 (15°≤ β ≤90°). Two equal edge cracks.

Materials Research Foundations **159** (2024) https://doi.org/10.21741/9781644902950

5.1 Two arbitrary inclined cracks

Figs. 2, 5 and 8 show the SIFs $k_{I,II}$, Figs. 3, 6 and 9 – the fracture angles, and Figs. 4, 7 and 10 – the critical loads as functions of inclination angles of the two edge cracks to the surface and for different distances *d*.

It is observed that for all angles β SIF k_I increases with increasing the distance *d* between the cracks and k_I tends to the value for a single edge crack, e.g. to $k_I = 1.12$ and $k_{II} = 0$ at $\beta = 90°$. Besides, for all parameters of the problem the values of k_I are smaller than the values of k_I for a single crack (Figs. 2, 5 and 8). That is, the shielding effect is observed, which is known for parallel cracks under tensile load normal to the crack lines.

Figs. 2–4 present results for two equal edge cracks inclined arbitrarily to the surface with the same angle $\beta=\beta_n$ (n=1, 2). Stress intensity factors k_I and k_{II} as functions of the inclination angle β are presented for the angles $60° \leq \beta \leq 120°$ in Figs. 2 a, b and for $15° \leq \beta \leq 90°$ in Figs. 2 c, d and for different distances *d* between the cracks.

In the interval $60° \leq \beta \leq 120°$ a small variation of the magnitude of k_I with β is observed (Fig. 2 a, b), but in the interval $15° \leq \beta \leq 60°$ for the small inclinations angles this variation is significant (Figs. 2 b, c). k_I is increased from 0.2 to 0.99 for $15° \leq \beta \leq 90°$ (for d=2) and then decreased for $90° \leq \beta \leq 120°$.

SIFs k_{II} are mostly nonzero, the absolute values of k_{II} are greater than k_I, and k_{II} is monotonically decreased for $60° \leq \beta \leq 120°$ (Fig. 2 a, b) and increased for $15° \leq \beta \leq 45°$ (Figs. 2 b, c).

The fracture angles ϕ for two edge cracks are presented in Figs. 3, strong influence of the inclination angles β on the fracture angles ϕ is observed. For all β fracture angles ϕ are increased and changed the sign from negative to positive at $\beta \approx 103°$ (crack 1) and at $\beta \approx 77°$ (crack 2) for d=2, for larger distances *d* these points are shifted towards $\beta \approx 99°$ (crack 1) and $\beta \approx 81°$ (crack 2) for d=4 (Fig. 3 a, b). These changes of sign mean the changes of direction of the crack propagation.

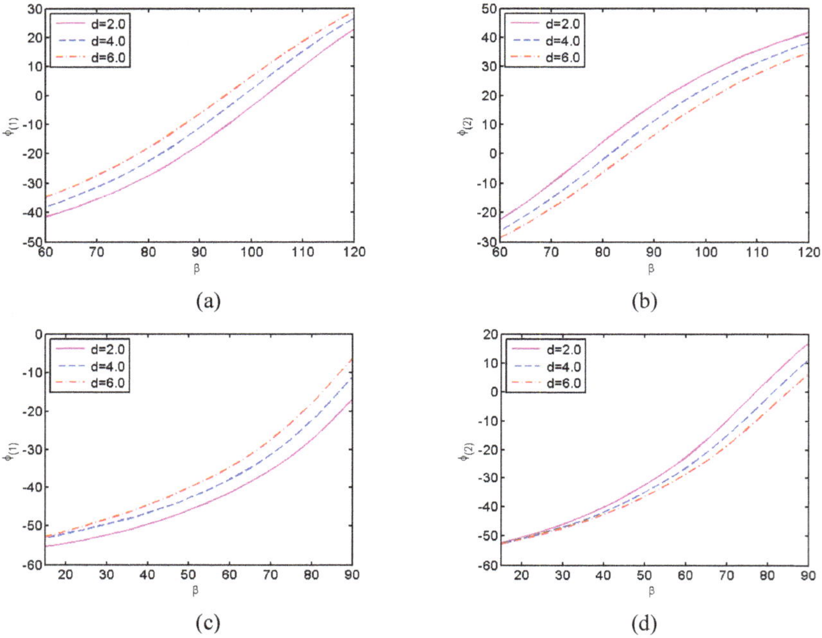

Figure 3. Fracture angles $\phi(1)$ and $\phi(2)$ as functions of the inclination angle $\beta=\beta n$ of the edge cracks to the surface for different distances d between the cracks: (a) for crack 1 ($60°\leq\beta\leq120°$), (b) for crack 2 ($60°\leq\beta\leq120°$), (c) for crack 1 ($15°\leq\beta\leq90°$), (d) for crack 2 ($15°\leq\beta\leq90°$). Two equal edge cracks.

Fig. 4 shows results for the non-dimensional critical loads for crack 1 and for both cracks in Fig. 4 b ($60°\leq \beta \leq 120°$) and 4 d ($15°\leq \beta \leq 90°$). The larger the distance between the cracks – the less the p_{cr}, i.e. the material becomes weaker with respect to fracture resistance. What crack starts to propagate first depends on the inclination angle, for $62°\leq \beta \leq 90°$ $p_{cr(1)}< p_{cr(2)}$ and the crack 1 propagates first and for $90°< \beta<118°$ the crack 2 will be starting first (Fig. 4 b). For small angles $15°\leq \beta \leq 62°$, where the crack 2 is close to the free surface, $p_{cr(2)}< p_{cr(1)}$ and the crack 2 will propagate first (Fig. 4 d). The opposite picture for critical loads and crack propagation is observed for other values of β.

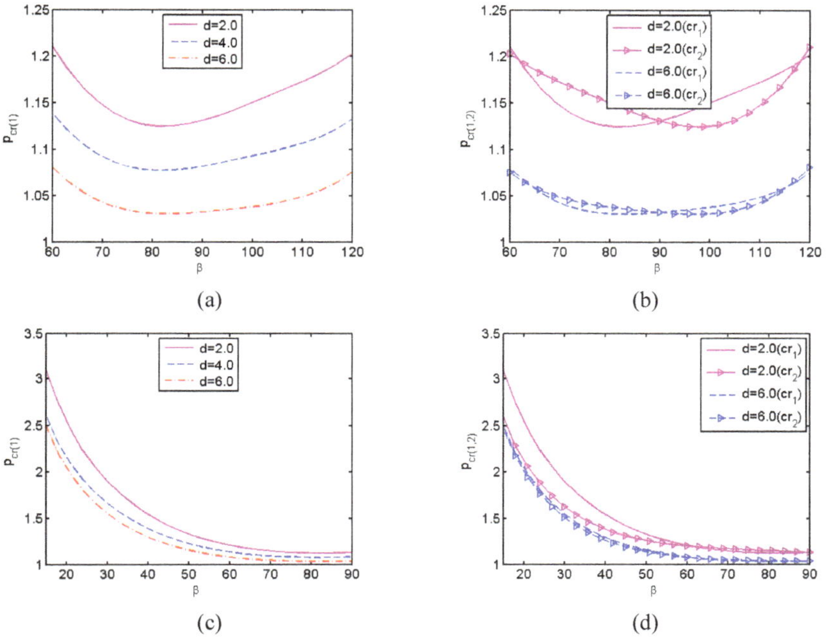

Figure 4. The non-dimensional critical load P_{cr} as function of the inclination angle $\beta=\beta n$ for two equal edge cracks for different distances d between the cracks: (a) for crack 1 ($60°\le\beta\le120°$), (b) for two cracks ($60°\le\beta\le120°$), (c) for crack 1 ($15°\le\beta\le90°$), (d) for two cracks ($15°\le\beta\le90°$). Two equal edge cracks.

Another case for two cracks with different sizes is presented in Figs. 5–7. It is assumed that non-dimensional crack sizes (a_k/a, $a=\max\limits_k a_k$) are a_1=1 and a_2=0.5, and as previously $\beta=\beta_n$ (n = 1, 2). It is observed, that distance d has small influence on the fracture parameters for the crack 1, i.e. SIFs, fracture angles and critical loads are nearly the same for the considered d-values d=2, 4 and 6 (Figs. 5 a, c, 6 a, c and 7 a, c). However, the influence of d on the crack 2 is strong (Figs. 5 b, d, 6 b, d and 7 b, d). Both cracks are sensitive to the inclination angle β. The influence of β on the SIFs and the fracture angles for the crack 1 are similar to the previous case and are differ from the previous for the crack 2 (see Figs. 5 and 6).

The critical loads are presented in Fig. 7. For two parallel cracks with different sizes the larger crack (crack 1) will propagate first because of $p_{cr(1)}< p_{cr(2)}$ for all angles β (Fig. 7 e, f). For close locate cracks the critical load $p_{cr(2)}$ for the smaller crack is much larger than $p_{cr(1)}$, Fig. 7. For far distances between the cracks (e.g. for $d=6$) the difference between the critical loads becomes less and the cracks have equal chances for propagation.

Figs. 8 – 10 show fracture parameters for two unequal cracks with sizes a_1=1 and a_2=0.5 (as previously), but with the fixed inclination angle β_1 =90° and varied β_2, for the distances d = 1, 2, 4, 6. For the fixed crack 1 the SIFs $k_{I(1)}$ and $k_{II(1)}$ are nearly the same as for the corresponding single edge crack, Fig. 8 a, and the fracture angle $\phi_{(1)}$ is small, Fig. 9 a. For the arbitrary inclined crack 2 the fracture angle $\phi_{(2)}$ is increased with increasing inclination angle β_2 and changed the sign from minus to plus. The critical loads are depicted in Fig. 10. The influence of the distance d on $p_{cr(1)}$ is not large, the maximal difference is 3.5% between the values of $p_{cr(1)}$ for d=1 and d=6, Fig. 10 a. A comparison of the critical loads for two cracks shows that the crack 1 will propagate first because of $p_{cr(1)} < p_{cr(2)}$, Fig. 10 b.

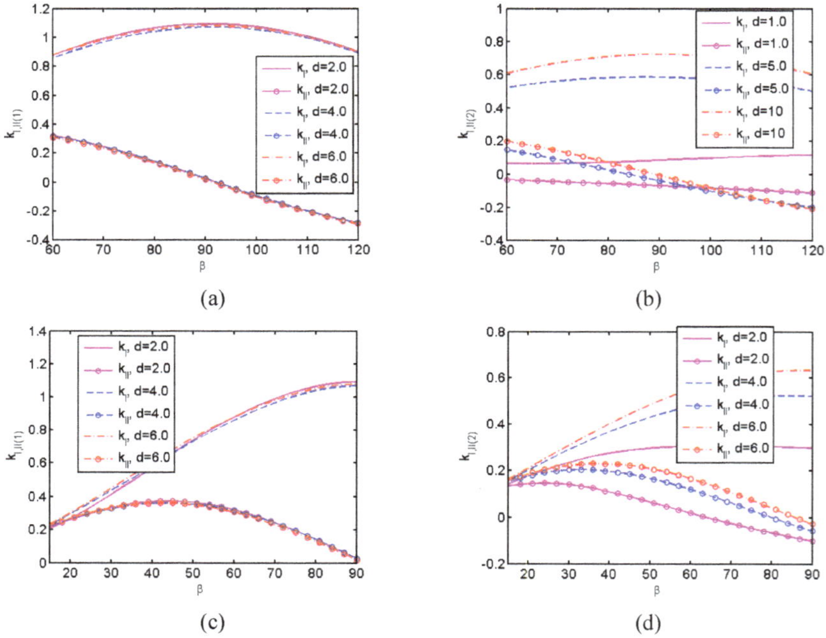

Figure 5. Stress intensity factors kI and kII as function of the inclination angle β=βn of the edge cracks to the surface for different distances d between the cracks: (a) for crack 1 (60°≤ β ≤120°), (b) for crack 2 (60°≤ β ≤120°), (c) for crack 1 (15°≤ β ≤90°), (d) for crack 2 (15°≤ β ≤90°). Two edge cracks with different sizes a1=1 and a2=0.5.

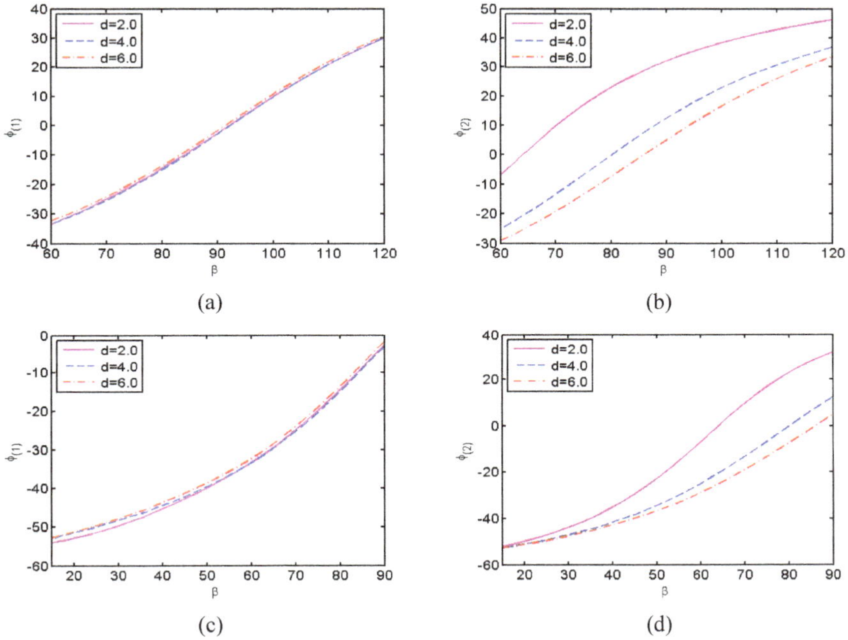

Figure 6. Fracture angles ϕ(1) and ϕ(2) as functions of the inclination angle β=βn of the edge cracks to the surface for different distances between the cracks: (a) for crack 1 (60°≤ β ≤120°), (b) for crack 2 (60°≤ β ≤120°), (c) for crack 1 (15°≤ β ≤90°), (d) for crack 2 (15°≤ β ≤90°). Two edge cracks with different sizes a1=1 and a2=0.5.

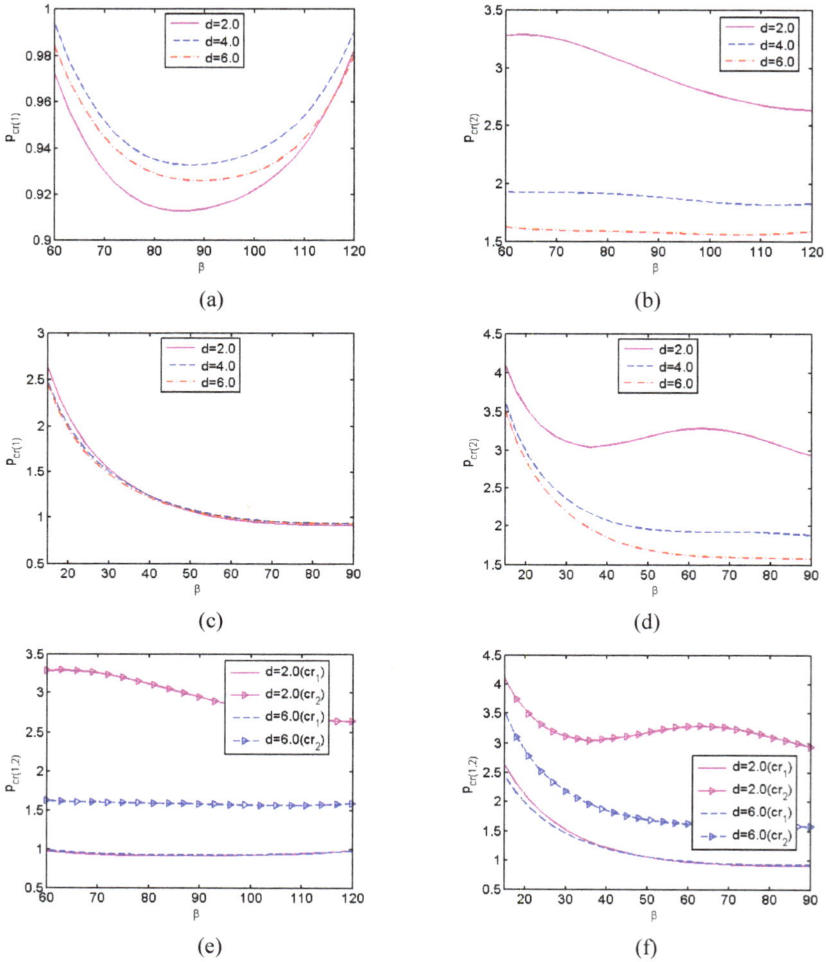

Figure 7. The non-dimensional critical load P_{cr} as function of the inclination angle $\beta=\beta n$ for two equal edge cracks for different distances d between the cracks: (a) for crack 1 ($60°\leq\beta\leq120°$), (b) for crack 2 ($60°\leq\beta\leq120°$), (c) for crack 1 ($15°\leq\beta\leq90°$), (d) for crack 2 ($15°\leq\beta\leq90°$), (e) for cracks 1 and 2 ($60°\leq\beta\leq120°$) and (f) for cracks 1 and 2 ($15°\leq\beta\leq90°$). Two edge cracks with different sizes a1=1 and a2=0.5.

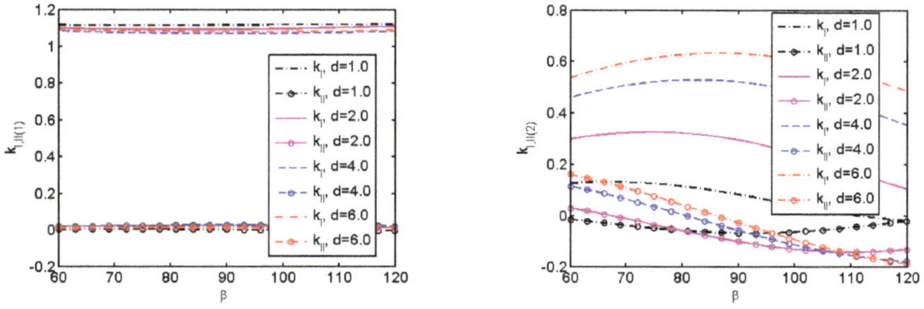

(a) (b)

Figure 8. Stress intensity factors kI and kII as function of the inclination angle β=β2 (60°≤ β ≤120°) and for β1=90° for different distances d between the cracks: (a) for crack 1, (b) for crack 2. Two edge cracks with different sizes a1=1 and a2=0.5.

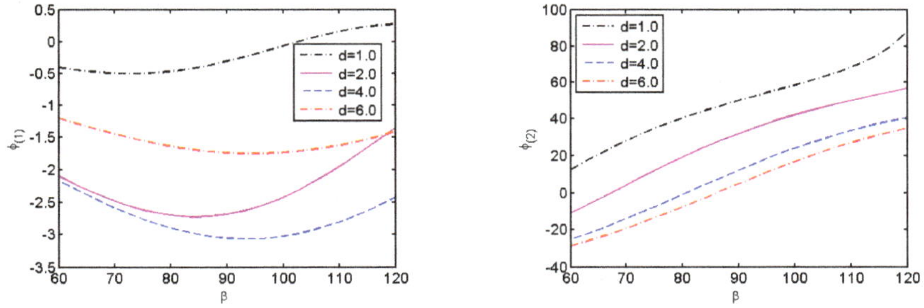

(a) (b)

Figure 9. Fracture angles ϕ(1) and ϕ(2) as functions of the inclination angle β=β2 (60°≤ β ≤120°) and for β1=90° for different distances d between the cracks: (a) for crack 1, (b) for crack 2. Two edge cracks with different sizes a1=1 and a2=0.5.

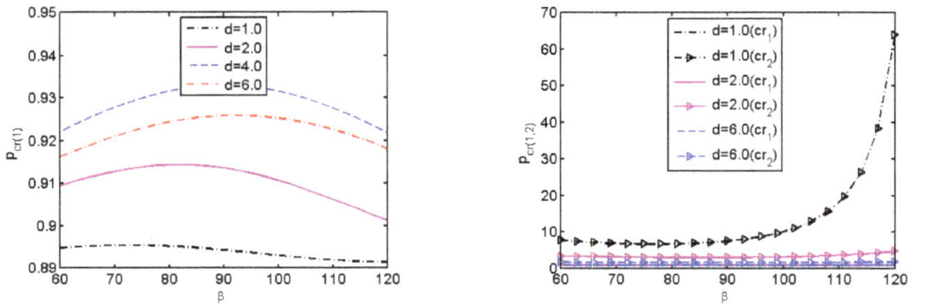

(a) (b)

Figure 10. The non-dimensional critical load P_{cr} as function of the inclination angle $\beta=\beta_2$ ($60°\leq \beta \leq 120°$) and for $\beta_1=90°$ for different distances d between the cracks: (a) for crack 1, (b) for cracks 1 and 2. Two edge cracks with different sizes $a_1=1$ and $a_2=0.5$.

The schemes of the direction of cracks propagation are shown in Fig. 11 for inclination angles β=90° and 60° and for different distances between the cracks. For a single edge crack with β=90° the fracture angle is equal to ϕ=0°. It is shown in Fig. 11 a both cracks will change the direction of the propagation and the fracture angles are larger for closely located cracks for *d*=2 than for the distances *d*=4 and 6 between the cracks where the fracture angles are ϕ=12° and 8°. The results for two unequal cracks are shown in Fig. 11 c. The crack 1 slightly deviates away from crack 2 if the distance is *d*=2 and will propagate straight for far distances between the cracks. At the same time the influence of the big crack on the propagation direction of the small is rather strong. Fig. 11 d show the fracture angles for the cracks inclined under the angle β=60°. For a single edge crack with β=60° the fracture angle is equal to 31°. The difference in the fracture angles due to their interaction is dependent on the distance between the cracks and the size of these cracks.

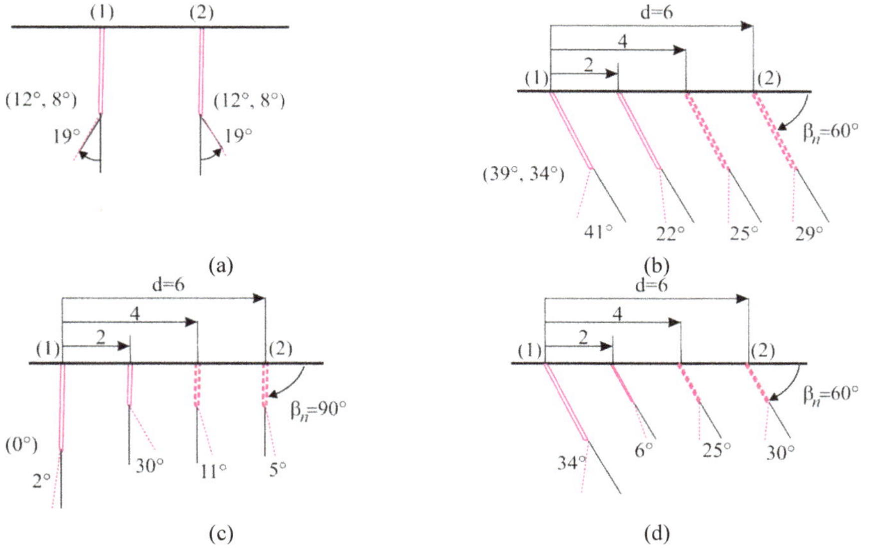

Figure 11. Schematic representation of the fracture angles: (a) and (b) for two equal cracks with inclined angles β=90° and 60° correspondingly; (c) and (d) for two crack of sizes a1=1 and a2=0.5 and with inclined angles β=90° and 60°.

As shown in Figs. 3, 6 and 9, there are crack configurations for which cracks are not deviating from their initial direction, i.e. ϕ =0. Table 1 presents such results, e.g. crack 1 has the fracture angle ϕ =0° for $\beta_{1,2}$=103°, the crack 2 for this case has ϕ =30°. For close located cracks (for d=1) with $\beta_{1,2}$=90° the crack interaction gives the fracture angles (ϕ_1, ϕ_2)=(0°, 50°), see column first in the Table 1. For a single edge crack for β=90° the crack has the zero fracture angle.

Table 1. Some cases for crack configurations for which cracks are not deviating from their initial direction.

d	d=1		d=2		d=4		d=6	
β, ϕ (degree)	$\beta_1 = \beta_2$	(ϕ_1, ϕ_2)	$\beta_1 = \beta_2$	(ϕ_1, ϕ_2)	$\beta_1 = \beta_2$	(ϕ_1, ϕ_2)	$\beta_1 = \beta_2$	(ϕ_1, ϕ_2)
(a_1, a_2)=(1.0, 1.0)	-	-	103	(0, 30)	99	(0, 20)	95	(0, 10)
	-	-	77	(−30, 0)	81	(−20, 0)	85	(−10, 0)
(a_1, a_2)=(1.0, 0.5)	90	(0, 50)	90	(0, 33)	90	(0, 10)	90	(0, 3)
	not exist	-	63	(−30, 0)	80	(−18, 0)	86	(−10, 0)

5.2 Three arbitrary inclined cracks

Consider the case for three arbitrary inclined edge cracks. Figs. (10)-(12) are for the cracks with same sizes and Figs. (15)-(22) for different sizes, i.e. a_1=1, a_2=0.5 and a_3=0.5.

Fig. 12 shows the SIFs k_I and k_{II} , Fig. 13 – the fracture angles, Figs. 14 – the critical loads as functions of inclination angle $\beta=\beta_n$ (n=1, 2, 3) for three interacting edge cracks of the same size. The results are presented for the crack 2 (the middle crack), for other cracks 1 and 3, the plots are similar to the previous case for two interacting cracks, Fig. 2. The distances between the cracks are equal (see Fig. 21 a) and the calculation is performed for d=2, 4 and 6.

The curves for k_I for all three cracks are similar, but the values for k_I are different $k_{I(2)} < k_{I(1,3)}$ for all β and d. The influence of the distance on the k_I is stronger for the crack 2, than for cracks 1 and 3. The SIF $k_{II(2)}$ is also less than k_{II} for cracks 1 and 3 and have small dependence on the distance d. The strong shielding effect is observed for the middle crack 2.

The fracture angles ϕ are presented in Fig. 13 only for the crack 2, for cracks 1 and 3 they are nearly the same as in Fig. 3 for two cracks. Some selected cases for directions of crack propagation are shown in Fig. 21 a, b. The influence of the inclination angle on the fracture angles is evident as well as the influence of interaction between the cracks.

Fig. 14 shows the non-dimensional critical loads for the three equal cracks, Fig. 14 a for $60° \leq \beta \leq 120°$ and Fig. 14 b for $15° \leq \beta \leq 90°$. The critical loads $p_{cr(1,3)} < p_{cr(2)}$ for all parameters, hence the outer cracks 1 and 3 will start to propagate first.

Some results are presented in Figs. 15 – 17 for cracks with different sizes a_1=1 and a_2= a_3=0.5 and equally inclined to the surface. Fig. 15 shows SIFs for cracks 2 and 3, the curves for the crack 1 (the bigger crack) are similar to the curves for the crack 1 in Fig. 5 a, c. $k_{I(2)} < k_{I(3)}$ and $k_{I(2,3)} < k_{I(1)}$ for all parameters, that is, the maximum shielding effect is observed for the crack 2. The SIFs k_{II} are small (close to zero) and their absolute values for cracks 2 and 3 are less than for crack 1.

The fracture angles ϕ are presented in Fig. 16 for the small cracks 2 and 3 and for the crack 1 the curves are similar to the curves in Fig. 6 a, c. Some schemes for the direction of the crack propagation are presented in Fig. 21 c, d.

The non-dimensional critical loads for the three unequal cracks are presented in Fig. 17. $p_{cr(2)} >> p_{cr(1,3)}$ for $60° \leq \beta \leq 120°$ and for $15° \leq \beta \leq 90°$ (d=2), besides $p_{cr(2)} > p_{cr(3)} > p_{cr(1)}$. The larger crack will propagate first.

The last case for the three edge cracks is presented in Figs. 18 – 22. The sizes of cracks are equal, the first crack is inclined with β_1 =90° and $\beta_{2,3}$=β are varied. The SIFs are shown in Fig. 18. The dependence of k_I and k_{II} with changing β are similar as for two interacting cracks, Fig. 8, but the values of k_I are smaller for the three crack case for all parameters. The fracture angles are presented in Fig. 19. The dependence of ϕ with β for crack 3 is similar to the two-crack case in Fig. 9 b. The fracture angles for the crack 1 (β_1 =90°) are larger than for the crack 1 interacting with only one crack, Fig. 9 a, and they are much smaller than the values ϕ for cracks 2 and 3, as expected, because of a single crack with β_1 =90° don't change the direction of propagation.

Fig. 20 shows the non-dimensional critical loads for this case of interacting cracks, Fig. 20 a for the crack 1 and Fig. 20 b is for three cracks. The critical load for the middle crack 2 is much larger than for other cracks for all β values. For inclinations $\beta_{2,3}$=β close to 90° ($88° \leq \beta \leq 102°$),

 Materials Research Forum LLC
https://doi.org/10.21741/9781644902950

when the three cracks nearly parallel, $p_{cr(1)} = p_{cr(3)}$, so that the cracks 1 and 3 will start to propagate first, for other inclination angles the weaker is the crack 1.

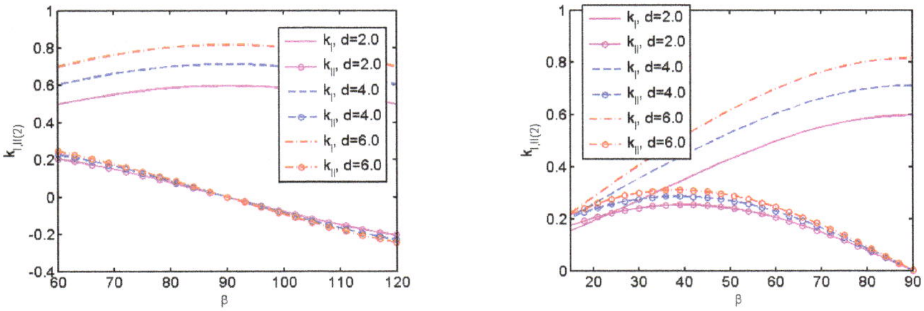

(a) (b)

Figure 12. Stress intensity factors kI and kII as functions of the inclination angle β=βn of the edge cracks to the surface for different distances d between the cracks: (a) for crack 2 (60°≤ β ≤120°) (b) for crack 2 (15°≤ β ≤90°) (SIFs for cracks 1 and 3 are similar as in Figs. 2 a, b for two crack). Three equal edge cracks.

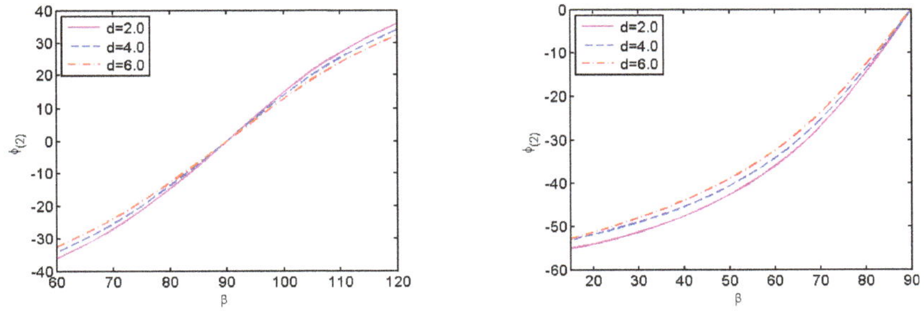

(a) (b)

Figure 13. Fracture angles ϕ(2) as functions of the inclination angle β=βn of the edge cracks to the surface for different distances d between the cracks: (a) for crack 2(60°≤ β ≤120°) and (b) for crack 2 (15°≤ β ≤90°). (The fracture angles ϕ for cracks 1 and 3 are similar as in Figs. 3 a, b for two crack.) Three equal edge cracks.

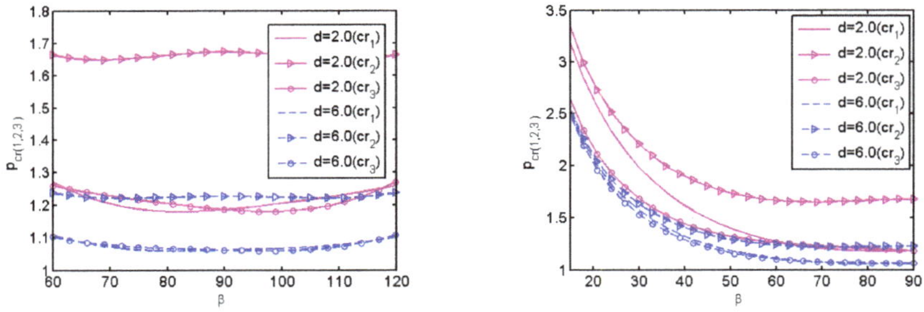

(a) (b)

Figure 14. The non-dimensional critical load P_{cr} as function of the inclination angle $\beta=\beta n$ for equal edge cracks for different distances d between the cracks: (a) for three cracks ($60° \leq \beta \leq 120°$), (b) for three cracks ($15° \leq \beta \leq 90°$).

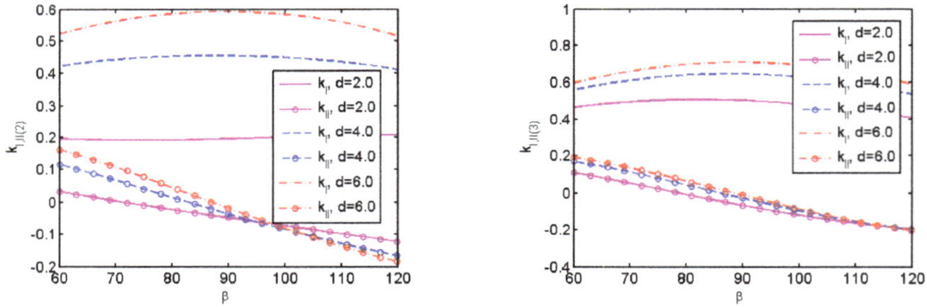

(a) (b)

Figure 15. Stress intensity factors kI and kII as function of the inclination angle $\beta=\beta n$ of the edge cracks to the surface for different distances d between the cracks: (a) for crack 2 ($60° \leq \beta \leq 120°$), (b) for crack 3 ($60° \leq \beta \leq 120°$). Three edge cracks with different sizes $a1=1$ and $a2=a3=0.5$.

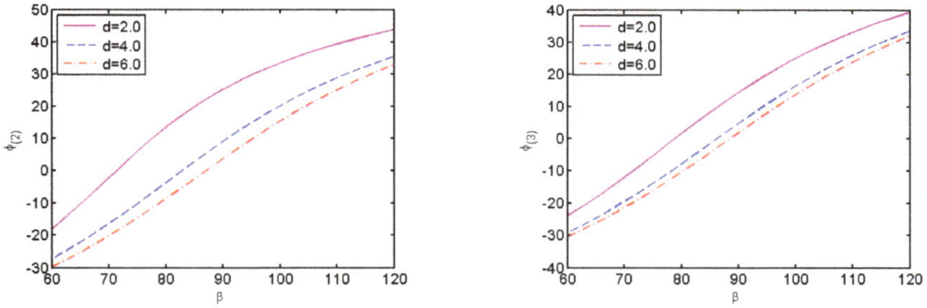

(a) (b)

Figure 16. Fracture angles $\phi(1)$ and $\phi(2)$ as functions of the inclination angle $\beta=\beta n$ of the edge cracks to the surface for different distances d between the cracks: (a) for crack 2 ($60°\leq\beta\leq120°$), (b) for crack 3 ($60°\leq\beta\leq120°$). Three edge cracks with different sizes a1=1 and a2= a3=0.5. (For small angles the figures are similar to the case of two cracks.)

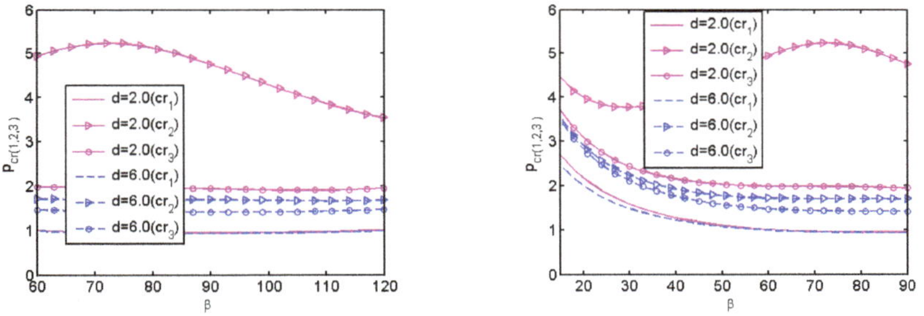

(a) (b)

Figure 17. The non-dimensional critical load P_{cr} as function of the inclination angle $\beta=\beta n$ for two equal edge cracks for different distances d between the cracks: (a) for 3 cracks ($60°\leq\beta\leq120°$), (b) for 3 cracks ($60°\leq\beta\leq120°$). Three edge cracks with different sizes a1=1 and a2= a3=0.5.

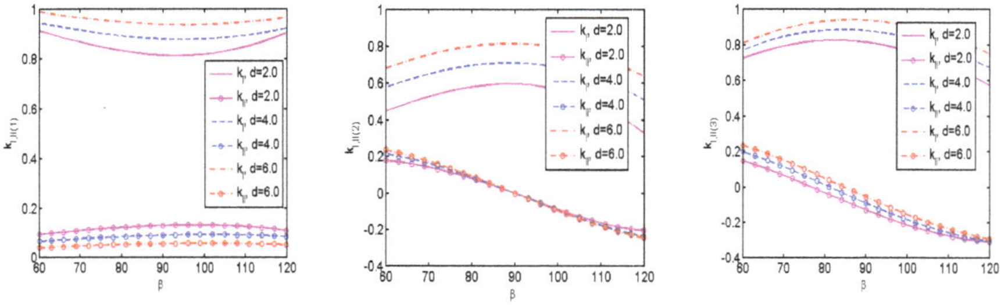

(a) (b) (c)

Figure 18. Stress intensity factors kI and kII as function of the inclination angle β=β2 (60°≤ β ≤120°) and for β1=90° for different distances d between the cracks: (a) for crack 1, (b) for crack 2, (c) for crack 3. Three equal edge cracks.

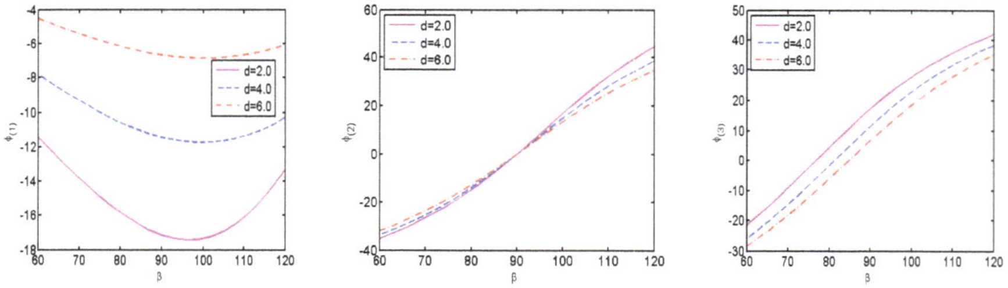

(a) (b) (c)

Figure 19. Fracture angles φ as functions of the inclination angle β=β2,3 (60°≤ β ≤120°) and for β1=90° for different distances d between the cracks: (a) for crack 1, (b) for crack 2, (c) for crack 3. Three equal edge cracks.

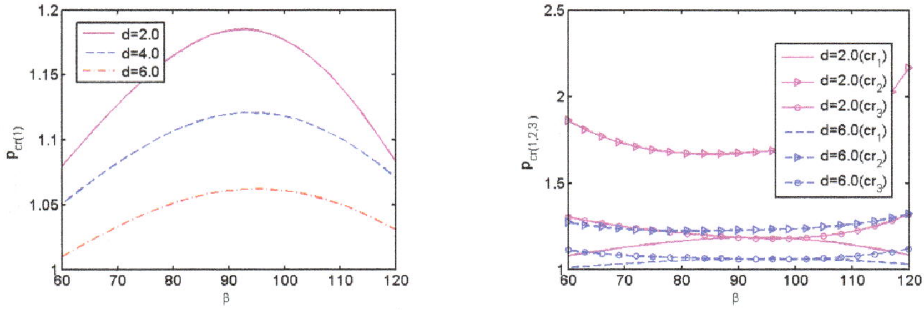

(a) (b)

Figure 20. The non-dimensional critical load P_{cr} as function of the inclination angle $\beta=\beta 2$, 3 ($60° \leq \beta \leq 120°$) and for $\beta 1=90°$ for different distances d between the cracks: (a) for crack 1, (b) for 3 cracks. Three edge cracks same size.

Some schemes of the direction of crack propagation for three cracks in four different arrangements are shown in Fig. 21. The schemes for fracture angles for other crack geometries can be built using the data in Figs. 13, 16 and 19.

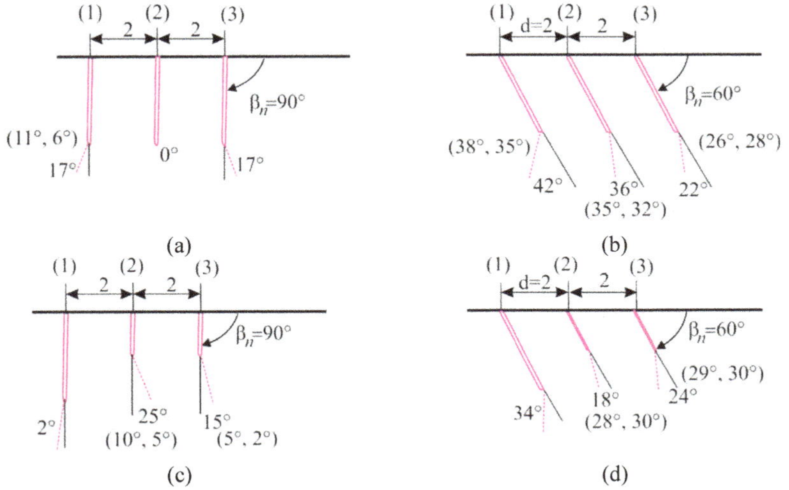

Figure 21. Schematic representation of the fracture angles: (a) and (b) for three equal cracks with inclined angles $\beta=90°$ and $60°$ correspondingly; (c) and (d) for three crack of sizes $a1=1$ and $a2= a3=0.5$ and with inclined angles $\beta=90°$ and $60°$.

5.3 Crack interaction effect

The shielding – amplification effects are observed in crack interaction problems, i.e. some geometries of the interacting cracks can enhance or suppress the propagation of each other. This problem has a long history of discussion especially for macro-microcrack interaction problems [21]. In a recent paper by Wang et al [22] the authors return to this problem. The shielding and amplification effects of transverse array of microcracks on a main crack were investigated using extended FEM with respect to SIFs. The SIFs at the main crack tip include not only the interaction between the main crack and each of the small cracks, but also include mutual interactions of microcracks. It was pointed out that the interaction between microcracks can weaken their amplification-shielding effect on the main crack and this effect was demonstrated in [22].

In the present problem for edge cracks we can check whether this interaction effect is present or not.

Consider the case of two edge cracks with sizes $(a_1, a_2) = (1.0, 0.5)$ and three edge cracks with sizes $(a_1, a_2, a_3)=(1.0, 0.5, 0.5)$, in both cases the cracks have same inclination angle β to the surface $\beta_n=\beta$.

Introduce the notation

$$\Delta_1 = k_I^{(1)} - k^0$$

for two cracks with distance between the cracks d=2 and for distance d=4

$$\Delta_2 = k_I^{(2)} - k^0$$

and

$$\Delta_3 = k_I^{(3)} - k^0$$

for 3 cracks with equal distances between the cracks d=2, where $k_I^{(n)}$ is the SIF at the tip of the large crack, and k^0 is the SIF for a single edge crack. These values of Δ determine the interaction effects between the cracks, i.e. the influence of small cracks on the large crack 1.

The value $\Delta_1 + \Delta_2 - \Delta_3$ will give the magnitude of the mutual interaction of small cracks (crack 2 and 3) and

$$f = [(\Delta_1 + \Delta_2 - \Delta_3) / k^0] \times 100\%$$

is the relative value of this crack interaction.

The data for SIFs are given in Table 2. SIF k^0=0.92 for a single crack with inclination angle β=60° can be found in [19] in Table 4 or in [11] in Table 1.

The SIF $k_I^{(1)}$ for crack 1 as influenced by a crack 2 at distance d=2 is 1.094 and this value is less than SIF k^0=1.12 on the value $|\Delta_1|$=0.026 (first column in Table 2), and for the cracks on the distance d=4 the SIF $k_I^{(2)}$ =1.071 is less than k^0 on $|\Delta_2|$=0.05 (second column in Table 2). The total interaction effect of the cracks 2 and 3 on the large crack 1 is $|\Delta_1+ \Delta_2|$=0.076 (if superposition of these interactions is assumed).

Materials Research Foundations **159** (2024) https://doi.org/10.21741/9781644902950

The SIF k_I for crack 1 under the influence of pair interacting cracks 2 and 3 is $k_I^{(3)}=1.061$ (all distances between the cracks are $d=2$) and $|\Delta_3|=0.06$ for this case. $|\Delta_3|$ is less than $|\Delta_1+\Delta_2|$ by 0.016. That is, the solution to the problem of three edge cracks gives interaction effect on 0.016 less than the superimposed effect of two separate cracks derived from the problems of the case of two-crack interaction. It was an example of calculations for $\beta=90°$, for $\beta=60°$ the calculations are similar and the results are presented in Table 2.

The last column of Table 2 shows that the interaction between small cracks (crack 2 and 3) weakens the shielding effect only by 1.5% for the cracks inclined at $\beta=90°$ and by 2.17% for the cracks with $\beta=60°$. These are rather small values, but in case of the interaction of multiple cracks the effect will be probably stronger.

Table 2. SIFs k_I at crack 1 interacting with small cracks and their interaction effects.

	d=2	d=4	d=2			
β	$k_I^{(1)}$, Δ_1	$k_I^{(2)}$, Δ_2	$k_I^{(3)}$, Δ_3	$(\Delta_1+\Delta_2)-\Delta_3$	k^0	f%
90°	1.094, –0.026	1.071, –0.05	1.061, –0.06	–0.016	1.12	–1.5
60°	0.88, –0.04	0.86, –0.06	0.84, –0.08	–0.02	0.92	–2.17

Conclusions

The effects of the interaction of edge cracks on further crack formation were studied with respect to main fracture characteristics, namely, stress intensity factors, fracture angles and critical loads. Some illustrative examples show the influence of inclination angles, distances between the cracks and lengths of the cracks on this interaction. The interaction of cracks leads to mixed mode conditions near the crack tips even for symmetric geometries and loading normal to the crack lines. As an illustration, a classical edge crack (i.e. the crack normal to the boundary and under tension normal to the crack line) in the presence of other cracks is under mixed mode conditions and deviates from its initial propagation direction, albeit the other cracks are small (Fig. 9 a and 11 c).

The crack shielding takes place (as expected for this problem) for most parameters of the problem. The maximum magnitude of the shielding effect is observed for closely located cracks and for a middle crack in the case of the interaction of three cracks. The influence of two interacting cracks on the third crack can weaken the shielding effect and it was shown for SIF Mode I which is dominant in this problem.

The applied method of singular integral equations (which have been solved by the well-known numerical method based on quadratic formulas for integrals) in combination with a fracture criteria (the maximum hoop stress criteria has been used) is an effective method for modeling of the crack interactions at the initial stage of their propagation.

Acknowledgement

The support of the German Research Foundation under the grant Schm 746/139-1 is greatly acknowledged.

References

[1] T.L. Anderson, Fracture Mechanics: Fundamentals and Applications, third ed., Taylor & Francis, 2005. https://doi.org/10.1201/9781420058215

[2] R. Daud, A.K. Ariffin, Sh. Abdullah, Al.E. Ismail, Interacting cracks analysis using finite element method, in: A. Belov (Ed.), Applied Fracture Mechanics, InTech, (2012) 359 – 380. https://doi.org/10.5772/54358

[3] https://en.wikipedia.org/wiki/Crocodile_cracking

[4] S. Rangaraj, K. Kokini, Multiple surface cracking and its effect on interface cracks in functionally graded thermal barrier coatings under thermal shock, Trans. ASME J. Appl. Mech. 70 (2003) 234-245. https://doi.org/10.1115/1.1533809

[5] A. Kawasaki, R. Watanabe, Thermal fracture behavior of metal/ceramic functionally graded materials, Eng. Fract. Mech. 69 (2002) 1713-1728. https://doi.org/10.1016/S0013-7944(02)00054-1

[6] A. Gilbert, K. Kokini, S. Sankarasubramanian, Thermal fracture of zirconia-mullite composite thermal barrier coatings under thermal shock: An experimental study, Surf. Coat. Technol. 202(10) (2008) 2152-2161. https://doi.org/10.1016/j.surfcoat.2007.09.001

[7] V. Petrova, S. Schmauder, Thermal fracture of a functionally graded/homogeneous bimaterial with a system of cracks, Theor. Appl. Fract. Mech. 55 (2011) 148-157. https://doi.org/10.1016/j.tafmec.2011.04.005

[8] V. Petrova, S. Schmauder, Mathematical modelling and thermal stress intensity factors evaluation for an interface crack in the presence of a system of cracks in functionally graded/ homogeneous bimaterials, Comp. Mater. Sci. 52 (2012) 171-177. https://doi.org/10.1016/j.commatsci.2011.02.028

[9] V. Petrova, S. Schmauder, Interaction of a system of cracks with an interface crack in functionally graded/ homogeneous bimaterials under thermo-mechanical loading, Comp. Mater. Sci. 64 (2012) 229-233. https://doi.org/10.1016/j.commatsci.2012.04.032

[10] V. Petrova, S. Schmauder, FGM/homogeneous bimaterials with systems of cracks under thermo-mechanical loading: Analysis by fracture criteria, Eng. Fract. Mech. 130 (2014) 12-20. https://doi.org/10.1016/j.engfracmech.2014.01.014

[11] V. Petrova, T. Sadowski, Theoretical modeling and analysis of thermal fracture of semi-infinite functionally graded materials with edge cracks, Meccanica 49 (2014) 2603-2615. https://doi.org/10.1007/s11012-014-9941-x

[12] B. Zhou, K. Kokini, Effect of surface pre-crack morphology on the fracture of thermal barrier coatings under thermal shock, Acta Mater. 52 (2004) 4189-4197. https://doi.org/10.1016/j.actamat.2004.05.035

[13] Y.Z. Feng, Z.-H. Jin, Thermal shock damage and residual strength behavior of a functionally graded plate with surface cracks alternating length, J. Therm. Stresses 35 (2012) 30-47. https://doi.org/10.1080/01495739.2012.637457

[14] A.M. Afsar, H. Sekine, Crack spacing effect on the brittle fracture characteristics of semi-infinite functionally graded materials with periodic edge cracks, Int. J. Fract. 102 (2000)

L61-L66.

[15] V.V. Panasyuk, M.P. Savruk, A.P. Datsyshin, Stress Distribution near Cracks in Plates and Shells (in Russian), Naukova Dumka, Kiev, 1976.

[16] M.P. Savruk, Two- Dimensional Problems of Elasticity for Body with Cracks (in Russian), Naukova Dumka, Kiev, 1981.

[17] F. Erdogan, G. Gupta, On the numerical solution of singular integral equations, Quart. Appl. Math., 29 (1972) 525-534. https://doi.org/10.1090/qam/408277

[18] F. Erdogan, G.C. Sih, On the crack extension in plates under plane loading and transverse shear, J. Basic. Eng. 85 (1963) 519-527. https://doi.org/10.1115/1.3656897

[19] N.-A. Noda, K. Oda, Numerical solution of the singular integral equations in the crack analysis using the body force method. Int. J. Fract. 58 (1992) 285-304. https://doi.org/10.1007/BF00048950

[20] C.E. Freese, Periodic edge cracks of unequal length in a semi-infinite tensile sheet. Int J. Fract. 12 (1976). 125-134. https://doi.org/10.1007/BF00036015

[21] V. Petrova, V. Tamuzs, N. Romalis, A survey of macro- microcrack interaction problems, ASME Appl. Mech. Rev. 53 (2000) 117-146. https://doi.org/10.1115/1.3097344

[22] H. Wang, Z. Liu, D. Xu, Q. Zeng, Z. Zhuang, Z. Chen, Extended finite element method analysis for shielding and amplification effect of a main crack interacted with a group of nearby parallel microcrack, Int. J. Damage Mech. 25 (1) (2016) 4-25. https://doi.org/10.1177/1056789514565933

Appendix

In Eqs. (2) the kernels $R_{nk}(t,x)$ and $S_{nk}(t,x)$ are written as

$$R_{nk}(t,x) = (1-\delta_{nk})K_{nk}(t,x) + \frac{e^{i\alpha_k}}{2}\left\{\frac{1}{X_n - \overline{T}_k} + \frac{e^{-2i\alpha_n}}{\overline{X}_n - T_k} + \right.$$
$$\left. + (\overline{T}_k - T_k)\left[\frac{1+e^{-2i\alpha_n}}{(\overline{X}_n - T_k)^2} - \frac{2e^{-2i\alpha_n}(X_n - T_k)}{(\overline{X}_n - T_k)^3}\right]\right\}, \quad (A.1)$$

$$S_{nk}(t,x) = (1-\delta_{nk})L_{nk}(t,x) + \frac{e^{-i\alpha_k}}{2}\left[\frac{T_k - \overline{T}_k}{(X_n - \overline{T}_k)^2} + \frac{1}{\overline{X}_n - T_k} - e^{-2i\alpha_n}\frac{X_n - T_k}{(\overline{X}_n - T_k)^2}\right], \quad (A.2)$$

and the kernels $K_{nk}(t,x)$ and $L_{nk}(t,x)$ are

$$K_{nk}(t,x) = \frac{e^{i\alpha_k}}{2}\left(\frac{1}{T_k - X_n} + \frac{e^{-2i\alpha_n}}{\overline{T}_k - \overline{X}_n}\right); \quad (A.3)$$

$$L_{nk}(t,x) = \frac{e^{-i\alpha_k}}{2}\left(\frac{1}{\overline{T}_k - \overline{X}_n} + \frac{T_k - X_n}{(\overline{T}_k - \overline{X}_n)^2} e^{-2i\alpha_n}\right), \tag{A.4}$$

where

$$T_k = te^{i\alpha_k} + z_k^0, \quad X_n = xe^{i\alpha_n} + z_n^0, \qquad n,k = 1,2,\ldots,N \tag{A.5}$$

and

$$\delta_{nk} = \begin{cases} 0 & for \;\; n \neq k; \\ 1 & for \;\; n = k. \end{cases}$$

The kernels $K_{nk}(t,x)$ and $L_{nk}(t,x)$ are the same as for the system of cracks in an infinite plane, and the additional terms in Eqs. (A.1) and (A.2) are responsible for the influence of the edge of the half plane. α_n is the inclination angle of n-th crack to the x-axis and $\alpha_n = -\beta_n$, Fig. 1; z_n^0 is the coordinate of the center of crack in global coordinate system (x,y).

Materials Research Foundations **159** (2024) https://doi.org/10.21741/9781644902950

CHAPTER 9

Modeling of Thermo-Mechanical Fracture of FGMs with Respect to Multiple Cracks Interaction

Vera Petrova [1,2]*, Siegfried Schmauder [1]

[1] IMWF, University of Stuttgart, Pfaffenwaldring 32, D-70569 Stuttgart, Germany

[2] Voronezh State University, University Sq.1, Voronezh 394006, Russia

veraep@gmail.com *, Siegfried.Schmauder@imwf.uni-stuttgart.de

Abstract

Different aspects of thermo-mechanical fracture of functionally graded materials (FGMs) are considered. Among them are the crack interaction problems in a functionally graded coating on a homogeneous substrate (FGM/H). The interaction between systems of edge cracks is investigated, as well as, how this mutual interaction influences the fracture process and the formation of crack patterns. The problem is formulated with respect to singular integral equations which are referred to the boundary equation methods. The FGM properties are modeled by exponential functions. The main fracture characteristics are calculated, namely, the stress intensity factors, the angles of deviation of the cracks from their initial propagation direction and the critical stresses when the crack starts to propagate. The last two characteristics are calculated using an appropriate fracture criterion. The problem contains different parameters, such as the geometry (location and orientation of cracks, their lengths, and the width of the FGM layer) and material parameters, i.e. the inhomogeneity parameters of elastic and thermal coefficients of the functionally graded material. The influence of these parameters on the thermo-mechanical fracture of FGM/H is investigated. As examples the following real material combinations are discussed: TiC/SiC, Al_2O_3/$MoSi_2$, $MoSi_2$/SiC, ZrO_2/nickel and ZrO_2/steel

Keywords

Edge And Internal Cracking, Thermal Fracture, Functionally Graded Coating, Stress Intensity Factors, Fracture Criteria, Fracture Angle

1. Introduction

Functionally graded coatings (FGCs) are applied in thermal, wear and corrosion barriers which are used in different fields, such as, nuclear energy (e.g., nuclear reactor components), aerospace (e.g., rocket engine components, space plane body), engineering (e.g., turbine blade, engine components), energy conversion (e.g., thermoelectric generator, fuel cell) as well as many other applications [1-3]. They are subjected to different thermal and mechanical loading and have to resist high temperature, wear and aggressive environments. However, cracks can initiate from initial defects or microcracks may appear during manufacturing or service. Therefore, the study of fracture of FGC structures is important for a better understanding of the fracture resistance of graded coatings.

Thermal fracture of FGCs is significantly affected by a complex crack interaction mechanism, e.g., interacting cracks can enhance or suppress the propagation of each other. The crack patterns strongly depend on the microstructure of the materials and type of loading. Numerous experimental results (e.g., [4-6]) showed that when FGCs are subjected to thermal shock, multiple cracks often occur at the ceramic surface. The studies from laser thermal shock tests [7, 8] indicated that multiple surface pre-cracks with a large density and short lengths reduce the crack driving force at the interface. At the same time very short pre-cracks were most vulnerable to further surface cracking which in turn induced the interface crack. In [8] a functionally graded plate with an array of periodically spaced surface cracks of alternating lengths subjected to thermal shocks was considered. The crack morphology was described by the ratio of the length of the short cracks to that of the long cracks (initial crack length ratio) as well as by the crack spacing. The thermal properties of the FGM plate were arbitrarily graded in the plate thickness direction, but the elastic properties were assumed to be constant.

In the present work the problem of thermo-mechanical fracture of FGCs on a homogeneous substrate (semi-infinite isotropic medium) is investigated in the frame of linear elastic fracture mechanics. It is supposed that the materials are brittle or quasi-brittle.

In previous (series of authors') works [9-13] the thermal fracture in the vicinity of an interface in a bimaterial compound which consists of an FGM and a homogeneous material was studied (an infinite FGM/H compound). The bimaterial contains an interface crack and internal arbitrary located cracks in the FGM and is subjected to a heat flux [9, 10] and/or a tensile load [11, 12] or a shear load [13]. For a special case when an interface crack is significantly larger in size than internal cracks in the FGM, the asymptotic analytical solution of the problem was obtained as a power series of the small parameter (the small parameter is equal to the ratio of the size of small internal cracks to the interface crack size). Approximate analytical formulas for the stress intensity factors (SIFs) at the interface crack tips were obtained. In [14] some results for edge cracks in FGM/homogeneous structures (a semi-infinite medium) were obtained in the frame of the approach used in [15, 16].

The paper is organized as follows. In the next section the description of the mathematical model is presented, it includes the geometry of the problem, modeling of properties of functionally graded material (FGM) with some examples of real FGMs, a description of thermo-mechanical loadings and the functions corresponding to these loadings, the main equations (singular integral equatios) for this problem and a method of their solution. In Section 3 a fracture criterion is described and the formulas for the calculation of the fracture angles (the direction of the crack propagation) and critical stresses will be presented. In Section 4 some results are discussed, which are followed by the conclusions in Section 5.

2. Model

2.1 Geometry of the problem

Structures consisting of FGCs on top of homogeneous substrates is considered with the presence of pre-existing systems of cracks in FGCs. The crack morphology in FGC systems depends on their composition profile, methods of producing and exploitation (loading) conditions (e.g., maximum-minimum temperature in the cycle of heating-cooling, number of cycles, additional

 Materials Research Forum LLC
https://doi.org/10.21741/9781644902950

mechanical loads). Different scenarios of propagation of cracks could be examined for typical crack patterns resulting from experiments available in literature [4, 6, 17].

At present the investigations devoted to the crack interactions in FGMs are restricted only to special cases of crack locations and the interaction of arbitrary located cracks has been not well examined, e.g. see [15]. In our previous works [9–13] an approximate approach for systems of arbitrarily located cracks in infinite FGM/H bimaterials was suggested and will be used partially in the present study of fracture of FGCs on a homogeneous substrate.

A functionally graded material on a homogeneous substrate is modeled by the geometry depicted in Fig. 1. The substrate is a homogeneous material 1 with subscribed properties and the upper material (on the top of the surface) is a pure material 2, at the interface the material has the same properties as the substrate, i.e. material 1. Through the thickness of the layer the properties are changed continuously from one material to the other.

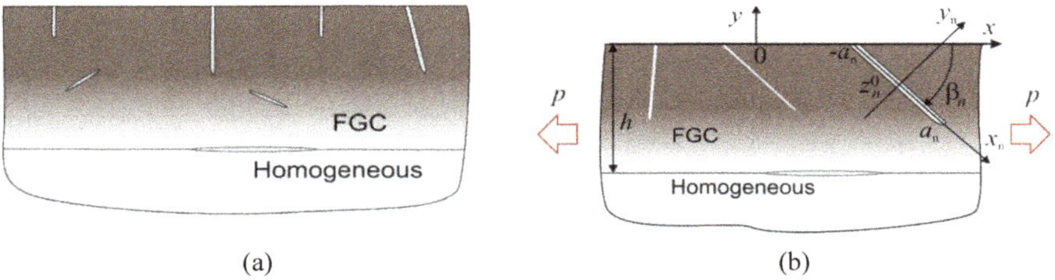

Figure 1. (a) FGC/H with system of cracks; (b) Edge cracks inclined arbitrarily with an angle β_n to the surface of the FGC: a_n – a half-length of n-th crack, $z_n^0 = x_n^0 + i\, y_n^0$ – the crack midpoint coordinate, h – the width of the FGC.

The layer contains a system of cracks and we consider some particular cases of the interaction of edge cracks arbitrarily inclined to the surface.

The positions of cracks are determined exactly by the midpoint coordinates $z_n^0 = x_n^0 + iy_n^0$ ($i^2 = -1$) and the inclination angles β_n to the surface. The cracks have sizes $2a_n$. The thickness of the FGM layer is h while the substrate is considered as a semi-infinite medium.

The coordinate systems are chosen as follows: the global coordinates (x, y) with the x-axis on the surface and the local coordinates (x_n, y_n) with centers in the midpoint of the n-th crack (Fig. 1b).

The relation between global coordinates (x, y) and local coordinates (x_n, y_n) systems is written as

$$z = z_n^0 + z_n e^{-i\beta_n}, \tag{1}$$

where z_n^0 is the origin coordinate of the local system in the global system and at the same time it is the midpoint coordinate of the n-th crack.

Materials Research Foundations **159** (2024) https://doi.org/10.21741/9781644902950

2.2 Modelling of functionally graded materials and examples of FGMs

It is assumed that thermo-mechanical properties of an FGC are continuous functions of the thickness coordinate y. As in previous works [9-14] the exponential form of this properties is used. Thus, the thermal expansion coefficient of the FGM layer is written as

$$\alpha_t(y) = \alpha_{t1}\exp(\varepsilon(y+h)),\ \ -h \le y \le 0, \tag{2}$$

and the Young's modulus is given as

$$E(y) = E_1\exp(\omega(y+h)),\ \ -h \le y \le 0, \tag{3}$$

with non-homogeneity parameters ε and ω. They are also called graded parameters.

The magnitudes of the dimensionless graded parameters εh and ωh (h is the thickness of the FGM layer) are obtained from Eqs. (2) and (3) as

$$\varepsilon h = \ln(\alpha_{t2} / \alpha_{t1}),\ \omega h = \ln(E_2 / E_1), \tag{4}$$

$$\alpha_{t2} = \alpha_t(y)_{|y=0}, \alpha_{t1} = \alpha_t(y)_{|y=-h},\ E_2 = E(y)_{|y=0}, E_1 = E(y)_{|y=-h}. \tag{5}$$

In the local coordinate system (x_n,y_n) connected with each crack the material parameters α_t and E (Eqs. (2) and (3)) take the form

$$\alpha_t(x_n, y_n) = \alpha_{t1}e^{\varepsilon(h+y_n^0)}e^{\varepsilon_1 x_n+\varepsilon_2 y_n},\ \varepsilon_1 = \varepsilon\sin(-\beta_n),\ \varepsilon_2 = \varepsilon\cos\beta_n,$$

$$E(x_n, y_n) = E_1e^{\omega(h+y_n^0)}e^{\omega_1 x_n+\omega_2 y_n},\ \omega_1 = \omega\sin(-\beta_n),\ \omega_2 = \omega\cos\beta_n$$

and on the crack lines, where $y_n = 0$, we will have

$$\alpha_t = \alpha_{t1}\exp(\varepsilon(h+y_n^0 - x_n\sin\beta_n)), \tag{6}$$

$$E = E_1\exp(\omega(h+y_n^0 - x_n\sin\beta_n)). \tag{7}$$

To derive these expressions Eq. (1) was used.

Functionally graded materials, as applied in thermal barrier coatings to protect substrates from high temperatures, should have a low thermal conductivity. At the same time they are desired to have a thermal expansion coefficient close to that of the material for the protected substrate. Some examples of real material combinations, which can be used in the present model, are given in Tables 3 and 4 [23]. In these tables the thermal expansion coefficients and Young's moduli of some ceramic/ceramic and ceramic/metal FGCs are included as well as the corresponding non-dimensional inhomogeneity coefficients of the FGMs. The thermal conductivity coefficients are also listed (for information about the better combination of FGMs), but the influence of this value on the fracture characteristics is not investigated in the present study.

Table 1. Material properties of ceramic/metal FGM/H (ZrO_2/Ti-6Al-4V)/Ti-6Al-4V

	Thermal expansion CTE ($Wm^{-1}K^{-1}$)	Inhomogeneity coefficient CTE	Young modulus E (GPa)	Inhomogeneity coefficient of E	Thermal conduct. ($*10^{-6}$ K^{-1})
ZrO_2 Ti-6Al-4V	$\alpha_{t2} = 10$ $\alpha_{t1} = 8.6$	$\alpha_{t2}/\alpha_{t1} > 1$ $\varepsilon h = 0.15 > 0$	$E_2 = 200$ $E_1 = 114$	$E_2/E_1 > 1$ $\omega h = 0.56 > 0$	$k_2 = 2$ $k_1 = 6.7$
(Ti-6Al-4V/ZrO_2)/ ZrO_2		$\varepsilon h = -0.15 < 0$		$\omega h = -0.56 < 0$	

Table 2. Material properties of some FGMs. The Young's moduli of these materials are similar

Material	Thermal expansion coeff. ($*10^{-6}$ K^{-1})		Inhomogeneity coefficient	Thermal conductivity ($Wm^{-1}K^{-1}$)	
FGM/H (Al_2O_3/ $MoSi_2$)/ $MoSi_2$ (Ceramic/Ceramic)					
Al_2O_3	α_{t2}	5	$\alpha_{t2}/\alpha_{t1} = 1$	k_2	25
$MoSi_2$	α_{t1}	5	$\varepsilon h = 0$	k_1	52
FGM/H ($MoSi_2$/ Al_2O_3)/ Al_2O_3			$\varepsilon h = 0$		
FGM/H ($MoSi_2$/ SiC)/ SiC (Ceramic/Ceramic)					
$MoSi_2$	α_{t2}	5	$\alpha_{t2}/\alpha_{t1} > 1$	k_2	52
SiC	α_{t1}	4	$\varepsilon h = 0.22 > 0$	k_1	60
FGM/H (SiC/ $MoSi_2$)/ $MoSi_2$			$\varepsilon h = -0.22 < 0$		
FGM/H (TiC/ SiC)/ SiC (Ceramic/Ceramic)					
TiC	α_{t2}	7	$\alpha_{t2}/\alpha_{t1} > 1$	k_2	20
SiC	α_{t1}	4	$\varepsilon h = 0.56 > 0$	k_1	60
FGM/H (SiC/ TiC)/ TiC			$\varepsilon h = -0.56 < 0$		
FGM/H (ZrO_2/ Ni)/ Ni (Ceramic/Metal)					
ZrO_2	α_{t2}	10	$\alpha_{t2}/\alpha_{t1} < 1$	k_2	2
Ni	α_{t1}	18	$\varepsilon h = -0.6 < 0$	k_1	90
FGM/H (Ni/ ZrO_2)/ ZrO_2			$\varepsilon h = 0.6 > 0$		
FGM/H (ZrO_2/ Steel)/ Steel (Ceramic/Metal)					
ZrO_2	α_{t2}	10	$\alpha_{t2}/\alpha_{t1} < 1$	k_2	2
Steel	α_{t1}	12	$\varepsilon h = -0.18 < 0$	k_1	20
FGM/H (Steel/ ZrO_2)/ ZrO_2			$\varepsilon h = 0.18 > 0$		

This exponential model describes the smooth variation of material properties of the FGM in the y-axis direction. For example, if α_{t1} is decreased with increasing y-coordinate (from $y = -h$ to y=0), then the inhomogeneity parameter ε is negative, and this case can correspond to a ceramic/metal FGM layer on a metal substrate with gradual transition from a metal at $y = -h$ to a ceramic at the upper part of the FGM layer ($\alpha_t^{ceramic} < \alpha_t^{metal}$).

2.3 Thermal and mechanical loadings

The FGM/H structure is subjected to tensile load p applied at infinity parallel to the free surface and is cooled by ΔT from the sintering temperature (Fig. 1b). In the case of cooling of the FGM/H tensile residual stresses are observed as shown in experimental investigations [6].

If an FGM/H structure is cooled, then residual stresses are arising due to mismatch in the coefficients of thermal expansion [15, 16]. The FGMs inhomogeneity will be accounted via the continuously varying residual stresses, and these stresses are the following [15]:

$$\sigma_{xx}^T(y) = [\alpha_t(y) - \alpha_{t1}]\Delta T E(y),\ \sigma_{xx}^e(y) = [E(y)/E_1 - 1]\sigma_{xx}^0,\ \sigma_{xx}^0 = p, \tag{8}$$

α_{t1} and E_1 are, respectively, the thermal expansion coefficient and the Young's modulus of a homogeneous material and at the interface, i.e. in the region $y \le -h$; $\alpha_t(y)$ and $E(y)$ are defined by formulas (2) and (3).

The method of linear superposition is used in the solution of this problem, so that the loads at infinity are reduced to the corresponding loads on the crack faces. Thus, the tensile load is reduced to the load p_n on the crack surfaces and written as

$$p_n = \sigma_n - i\tau_n = p(1 - \exp(2i\beta_n))/2 \quad (n = 1, 2, .., N). \tag{9}$$

In the common case of FGMs, the full load on the n-th crack consists of p_n, σ_n^T and σ_n^e, where the index "n" denotes that the functions are written in the local coordinate systems (x_n, y_n) connected with cracks. If the materials are elastically homogeneous, then $E_1 = E(y)$ and consequently $\sigma_{xx}^e = 0$ in Eq. (8) – see Table 2 for examples of such materials for which the graded parameter of the Young's modulus is $\omega=0$.

2.4 Main equations

The problem is solved by using the method of singular integral equations. For arbitrary located cracks in a half-plane the system of singular integral equations is written as [20, 21]

$$\int_{-a_n}^{a_n} \frac{g_n'(t)dt}{t-x} + \sum_{\substack{k=1 \\ k \ne n}}^{N} \int_{-a_k}^{a_k} [g_k'(t)R_{nk}(t,x) + \overline{g_k'(t)}S_{nk}(t,x)]dt = \pi p_n(x),\ |x| < a_n,$$

$$n = 1,2,\ \dots,\ N, \tag{10}$$

and for internal cracks the following condition is fulfilled

$$\int_{-a_n}^{a_n} g_n'(t)dt = 0. \tag{11}$$

It is the condition of displacement continuity at crack tips. An overbar $(\overline{\dots})$ denote the complex conjugate. N is the number of cracks.

Materials Research Forum LLC
https://doi.org/10.21741/9781644902950

The unknown functions in this formulation are the derivatives of displacement jumps on the crack lines

$$g_n'(x) = \frac{2\mu}{i(\kappa+1)}\frac{\partial}{\partial x}\left([u_n]+i[v_n]\right).$$

Here $[u_n]$ and $[v_n]$ are shear and vertical displacement jumps, respectively, on the n-th crack line, $\mu = E/2(1+\nu)$ is the shear modulus, E - Young's modulus, ν - Poisson's ratio, $\kappa = 3-4\nu$ for the plane strain state and $\kappa = (3-\nu)/(1+\nu)$ for the plane stress state. The regular kernels contain geometry of the problem and they can be found in [14, 20, 21].

The functions p_n in the right side of Eq. (10) are known loadings and they consist of the loads determined by Eqs. (8) and (9).

The stress intensity factors (SIFs) are found as [20, 21]

$$K_{nI}^{\pm} - iK_{nII}^{\pm} = \mp \lim_{x_n \to \pm a_n} \sqrt{(a_n^2 - x_n^2)/a_n}\, g_n'(x_n),$$

where the upper part of the "±" or "∓" signs refers to the right tip and the lower to the left tip of the n-th crack.

2. Numerical solution and determination of stress intensity factors

The solution of singular integral equations (Eq. 2) is obtained by a numerical method which is based on Gauss-Chebyshev quadrature, the description of this method can be found in [18-21].

The equations (10), (11) are rewritten in dimensionless form with the non-dimensionless coordinates $\xi = t/a_k$ and $\eta = x/a_n$, where $2a_k$ is the length of the k-th crack. The unknown function $g_n'(\eta)$ consists of a function $u_n(\eta)$ (a bounded continuous function in the segment [-1,1]) and the weight function $1/\sqrt{1-\eta^2}$, that is,

$$g_n'(\eta) = u_n(\eta)/\sqrt{1-\eta^2}\,. \tag{12}$$

For edge cracks the function $g_n'(\eta)$ possess a singularity less than $1/\sqrt{1+\eta}$ at the edge point $\eta = -1$ and this condition is accounted as [15, 16]

$$u_n(-1) = 0\,. \tag{13}$$

In spite of that the exact singularity at the edge points is not taking into account, the numerical results have shown good accuracy [15, 16] when the SIFs at the internal crack tip are calculated. If the stress-strain state in the vicinity of the edge crack tip is examined, then the exact order of singularity at this tip should be taking into account.

Using Gauss's quadrature formulae for the regular and singular integrals the integral equations are reduced to the following system of *N*x*M* (*N* – number of cracks, *M* – number of nodes) algebraic equations

$$\frac{1}{M}\sum_{m=1}^{M}\sum_{k=1}^{N}\left[u_k(\xi_m)R_{nk}(\xi_m,\eta_r)+\overline{u_k(\xi_m)}S_{nk}(\xi_m,\eta_r)\right]=\pi p_n(\eta_r), \tag{14}$$

$$\sum_{m=1}^{M}(-1)^m u_n(\xi_m)\tan\frac{2m-1}{4M}\pi=0 \quad (n=1,2,\ldots,N,\ r=1,2,\ldots,M\text{-}1), \tag{15}$$

with

$$\xi_m=\cos\frac{2m-1}{2M}\pi \quad (m=1,2,\ldots,M),\quad \eta_r=\cos\frac{\pi r}{M} \quad (r=1,2,\ldots,M\text{-}1).$$

M is the total number of discrete points of the unknown functions $u_n(\eta)$ within the interval (-1,1). Applying the conjugate operation to the system (14) and (15) additional *N*x*M* equations are obtained, i.e. (2*N*)x*M* equations should be solved, where *N* is the number of cracks.

If internal cracks are considered then instead of Eq. (15) the following equation should be used

$$\sum_{m=1}^{M}u_n(\xi_m)=0.$$

After solution of the algebraic system (14) and (15) the functions $u_n(\eta)$ are calculated by the interpolation formula:

$$u_n(\eta)=\frac{2}{M}\sum_{m=1}^{M}u_n(\xi_m)\sum_{r=0}^{M-1}T_r(\xi_m)T_r(\eta)-\frac{1}{M}\sum_{m=0}^{M}u_n(\xi_m). \tag{16}$$

The functions T_r are Chebyshev polynomials of the first kind.

The stress intensity factors (SIFs) are obtained from the following formulas [15, 16]

$$K_{nI}^{\pm}-iK_{nII}^{\pm}=\mp\lim_{\eta\to\pm1}\sqrt{a_n}\sqrt{1-\eta^2}\,g_n'(\eta),$$

$$K_{In}^{+}-iK_{IIn}^{+}=-\sqrt{a_n}u_n(+1)$$

$$=p_n\sqrt{a_n}\frac{1}{M}\sum_{m=1}^{M}(-1)^m u_n(\xi_m)\cot\frac{2m-1}{4M}\pi, \tag{17}$$

$$K_{In}^{-}-iK_{IIn}^{-}=\sqrt{a_n}u_n(-1)$$

$$= p_n \sqrt{a_n} \frac{1}{M} \sum_{m=1}^{M} (-1)^{M+m} u_n(\xi_m) \tan \frac{2m-1}{4M} \pi \qquad (n = 1, 2, \ldots, N). \tag{18}$$

The functions (16) are used in this calculation.

3. Fracture criteria, Fracture angles, Critical loads

For general crack problems the stress intensity factors are both nonzero, i.e. mixed-mode conditions hold in the vicinity of cracks. For this mixed-mode case the cracks deviate from their initial propagation direction. For the prediction of the crack growth and direction of this growth a fracture criterion should be applied. Using the maximum circumferential stress criterion, see [22], the direction of the initial crack propagation (Fig. 2) is evaluated as

$$\varphi = 2\arctan\left[\left(K_I - \sqrt{K_I^2 + 8K_{II}^2}\right) / 4K_{II}\right] \tag{19}$$

and the critical stresses can be calculated from the expression

$$\cos^3(\varphi/2)\left(K_I - 3K_{II}\tan(\varphi/2)\right) = K_{Ic} / \sqrt{\pi} . \tag{20}$$

Here K_{Ic} is the fracture toughness of the material. The critical stresses are given as

$$p_{cr} = P_{cr} / p_0 = P_{cr} / (K_{Ic} / \sqrt{2\pi a}) = 1 / [\cos^3(\varphi/2)\left(k_I - 3k_{II}\tan(\varphi/2)\right)]. \tag{21}$$

Here $k_{I,II}$ are non-dimensional SIFs

$$k_{I,II} = K_{I,II} / K^0, \quad K^0 = p\sqrt{2a}, \tag{22}$$

and $p_0 = K_{Ic} / \sqrt{2\pi a}$ is the critical load for a single crack in a material with the fracture toughness K_{Ic}.

For the system of cracks the fracture starts from the crack tip where P_{cr} is minimal, i.e. $\min_k [P_{cr(k)} / p_0]$.

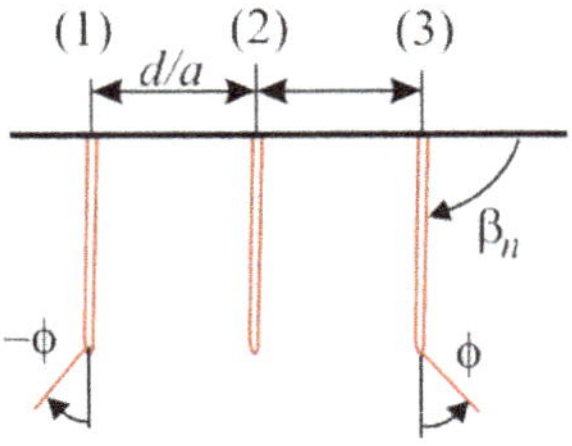

Figure 2. Schematic representation of three edge cracks with fracture angles ϕ.

Materials Research Foundations **159** (2024) https://doi.org/10.21741/9781644902950

4. Results and parametric analysis

The method described in Sections 2 and 3 can be applied to different systems of cracks. The verification of the method has been done in [14], where the results for some special cases were compared with the results for SIFs for a single inclined edge crack and with SIFs for periodic edge cracks. In [24] some results with respect to edge crack interaction in a homogeneous medium were presented.

In order to investigate the mutual interaction of cracks consider systems of two and three arbitrary inclined edge cracks. The geometry of cracks is considered with respect to experimental data available in literature, e.g. the experimental results in Refs. [4-6] showed that, when FGCs are subjected to thermal shock, multiple cracks occur at the ceramic surface during cooling-heating processes.

The following non-dimensional parameters are used in the calculations: $h/a = 4$, $d/a = (2, 4, 6)$, $a = \max_k a_k$, $a = 1mm$, $\varepsilon a = -1$ and $\omega a = 0$. The non-dimensional parameter $\varepsilon a = -1$ corresponds to smaller values of the thermal expansion coefficient in the upper part of the FGM layer. It should be mentioned that the results for fracture angles ϕ for the cracks in a material with $\varepsilon a = 0$ (homogeneous material) and with $\varepsilon a = -1$ are nearly the same. The designations for the distances d/a, the fracture angles ϕ and crack numbering can be seen in Fig. 2. The loading is described in Section 2.

Figs. 3-6 schematically represents the crack patterns due to the interaction of the cracks. Fig. 3 shows the influence of the interaction of two and three equal edge cracks with $\beta = 90°$ on fracture angles ϕ of the direction of crack propagation for different distances between cracks. For a single edge crack with $\beta = 90°$ the fracture angle is equal to $\phi = 0°$. In the case of two cracks (Fig. 3a), both cracks change the direction of their propagation. The direction of the angles ϕ_1 and ϕ_2 is opposite, i.e. the cracks repulse each other. Besides, the less the distance d/a – the larger the fracture angle ϕ (as it could be expected). For three edge cracks (Fig. 3b) the behavior of the outer cracks is similar to the behavior of the two interacting cracks (Fig. 3a). The middle crack is in the neutral position $\phi = 0$, i.e. the crack will not change the direction of the propagation.

Fig. 4 shows the nonsymmetrical cases of the interaction of two and three unequal edge cracks. The crack 1 slightly deviates away from the small crack 2 if the distance is $d/a=2$ and will propagate straight for far distances between the cracks (Fig. 4a). At the same time the influence of the large crack on the propagation direction of the small one is rather strong (Fig. 4a, left). The small crack will propagate away and a repulsion of these two cracks is observed.

In the case of three non-equal cracks the propagation of the large crack is similar to the case of two cracks (Fig. 4b). The small cracks have fracture angles with the same sign, and both small cracks deviate away from the large crack. In contrast to the case of the three equal cracks interaction, where the middle crack was in a neutral position (Fig. 3b), in this case the behavior of the middle crack is similar to the behavior of the outer small crack. The influence of the large crack on the propagation direction of the small middle crack is rather strong and stronger than on the outer small crack (Fig. 4b).

Figs. 5 and 6 show the results for systems of edge cracks with inclination angles $\beta = 60°$. For a single edge crack with $\beta=60°$ the fracture angle is equal to $\phi_0 = 31°$. The difference in the fracture angles due to their interaction is dependent on the distance between the cracks and the size of these cracks. The influence of this interaction on the fracture angles is more complicated than for the previous case for cracks with β=90°. For two equal cracks the crack 1 has the fracture angle $\phi_1 > \phi_0$ in contrast to the crack 2 with $\phi_2 < \phi_0$, Fig. 7. The difference between $\phi_{1,2}$ and ϕ_0 becomes smaller with larger distance between the cracks. For three cracks we have the following: $\phi_1 > \phi_0$, $\phi_2 > \phi_0$, and $\phi_3 < \phi_0$. The direction of the crack propagation for all cracks in Figs. 5 and 6 is the same, the fracture angles have the same sign.

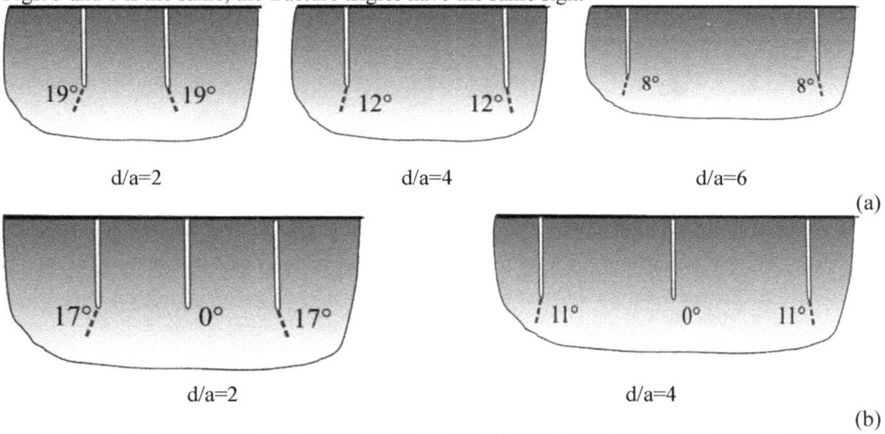

Figure 3. Fracture angles for two (a) and three (b) equal cracks with different distances between them, $\beta = 90^0$.

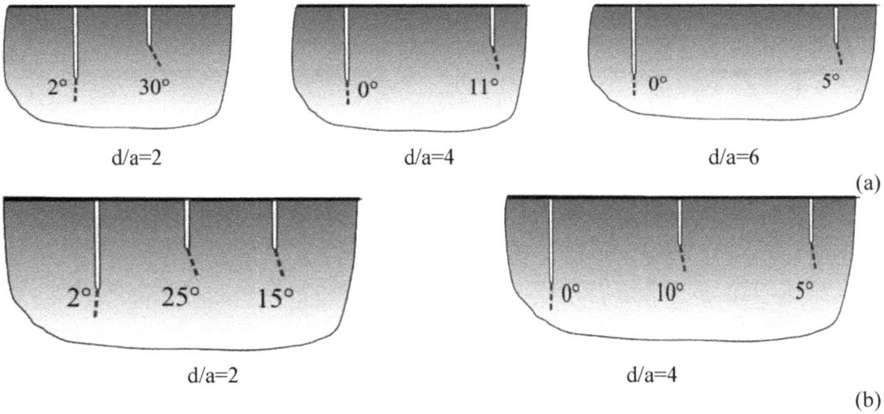

Figure 4. Fracture angles for two (a) and three (b) cracks with different distances between them, $a_2=0.5a_1$ (a), $a_2=a_3=0.5a_1$ (b), $\beta = 90^0$.

Materials Research Forum LLC
https://doi.org/10.21741/9781644902950

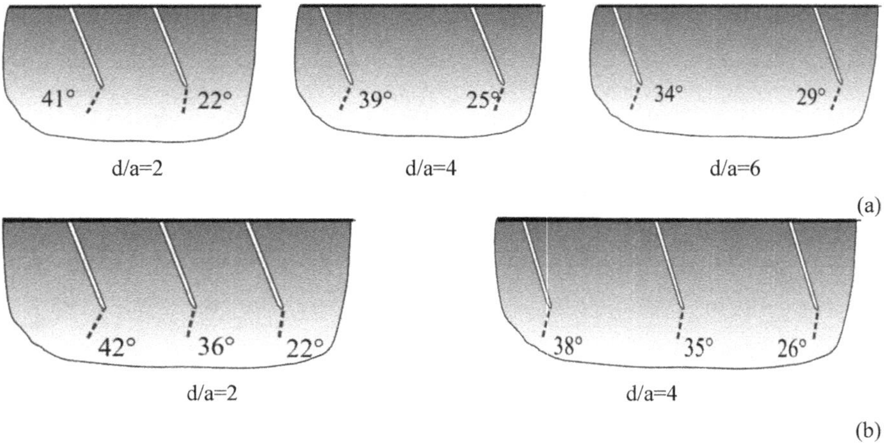

Figure 5. Fracture angles for two (a) and three (b) equal cracks with different distances between them, $\beta = 60^0$.

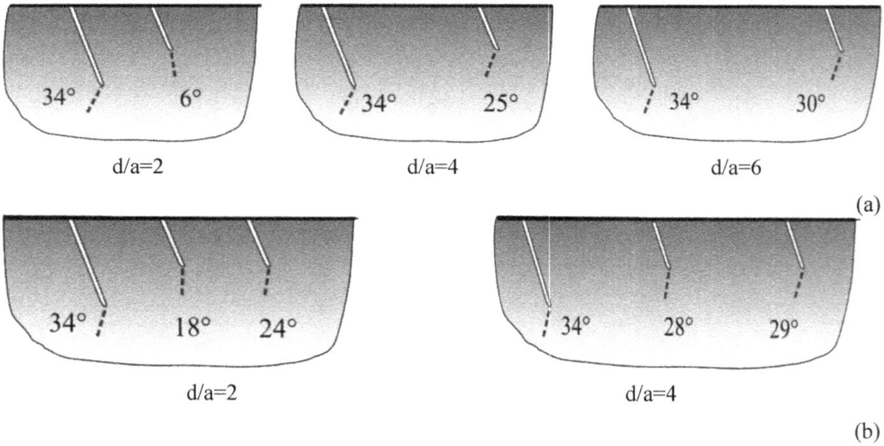

Figure 6. Fracture angles for two (a) and three (b) cracks with different distances between them, $a_2=0.5a_1$ (a), $a_2=a_3=0.5a_1$ (b), $\beta = 60^0$.

Table 3 presents the results for fracture angles for two edge cracks with slightly perturbed inclination angle β to the surface, i.e. for β =85° and 95°. Small changes in the inclination angle β cause strong influences on the fracture angle ϕ.

The perturbation effect is known for the crack interaction problem, for example, when two collinear Mode I cracks are growing towards each other, they do not merge tip to tip, but instead repeal each other [25]. In [25] the stability of crack paths to small perturbations was investigated experimentally for arrays of cracks in heterogeneous plates under tension. The main result of their investigation was the analysis of the geometrical conditions for which cracks are attracted towards another and when they are repelled.

Table 3. The perturbation effect for two edge cracks

β	85°			95°			90°		
d/a	2	4	6	2	4	6	2	4	6
ϕ_1 ϕ_2 (a_1=a_2)	−25° 10°	−18° 4°	−12° 0°	−10° 25°	−4° 18°	0° 12°	−19° 19°	−12° 12°	−8° 8°
ϕ_1 ϕ_2 (a_2=0.5 a_1)	−10° 25°	− 10° 5°	−10° 0°	5° 35°	5° 17°	5° 10°	−2° 31°	−2° 12°	−2° 5°

Conclusions

A semi-analytical model for fracture analysis of FGCs on a homogeneous substrate subjected to thermo-mechanical loadings is described. A system of pre-existing cracks in the FGC is studied in detail. This typical crack system is observed in experiments and the details are available in the literature [4-6]. The influence of the geometry of the problem, i.e. the crack sizes, the inclination angles and distances between the cracks on the fracture angles of the cracks was studied. It was shown that the directions of crack propagation depend mainly on the geometry of the problem, while the influence of inhomogeneity parameters of the thermal expansion coefficients of FGM is negligible.

Acknowledgement

The support of the German Research Foundation under the grant Schm 746/139-2 is greatly acknowledged.

References

[1] D.K. Jha, T. Kant, R.K. Singh, A critical review of recent research on functionally graded plates, Compos. Struct. 96 (2013) 833-849. https://doi.org/10.1016/j.compstruct.2012.09.001

[2] Y. Miyamoto, W.A. Kaysser, B.H. Rabin, A. Kawasaki, R.G. Ford, Functionally Graded Materials: Design, Processing and Applications, Kluwer Academic, Dordrecht, 1999. https://doi.org/10.1007/978-1-4615-5301-4

[3] U. Schulz, M. Peters, Fr.-W. Bach, G. Tegeder, Graded coatings for thermal, wear and corrosion barriers, Mater. Sci. Eng. A362 (2003) 61-80. https://doi.org/10.1016/S0921-5093(03)00579-3

[4] A. Kawasaki, R. Watanabe, Thermal fracture behavior of metal/ceramic functionally

graded materials, Eng. Fract. Mech. 69 (2002) 1713-1728. https://doi.org/10.1016/S0013-7944(02)00054-1

[5] S. Rangaraj, K. Kokini, Multiple surface cracking and its effect on interface cracks in functionally graded thermal barrier coatings under thermal shock, Trans. ASME J. Appl. Mech. 70 (2003) 234-245. https://doi.org/10.1115/1.1533809

[6] A. Gilbert, K. Kokini, S. Sankarasubramanian, Thermal fracture of zirconia-mullite composite thermal barrier coatings under thermal shock: An experimental study, Surf. Coat. Technol. 202(10) (2008) 2152-2161. https://doi.org/10.1016/j.surfcoat.2007.09.001

[7] B. Zhou, K. Kokini, Effect of surface pre-crack morphology on the fracture of thermal barrier coatings under thermal shock, Acta Mater. 52 (2004) 4189–4197. https://doi.org/10.1016/j.actamat.2004.05.035

[8] Y. Z. Feng, Z.-H. Jin, Thermal shock damage and residual strength behavior of a functionally graded plate with surface cracks alternating length, J. Therm. Stresses 35 (2012) 30-47. https://doi.org/10.1080/01495739.2012.637457

[9] V. Petrova, S. Schmauder, Thermal fracture of a functionally graded/homogeneous bimaterial with a system of cracks, Theor. Appl. Fract. Mech. 55 (2011) 148-157. https://doi.org/10.1016/j.tafmec.2011.04.005

[10] V. Petrova, S. Schmauder, Mathematical modelling and thermal stress intensity factors evaluation for an interface crack in the presence of a system of cracks in functionally graded/ homogeneous bimaterials, Comp. Mater. Sci. 52 (2012) 171-177. https://doi.org/10.1016/j.commatsci.2011.02.028

[11] V. Petrova, S. Schmauder, Interaction of a system of cracks with an interface crack in functionally graded/ homogeneous bimaterials under thermo-mechanical loading, Comp. Mater. Sci. 64 (2012) 229-233. https://doi.org/10.1016/j.commatsci.2012.04.032

[12] V. Petrova, S. Schmauder, FGM/homogeneous bimaterials with systems of cracks under thermo-mechanical loading: Analysis by fracture criteria, Eng. Fract. Mech. 130 (2014) 12-20. https://doi.org/10.1016/j.engfracmech.2014.01.014

[13] V. Petrova, S. Schmauder, Crack closure effects in thermal fracture of functionally graded/ho-mogeneous bimaterials with systems of cracks, ZAMM 95(10) (2015) 1027-1036. https://doi.org/10.1002/zamm.201400294

[14] V. Petrova, T. Sadowski, Theoretical modeling and analysis of thermal fracture of semi-infinite functionally graded materials with edge cracks, Meccanica, 49 (2014) 2603-2615. https://doi.org/10.1007/s11012-014-9941-x

[15] A.M. Afsar, H. Sekine, Crack spacing effect on the brittle fracture characteristics of semi-infinite functionally graded materials with periodic edge cracks, Int. J. Fract. 102 (2000) L61-L66.

[16] H. Sekine, A.M. Afsar, Composition profile for improving the brittle fracture characteristics in semi-infinite functionally graded materials, JSME International Journal, Ser. A 42(4) (1999) 592-600. https://doi.org/10.1299/jsmea.42.592

[17] K. Kokini, J. DeJonge, S. Rangaraj, B. Beardsley, Thermal shock of functionally graded thermal barrier coatings with similar thermal resistance, Surf. Coat. Technol. 154 (2002)

223-231. https://doi.org/10.1016/S0257-8972(02)00031-2

[18] F. Erdogan, G. Gupta, On the numerical solution of singular integral equations, Quart. Appl. Math. 29 (1972) 525-534. https://doi.org/10.1090/qam/408277

[19] A.I. Kalandia, Mathematical Methods of Two-Dimensional Elasticity (in Russian), Nauka, Moscow, 1973.

[20] V.V. Panasyuk, M.P. Savruk, A.P. Datsyshin, Stress Distribution near Cracks in Plates and Shells (in Russian), Naukova Dumka, Kiev, 1976.

[21] M.P. Savruk, Two- Dimensional Problems of Elasticity for Body with Cracks (in Russian), Naukova Dumka, Kiev, 1981.

[22] F. Erdogan, G.C. Sih, On the crack extension in plates under plane loading and transverse shear, J. Basic. Eng. 85 (1963) 519-527. https://doi.org/10.1115/1.3656897

[23] J.F. Shackelford, W. Alexander, CRC Materials Science and Engineering Handbook, CRC Press, Boca Raton, 2001. https://doi.org/10.1201/9781420038408

[24] V. Petrova, S. Schmauder, A. Shashkin, Modeling of edge cracks interaction, Frat. ed Integrita Strutt. 36 (2016) 8-26. https://doi.org/10.3221/IGF-ESIS.36.02

[25] P.-P. Cortet, G. Huillard, L. Vanel, S. Ciliberto, Attractive and repulsive cracks in a heterogeneous material, J. Stat. Mech.: Theory and Experiment, 10 (P10022) 2008 12 p. https://doi.org/10.1088/1742-5468/2008/10/P10022

Materials Research Foundations **159** (2024) https://doi.org/10.21741/9781644902950

CHAPTER 10

Fracture of Functionally Graded Thermal Barrier Coating on a Homogeneous Substrate: Models, Methods, Analysis

Vera Petrova [1,2]*, Siegfried Schmauder [1]

[1] IMWF, University of Stuttgart, Pfaffenwaldring 32, D-70569 Stuttgart, Germany

[2] Voronezh State University, University Sq.1, Voronezh 394006, Russia

veraep@gmail.com *, Siegfried.Schmauder@imwf.uni-stuttgart.de

Abstract

Fracture of functionally graded thermal barrier coatings on a homogeneous semi-infinite substrate (FGC/H) is studied under the influence of thermal loadings. A mathematical model for the FGC/H with pre-existing systems of cracks is presented in point of view of the formulation of the boundary conditions and assumptions, and afterwards the integral equations. Methods for solving the problem are described, and then demonstrated for some special cases of the arrangement of cracks. The effect of geometry (the thickness of the coating and location – orientation of multiple cracks) and inhomogeneity parameters of FGCs on the main fracture characteristics of the material is analyzed. The study can be used to improve the thermal fracture resistance of FGC/H systems.

1. Introduction

Thermal barrier coatings are widely used in industry, for example, in aerospace and airplane construction, and in power engineering. The effectiveness of rocket equipment, power turbines and aircraft gas turbine engines depends on the operating temperature. High temperatures and abrupt temperature changes lead to high-temperature cracking and melting of metal parts. Thermal barrier coatings create a layer with low thermal conductivity on the surface of the workpiece, which reduces the negative impact of high temperatures. For this purpose, various ceramics are used. Ceramics have a low thermal conductivity, but they are brittle. In addition, as a result of the difference in coefficients of thermal expansion of the coating and the substrate, high thermal stresses at the bond interface occur, which leads to cracking and debonding of the coatings. To overcome this problem, new materials are being sought, and new coatings are designed, namely, layered coatings and coatings from functionally gradient materials [1]. Functionally graded materials (FGMs) are a special class of composites with properties that vary along one direction (one spatial coordinate). Generally, the change in the properties of FGMs is connected with a corresponding variation in the chemical composition and/or physical structure of the material along this direction.

The study of the strength and fracture toughness of such FGMs is of great importance, especially under the combined effect of temperature, mechanical stresses and aggressive environment. Nowadays, a lot of studies has been devoted to these problems, reviews can be found in [1, 2]. However, the interaction of cracks and defects in FGMs under thermal loading is not sufficiently developed.

Investigation of the fracture of the functionally graded coating on a homogeneous substrate (FGC/H) under the influence of thermal loads includes the following tasks:

Modeling of inhomogeneous physical and mechanical properties of FGMs and the determination of the stress-strain state in FGC/H structures.

The problem of edge, internal and interface cracks and their interaction.

The influence of material heterogeneity on the interaction of cracks, as well as the mutual effect of thermal and mechanical loads on cracks and on the interaction of cracks.

In the present paper, the problem of the interaction of systems of cracks (edge and internal) in a functionally graded coating on a homogeneous semi-infinite substrate subjected to thermal loads is considered. The applied methods are similar to those used in reference [3] for an infinite bimaterial formed from functionally gradient and homogeneous materials, and in [4] for a FGM coating on a semi-infinite substrate with a periodic system of edge cracks in the coating.

The paper is organized as follows. The statement of the problem is presented in Section 2, where the geometry of the problem and the loading are described, and also the boundary conditions and assumptions are formulated. Then, in Section 3 models for functionally graded materials (FGMs) and for cracks in FGMs are described. In Section 4 the formulation of integral equations is followed by methods for solving these equations. Afterwards, in Section 5 the main fracture characteristics are calculated, namely, the stress intensity factors, as well as the fracture angles and the critical loads. In Section 6 a few illustrative examples are presented and parametric analysis is performed. Some concluding remarks are made in the final Section 6.

2. Statement of the problem

2.1 Geometry of the problem

The general case of the geometry of the problem is depicted in figure 1a. The upper layer of thickness h is made from a functionally graded material (FGM), and the semi-infinite substrate consists of a homogeneous material. The functionally graded coating (FGC) and the substrate are perfectly bonded with the exception of an interface crack disposed at the interface between the two materials. The FGC contains arbitrary located cracks of length $2a_k$ (k=1,…,N), which can be internal and/or edge cracks. The coordinate systems are chosen as follows: the global coordinates (x, y) with the x-axis located on the surface of the FGC/H structure and the local coordinates (x_k, y_k) connected with cracks and with centers in the midpoint of the k-th crack. The crack positions are determined explicitly by the midpoint coordinates z_k^0 and the inclination angles α_k to the x-axis (or β_k for edge cracks, $\beta_k = -\alpha_k$) (figure 1b).

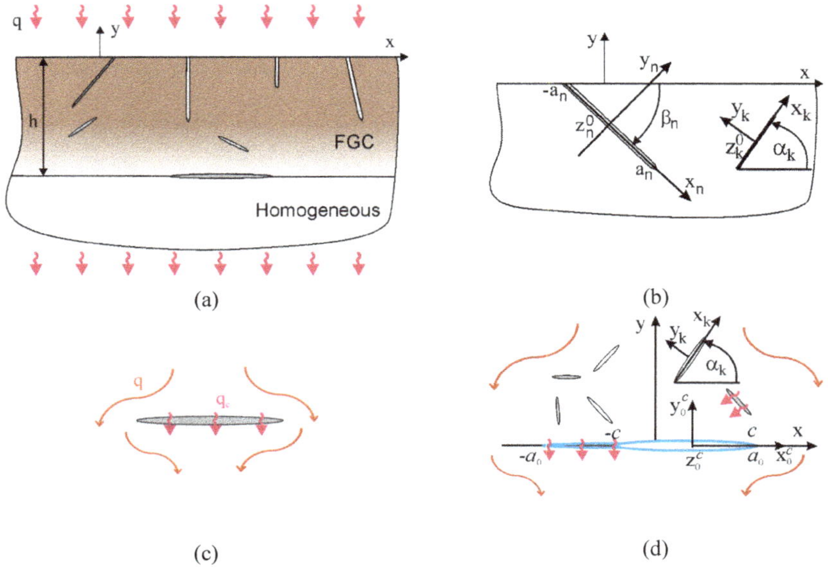

Figure 1. (a) An FGC/H structure with a system of cracks. (b) Coordinate systems connected with cracks. (c) A partially thermal conducting crack. (d) Partially closed crack with an open zone of length 2c and a system of open or closed cracks.

2.2 Thermal and mechanical loads

The following loads are considered: a steady state thermal flux applied normal to the surface of the FGC/H structure (figure 1a), a tension parallel to this surface (in the *x*-direction), and cooling by ΔT. In the case of cooling of the FGC/H structure tensile residual stresses are observed as shown in the experimental investigations [5]. Since the problem is linear, the results for each load can be superimposed. In the case of FGC/H under a heat flux the problem is solved in two steps, first, the thermal problem for the FGC/H structure with cracks, and then the thermo-elastic problem for the same geometry.

2.3 Assumptions

The following assumptions are used for this problem:

The uncoupled, quasi-static thermo-elasticity theory is applied. In this theory the temperature distribution is independent of the mechanical field, and the solution consists of the determination of the temperature field and then the determination of the thermal stresses.

The thermal and mechanical properties of an FGC are continuous functions of the thickness coordinate *y*.

The non-homogeneity of the functionally graded material is revealed in the form of the corresponding inhomogeneous stress distributions on the surfaces of the cracks [3, 4, 6]. In this case, the properties of the FGM should vary slightly with the depth of the layer.

2.4 Boundary conditions

The thermal problem for a system of thermally insulated cracks in an infinite bimaterial consisting of a functionally graded material and a homogeneous substrate has been formulated in [3]. In the present problem the partially thermal insulated cracks in the FGC are considered with the insulation coefficient η_k ($0 \le \eta_k \le 1$). The description of the model is given below in Section 3. Cracks are free of stresses, unless other conditions are mentioned. The FGC and the substrate are perfectly bonded, i.e. ideal thermal and mechanical conditions are fulfilled outside an interface crack, namely, the temperatures and the thermal fluxes are equal at the interface, and the stresses and displacements are also equal.

2.5 Methods

For the formulation of the problem the method of complex variables is used, and the method of superposition. Due to the superposition principle the common problem is decomposed into sub-problems, each of which contains one crack. Besides, the loads (thermal and mechanical) at infinity are reduced to the corresponding loads on the crack faces. The solution of each sub-problem can be obtained in explicit form (and can be analyzed separately) or in an integral form, and this integral form is used for constructing the equations of the complete problem. The methods for constructing integral equations for systems of cracks in a homogeneous material can be found in [7].

3. Modeling of functionally graded materials and cracks in FGMs

3.1 Models for functionally graded materials

The thermal and mechanical properties of an FGC are continuous functions of the thickness coordinate y. As in the previous works [3, 8, 9] the exponential form of these properties is used:

$$f(y) = f_1 \exp(\zeta_n (y+h)), \quad -h \le y \le 0 \tag{1}$$

here

$$f = \{k,\ \alpha_t,\ E\},\ f_1 = \{k_1,\ \alpha_{t1},\ E_1\},\ \zeta_n = \{\delta,\ \varepsilon,\ \omega\},$$

k is thermal conductivity, α_t – thermal expansion coefficient and E – the Young's modulus with non-homogeneity parameters δ, ε and ω, respectively. f_1 are thermal and mechanical properties of the homogeneous substrate.

The Poisson's ratio is assumed to be constant and is equal to the value of the homogeneous substrate. The previous studies show that the effect of Poison's ratio on the stress intensity factors is negligibly small (see, for example, [10]).

The values of the dimensionless graded parameters $\zeta_n h$ (h is the thickness of the FGC) are obtained from Eq. (1) as

$$\zeta_n h = \ln(f_2 / f_1),\quad f_2 = f(y)_{|y=0}\quad f_1 = f(y)_{|y=-h}$$

An example of ceramic/metal FGC/H is (ZrO_2/Ti-6Al-4V)/Ti-6Al-4V with f_2 / f_1 ={2/6.7, 10/8.6, 200/114} and, respectively, with the inhomogeneity parameters $\{\delta h,\ \varepsilon h,\ \omega h\}$={−1.2, 0.15, 0.56} (see [8, 11]). The variation of non-dimensional properties $f / f_1 = \exp(\zeta_n h(y / h+1))$ with dimensionless coordinate *y/h* is displayed in figure 2a. The coefficient of thermal expansion and the Young's modulus increase toward the upper part of the FGC/H structure, while the coefficient of thermal conductivity decreases (figure 2a).

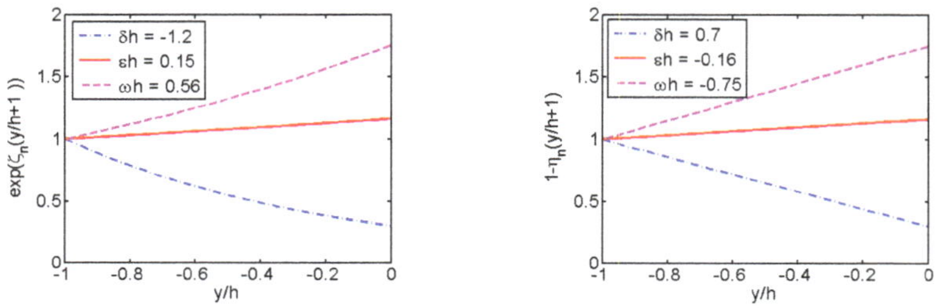

Figure 2. The variation of the exponential function $f / f_1 = \exp(\zeta_n h(y / h+1))$ *(a) and the linear function* $f(y) = f_1(1-\eta_n h(y / h+1))$ *(b) with the coating depth y/h for the FGC/H structure.*

The exponential form of properties (1) is convenient for the mathematical formulation of equations. Other types of variation laws can be a linear function, a power law or other continuous functions [12]. An example of the linear law for our case is written as

$$f(y) = f_1(1-\eta_n(y+h))\ \text{with}\ \eta_n = \{\delta,\ \varepsilon,\ \omega\}. \tag{2}$$

The inhomogeneity parameters are defined as $\eta_n h = 1 - f_2 / f_1$, and for the above mentioned ceramic/metal FGC/H structure are the following: $\eta_n h = \{\delta h,\ \varepsilon h,\ \omega h\}=\{0.7, -0.16, -0.75\}$. Figure 2b shows the variation of this linear function with coordinate *y/h*.

For practical application, if the spatial composition profile is known, it can be useful to apply a formula based on the rule of mixtures for determining the effective properties of composites. Theoretical mixing laws have been considered extensively for different types of composites, and have been applied to FGMs [13, 14, 15]. For example, the thermal conductivity is defined by

$$k(y) = k_m\left[1+\frac{V_i(y)(k_i-k_m)}{k_m+(k_i-k_m)(1-V_i(y))/3}\right],$$

where the subscripts i and m stand for the inclusion and matrix properties, respectively. The volume fraction of the inclusion phase $V_i(y)$ is assumed in the form of a power-law function with a non-homogeneity parameter p

$$V_i(y) = ((y+h)/h)^p,$$

and the matrix phase is calculated as $V_m = 1 - V_i$.

3.2 Models for cracks

Partially thermally permeable cracks: A model of partially thermally permeable (insulated) cracks is used, figure 1c. The simplest model of partially insulated interfacial crack was used, for example, in [9, 16]. The insulation coefficient η_k $(0 \le \eta_k \le 1)$ is introduced, here the limiting value $\eta_k = 0$ represents the fully insulating crack and $\eta_k = 1$ – the fully conducting crack. In common case the coefficient $\eta_k(x)$ can be the function of the crack coordinate.

Cracks with contact zones: In the present work possible crack closure and contact of the crack surfaces is taken into account in the model (figure 1d), and its contribution is assessed. Previously, the model of a crack with contact zones was considered in the problem of the interaction between an interface crack and internal cracks in FGM/homogeneous bimaterials under the influence of thermal flux in [17] and under the combination of Mode II load and thermal flux in [18]. It was supposed that the closure of the crack does not affect the temperature distribution, and therefore only the thermoelastic problem was reformulated and solved.

Cracks with plastic zones: The plasticity is considered through a special mode, e.g. Dugdale model [19]. In [20] this model was applied for the problem of the interaction between a system of microcracks and a main crack with plastic zones. This plasticity model allows to formulate the equations within the framework of linear fracture mechanics.

4. Formulation of the integral equations of the problem and solution

4.1 System of singular integral equations

As it was mentioned in Section 2, assuming that the inhomogeneity of the FGM is revealed in the form of the corresponding inhomogeneous stress distributions on the surfaces of cracks, the solution of the boundary value problem of elasticity is reduced to the solution of the system of singular integral equations [3, 6, 7]:

$$\int_{-a_n}^{a_n} \frac{g_n'(t)\mathrm{d}t}{t-x} + \sum_{\substack{k=1 \\ k\neq n}}^{N} \int_{-a_k}^{a_k} [g_k'(t)R_{nk}(t,x) + \overline{g_k'(t)}S_{nk}(t,x)]\mathrm{d}t = \pi p_n(x), \quad |x| < a_n,$$

$$(n = 1, 2, \ldots, N). \qquad (3)$$

Here the unknown functions $g_n'(x)$ are the derivatives of displacement jumps on the crack lines

Materials Research Foundations **159** (2024) https://doi.org/10.21741/9781644902950

$$g_n'(x) = \frac{2\mu}{i(\kappa+1)}\frac{\partial}{\partial x}([u_n]+i[v_n]),$$

where $[u_n]$ and $[v_n]$ are shear and vertical displacement jumps, respectively, on the n-th crack line, $\mu = E/2(1+\nu)$ is the shear modulus, E - Young's modulus, ν - Poisson's ratio, $\kappa = 3-4\nu$ for the plane strain state and $\kappa = (3-\nu)/(1+\nu)$ for the plane stress state. In equations (3) an overbar $\left(\overline{...}\right)$ denotes the complex conjugate.

The regular kernels R_{nk} and S_{nk} in equations (3) determine the geometry of the problem, and on the right side of equations (3) the functions p_n are known functions determined by the load on the crack lines.

For internal cracks, the condition of displacement continuity at crack tips has to be taken into account. The system of equations for the heat conduction problem has a similar form, where the unknowns are the derivatives of temperature jumps on the crack lines, and on the right side of the equations – the known functions of the heat fluxes on the cracks.

4.2 Solution of the equations

Many results concerning the interaction of cracks were obtained using methods of integral equations [3, 4, 6-9]. Various numerical methods such as series expansions, boundary element methods and others can be used to obtain an approximate solution. The solutions presented below are obtained for FGM/H structures.

4.2.1 Approximate analytical method

In the previous works [3, 17, 18] results for an infinite bimaterial consisting of an FGM and a homogeneous material with an interface crack and a system of internal cracks in the FGM were obtained by the small parameter method. The solution of equations (3) was presented in explicit approximate form and was written for the interface crack as

$$g_0'(x) = g_{00}'(x) + \lambda^2 g_{02}'(x),$$

and the stress intensity factors (SIFs) at the interface crack tips were presented as

$$k_I^{\pm} - ik_{II}^{\pm} = k_0\left\{1+\lambda^2\sum_{n=1}^{N} g_n(\alpha_n, z_n^0/a_0, \delta a_0, \varepsilon a_0)\right\}.$$

The stress intensity factors are found by the formula [4, 7]

$$k_{nI}^{\pm} - ik_{nII}^{\pm} = \mp \lim_{x_n \to \pm a_n} \sqrt{(a_n^2 - x_n^2)/a_n}\, g_n'(x_n), \tag{4}$$

where the upper part of the "$\pm$" or "$\mp$" signs concerns to the right tip and the lower to the left tip of the crack.

It was assumed that an interface crack is significantly larger in size than internal cracks in the FGM. For this special case the small parameter λ is equal to the ratio of the size of small internal cracks to the interface crack size, i.e. $\lambda = a / a_0$ with $a = a_n$. The solution takes into account the interaction of the interface crack and each of the internal cracks (the interaction between small cracks is not accounted). These SIF functions contain parameters of the geometry of the problem (the midpoint coordinates z_n^0 of cracks, the orientation angles α_n of cracks and the crack sizes) and the parameters of the non-homogeneity of the FGM (δ and ε). It was shown that the non-homogeneity parameters δ and ε of thermo-conductivity and of thermal expansion coefficients, correspondingly, notably affect the SIFs of the interface crack. The SIFs can be amplified or shielded by the system of microcracks.

4.2.2 Numerical solution.

The solution of the equations is obtained by the method of mechanical quadratures [7, 21]. Applying quadrature formulas based on Chebyshev polynomials, the integral equations (3) reduce to systems of algebraic equations. In [8], a scheme for solving this method is given. After determining the unknowns, the main characteristics of the fracture mechanics are calculated, namely, the stress intensity factors at the crack tips (4) or crack tip opening displacements.

5. Fracture criteria

Experimental and theoretical studies show that with a mixed-mode type of loading, the crack propagation deviates from its original one. In FGMs, a mixed stress-strain state can also arise due to the heterogeneity of the material. Applying a fracture criterion, it is possible to determine the angles of the deflection of the propagation of cracks from their original direction and the critical loads at which this propagation occurs. For example, according to the criterion of maximum normal stresses [22] (the formulation of this criterion and the method of application can be also found in [7]), the crack deflection angle (or the fracture angle) is calculated by the formula

$$\varphi_0 = 2\,\mathrm{arctg}\left[\left(k_I - \sqrt{k_I^2 + 8k_{II}^2}\right) / 4k_{II}\right], \tag{5}$$

where k_I and k_{II} are the stress intensity factors.

The critical stresses are calculated from the expression

$$\cos^3(\varphi / 2)\left(k_I - 3k_{II}\tan(\varphi / 2)\right) = K_{ICtip} / \sqrt{\pi}\,.$$

Here K_{ICtip} is the fracture toughness of the material near the crack tip. First, it is obtained the angle of the crack propagation using the results for the calculation of SIFs. Then, the local fracture stability is evaluated. This criterion is the simplest one to employ.

In the minimum strain energy density criterion [23] the strain energy density factor (SDF) is introduced so that it is a quadratic form of SIFs (with coefficients in this form presented as functions of fracture angles). Consequently, it is postulated that the crack will grow in the direction where the SDF is minimal and when this minimum reaches a critical value. This criterion depends on Poisson's ratio.

The maximum strain energy release rate (SERR) criterion [24] posts that the crack will grow along the direction for which the strain energy release rate reaches a critical value. Note, this critical value is material dependent.

As a result (by using these criteria) the fracture angles are obtained as functions of the geometry of the problem and of the non-homogeneity parameters of the FGC with additional parameters due to thermal loadings. Then, the critical stresses and critical heat fluxes are obtained near cracks (for different crack locations).

In [25], the results for these three criteria were compared for the problem for an infinite FGM/H structure and it was shown that they are nearly the same. In the present work the criterion of maximum normal stresses [22] is applied.

6. Results and analysis

The problem contains geometric parameters, such as the size of the cracks, the coordinates of their centers and the inclination angles, and the thickness of the FGM layer, and the parameters of material inhomogeneity, determined by formulas (1) or (2). The presented model allows analyzing the influence of these parameters on the main characteristics of the problem.

Fig. 3 schematically shows three edge cracks and their deviation angles ϕ, in Fig. 3a the cracks are of the same size, and in Fig. 3b the length of the first crack is twice the length of the second and third ones. The following parameters were used for the calculation: $h/a = 4$, $d/a = 2$, $a = \max a_k$, $a = 1mm$, $\varepsilon a = -1$ and $\omega a = 0$. The distance between the cracks is equal to the length of the large crack (d/a=2), β = 90^0. At $\varepsilon a = -1$ the coefficient of thermal expansion increases with the depth of the layer. The FGC/H structure is cooled by ΔT, in this case tensile residual stresses parallel to the boundary ($y = 0$) arise. With this load, for one edge crack the value of the angle is equal to $\phi = 0°$. The change in the angles ϕ due to the interaction of the cracks is shown in Fig. 3. It should be noted that the thermal expansion inhomogeneity coefficient does not greatly affect the deflection angles ϕ. Earlier it was shown that the coefficient of inhomogeneity of thermal conductivity significantly affects the interaction of cracks [3, 17].

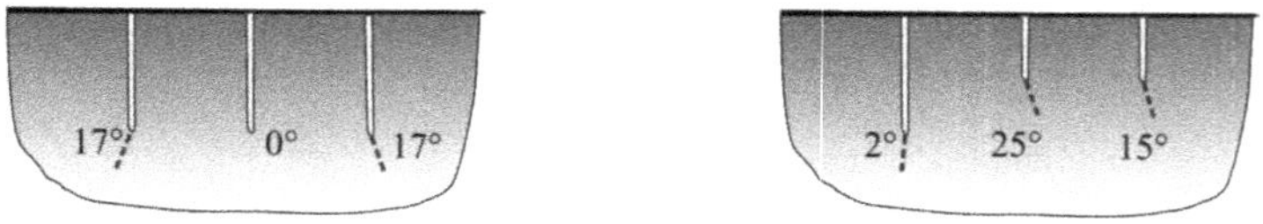

(a) (b)

Figure 3. Angles of deflection of the propagation of cracks under thermal loading. (a) Three edge cracks of equal length. (b) Edge cracks of different lengths a=a_2=a_3=0.5a_1. The distance between the cracks is equal to the length of the large crack d/a=2, β = 90^0.

The influence of the remaining parameters, geometrical and physical, will be considered later separately for the corresponding problems for thermal and mechanical loads.

Conclusions

A general theoretical formulation of the model for the thermal fracture analysis of functionally graded coatings on a homogeneous substrate (FGC/H) has been performed by means of integral equations. The FGC/H structure with pre-existing systems of interacting cracks in the FGC is subjected to thermal and mechanical loadings. The FGM properties are modeled by continuous functions of the spatial coordinate (the thickness of the coating). This approximate model allows taking into account some special crack models such as the partially thermally permeable cracks and the cracks with contact and plastic zones. The solution of equations is obtained numerically, using the special quadrature formulae for the singular and regular integrals in the integral equations. Besides, an approximate analytical solution previously obtained for a special case, when an interface crack is significantly larger in size than internal cracks in the functionally graded material for an infinite FGM/H structures, is given. In the present study the main emphasis is done on the fracture with multiple crack interactions. At first, the stress intensity factors are calculated, then, using the fracture criterion of maximum normal stresses [22], the crack deflection angles (or the fracture angles) are obtained. An illustrative example is presented to show the influence of the parameters of the problem on the edge crack interaction and their deviation from the initial propagation. Other crack models in the FGC/H structures will be considered in future works. The study of fracture of the FGC/H structures is important for a better understanding of the thermal fracture of graded coatings and for improving the fracture resistance of these systems.

Acknowledgments

The authors would like to acknowledge the financial support of the German Research Foundation under Grant Schm 746/139-2.

References

[1] Y. Miyamoto, W.A. Kaysser, B.H. Rabin, A. Kawasaki, R.G. Ford, Functionally Graded Materials: Design, Processing and Applications, Kluwer Academic, Dordrecht, 1999. https://doi.org/10.1007/978-1-4615-5301-4

[2] V. Birman, L.W. Byrd, Modeling and analysis of functionally graded materials and structures, ASME Appl. Mech. Rev. 60 (2007) 195-216. https://doi.org/10.1115/1.2777164

[3] V. Petrova, S. Schmauder, Thermal fracture of a functionally graded/homogeneous bimaterial with a system of cracks, Theor. Appl. Fract. Mech. 55 (2011) 148-157. https://doi.org/10.1016/j.tafmec.2011.04.005

[4] A.M. Afsar, H. Sekine, Crack spacing effect on the brittle fracture characteristics of semi-infinite functionally graded materials with periodic edge cracks Int. J. Fract. 102 (2000) L61-L66.

[5] A. Gilbert, K. Kokini, S. Sankarasubramanian, Thermal fracture of zirconia-mullite composite thermal barrier coatings under thermal shock: An experimental study, Surf. Coat. Technol. 202 (2008) 2152-2161. https://doi.org/10.1016/j.surfcoat.2007.09.001

[6] V. Petrova, T. Sadowski, Theoretical modeling and analysis of thermal fracture of semi-

infinite functionally graded materials with edge cracks, Meccanica 49 (2014) 2603-2615. https://doi.org/10.1007/s11012-014-9941-x

[7] V.V. Panasyuk, M.P. Savruk, A.P. Datsyshin, Stress Distribution near Cracks in Plates and Shells (in Russian), Naukova Dumka, Kiev, 1976.

[8] V. Petrova, S. Schmauder, Modeling of thermo-mechanical fracture of FGMs with respect to multiple cracks interaction, Phys. Mesomech. 20 (2017) 241-249. https://doi.org/10.1134/S1029959917030018

[9] S. El-Borgi, L. Hidri, R. Abdelmoula, An embedded crack in a graded coating bonded to a homogeneous substrate under thermo-mechanical loading, J. Therm. Stresses 29 (2006) 439-466. https://doi.org/10.1080/01495730500360591

[10] F. Delale, F. Erdogan, The crack problem for a nonhomogeneous plane, ASME J. Appl. Mech. 50 (1983) 609-614. https://doi.org/10.1115/1.3167098

[11] J.F. Shackelford, W. Alexander, CRC Materials Science and Engineering Handbook, CRC Press, Boca Raton, 2001. https://doi.org/10.1201/9781420038408

[12] M. Sevcik, P. Hutar, L. Nahlik, Z. Knesl, An evaluation of the stress intensity factor in functionally graded materials, Appl. Comput. Mech. 3 (2009) 401-410.

[13] M. Tilbrook, R. Moon, M. Hoffman, Crack propagation in graded composites, Composites Science and Technology 65 (2005) 201-220. https://doi.org/10.1016/j.compscitech.2004.07.004

[14] T. Fujimoto, N. Noda, Influence of the Compositional Profile of Functionally Graded Material on the Crack Path under Thermal Shock, J. Am. Ceram. Soc. 84 (2001) 1480-1486. https://doi.org/10.1111/j.1151-2916.2001.tb00864.x

[15] V.N. Burlayenko, H. Altenbach, T. Sadowski, S.D. Dimitrova, A. Bhaskar, Modelling functionally graded materials in heat transfer and thermal stress analysis by means of graded finite elements, Appl. Math. Model. 45 (2017) 422-438. https://doi.org/10.1016/j.apm.2017.01.005

[16] K.Y. Lee, S.J. Park, Thermal stress intensity factors for partially insulated interface crack under uniform heat flow, Eng. Fract. Mech. 50 (1995) 475-482. https://doi.org/10.1016/0013-7944(94)00243-B

[17] V. Petrova, S. Schmauder, Mathematical modelling and thermal stress intensity factors evaluation for an interface crack in the presence of a system of cracks in functionally graded, Comput. Mater. Sci. 52 (2012) 171-177. https://doi.org/10.1016/j.commatsci.2011.02.028

[18] V. Petrova, S. Schmauder, Crack closure effects in thermal fracture of functionally graded/homogeneous bimaterials with systems of cracks, ZAMM 195 (2015) 1027-1036. https://doi.org/10.1002/zamm.201400294

[19] D.S. Dugdale, Yielding in steel sheets containing slits, J. Mech. Phys. Solid. 8 (1960) 100-104. https://doi.org/10.1016/0022-5096(60)90013-2

[20] V. Tamuzs, V. Petrova, S. Tarasovs, Interaction of micro-cracks with a macro-crack yielded in a narrow strip, Theor. Appl. Fract. Mech. 41 (2004) 291-299.

https://doi.org/10.1016/j.tafmec.2003.11.016

[21] F. Erdogan, G. Gupta, On the numerical solution of singular integral equations, Quart. Appl. Math. 29 (1972) 525-534. https://doi.org/10.1090/qam/408277

[22] F. Erdogan, G.C. Sih, On the crack extension in plates under plane loading and transverse shear, J. Basic. Eng. 85 (1963) 519-527. https://doi.org/10.1115/1.3656897

[23] G.C. Sih, Mechanics of Fracture Initiation and Propagation: Surface and Volume Energy Density Applied as Failure Criterion, Kluwer Academic Publishers, Dordrecht, The Netherlands, 1991.

[24] M.A. Hussain, S.L. Pu, J. Underwood, Strain energy release rate for a crack under combined mode I and mode II, ASTM STP 560 (1974) 2-28. https://doi.org/10.1520/STP33130S

[25] V. Petrova, S. Schmauder, FGM/homogeneous bimaterials with systems of cracks under thermo-mechanical loading: Analysis by fracture criteria, Eng. Fract. Mech. 130 (2014) 12-20. https://doi.org/10.1016/j.engfracmech.2014.01.014

CHAPTER 11

Analysis of Interacting Cracks in Functionally Graded Thermal Barrier Coatings

Vera Petrova [1,2]*, Siegfried Schmauder [1]

[1] IMWF, University of Stuttgart, Pfaffenwaldring 32, D-70569 Stuttgart, Germany

[2] Voronezh State University, University Sq.1, Voronezh 394006, Russia

veraep@gmail.com *, Siegfried.Schmauder@imwf.uni-stuttgart.de

Abstract

The problem of thermal fracture of a structure consisting of a functionally graded coating (FGC) on a homogeneous substrate (FGC/H) subjected to a thermal loading is investigated. The main focus is laid on the application of fracture criteria for an FGC and the determination of critical mechanical and thermal loads. The application of fracture criteria requires knowledge of the fracture toughness near the crack tips. Thus, it is assumed that the fracture toughness of an FGC, as well as other material properties, continuously varies through the thickness of the coating. Besides, having a system of cracks, it is important to determine which crack is most dangerous with respect to possible propagation. The proposed model, combined with a detailed parametric analysis, provides a reliable basis for optimizing FGCs in order to improve the fracture resistance of FGC/homogeneous systems operating under high temperatures. An illustrative example for a real combination of materials (ceramic/metal)/metal is provided, which demonstrates that it is important to consider the variation of fracture toughness in FGCs in determining critical loads at which cracks can lead to complete structural fracture.

Keywords

Thermal Fracture; System of Cracks; Functionally Graded Coatings; Fracture Criteria

1. Introduction

Functionally graded materials (FGMs) represent a concept for composites that consist of a graded pattern of material composition and/or microstructures and, accordingly, with properties continuously graded in the same spatial direction. FGMs have been developed as ultrahigh temperature resistant materials for the aerospace industry, but they currently have a wide range of applications, e.g. in the automotive and aeronautic industries, in power engineering, as well as in the biomedical industry. For thermal barrier coatings, functionally graded materials consisting of ceramics on the top and metals underneath of the structure are used. At elevated temperatures and changes of temperatures in combination with mechanical loadings, different crack patterns are observed in functionally graded coatings (FGCs) [1]. To analyze the complex fracture behavior in FGMs and the fracture process, one needs to know the distribution of fracture toughness in FGMs.

The present paper deals with the problem of thermal fracture of a structure consisting of a functionally graded coating on a homogeneous substrate (FGC/H) subjected to thermal loading. The main focus is laid on the application of fracture criteria for FGMs and the determination of critical mechanical and thermal loads. In order to apply fracture criteria for an FGM, it is required to introduce a model how to determine the fracture toughness near the crack tips on the basis of available values of the fracture toughness of the constituent materials of FGCs. An experimental investigation of FGMs consisting of partially stabilized zirconia (PSZ) and stainless steel (see [2]) demonstrates the influence of microstructure on fracture toughness distribution, namely, the fracture toughness in the FGMs is higher in the FGM with a finer microstructure than in the FGM with a coarse (rough) microstructure; in addition, the fracture toughness in the FGMs is higher than in non-FGM. The fracture toughness distribution in FGMs is in good agreement with changes in the morphology and components of FGM, e.g. the fracture toughness increases with increasing steel content [2,3]. Theoretical models for estimating the fracture toughness of FGMs have been discussed in many papers, e.g. [4,5,6]. It is physically reasonable to assume that the fracture toughness is a function of a spatial coordinate, like other material properties. In [6], an exponential function and power-law function were chosen to describe the fracture toughness of the FGM, and in [4], the rule of mixtures was used to determine the fracture toughness of a ceramic-metal FGM. These functions are convenient for implementation in FGM models, but, generally, in these cases, the fracture toughness can be overestimated, since the fracture toughness of a metal in bulk form is much higher than that of metal particles dispersed in a brittle matrix [4,6].

The present work is devoted to the study of the interaction of multiple cracks in FGCs with respect to the main characteristics of fracture, such as stress intensity factors and critical loads. The theoretical basis for this problem has been described in authors' previous works [7,8], but the main parts are repeated here for completeness. The fracture angles (the deviation of cracks from their initial direction of propagation) were investigated previously in [7,8], so that here the main attention is paid on the critical loads for the pre-existing system of cracks, edge and internal cracks, in the FGC.

This investigation can provide important knowledge for choosing appropriate material combinations and the gradation profile of the FGCs in order to optimize the fracture resistance of FGC/homogeneous structures operating under high temperature.

2. Statement of the problem

2.1 Geometry of the problem and loading

Consider a structure consisting of a functionally graded layer of thickness h on a semi-infinite homogeneous substrate, as shown in Fig. 1a. For thermal barrier coatings (TBCs), the top of the layer should be made of ceramic, and the homogeneous substrate is made of metal. TBCs operate at high temperatures at which oxidation processes can occur, resulting in the formation of an oxide layer between the coating and the underlying alloy. That is, the structure under consideration is a functionally graded coating (FGC) with a material gradient perpendicular to the interface on a homogeneous substrate with the thermally grown oxide (TGO) layer between them. The structure will be called FGC/TGO/H.

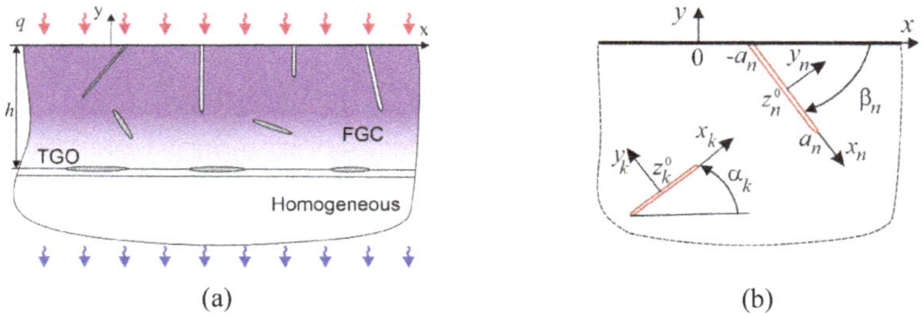

Figure 1. Geometry of the problem: (a) FGC/TGO/H with a system of cracks, and (b) coordinate systems connected with cracks.

In the functionally graded coating a system of N cracks of length $2a_k$ ($k = 1, 2, \ldots, N$) is located, which can be edge and/or internal cracks. The oxide layer is modeled as a weak layer and is represented by cracks located in this thin layer, Fig. 1a. The global coordinate system (x, y) is connected to the FGC surface, and the local coordinates (x_k, y_k) are referring to each crack with the x-axis on the crack lines as shown in Fig. 1b. Using these coordinate systems, the positions of the cracks are determined by their midpoint coordinates (x_k^0, y_k^0) and the inclination angles α_k to the x-axis, or β_k for edge cracks, $\beta_k = -\alpha_k$, Fig. 1b. Further, the complex variables method will be used in the formulation of the equations of the problem; therefore, it is convenient to represent the midpoint coordinates of the cracks in complex form as $z_k^0 = x_k^0 + iy_k^0$ (i is the imaginary unit).

The FGC/TGO/H structure is cooled by ΔT, $\Delta T > 0$ (this can be cooling from operating temperatures) and an additional tensile load p is applied parallel to the surface.

2.2 Key considerations

The problem of interacting cracks in functionally graded coatings under thermo-mechanical loading is studied in the frame of linear fracture mechanics. Numerous investigations (e.g. [4]) reveal that the distribution of stresses near the crack tips in functionally graded materials (FGMs) has the same inverse square root singularity as in homogeneous materials, if the material properties are continuous functions of a spatial variable. Hence, the stress intensity factors (SIFs) can be determined near the crack tips. For interacting cracks in an FGC, the mixed-mode fracture conditions is realized, that is, both stress intensity factors mode I (K_I) and mode II (K_{II}) are non-zero. Further, fracture analysis is possible by applying an appropriate fracture criterion for mixed-mode cracks and determining the direction of the crack propagation and the critical loads when this propagation starts. To obtain the critical loadings, it is important to know the fracture toughness near the crack tips [5,6,9]. Therefore, a rule should be established for determining fracture toughness for a functionally graded material. An experimental evaluation of fracture toughness and residual stresses in ceramic-metal functionally graded materials can be found in [2,3,10]. The thermal and mechanical properties of the FGC are assumed to be continuous functions of the thickness coordinate. It is also expected that, like other material properties of functionally graded materials (FGMs), the fracture toughness is a continuous function of the same spatial coordinate. Models for material properties for FGMs, including a model for the fracture toughness, are presented in the next section.

If the coating contains many cracks, fracture is expected to start from a weakest crack. The weakest crack can be determined by the highest value of SIFs (which is applicable for homogeneous materials, but not for FGMs), or by a smallest value of critical load at the crack. In determining critical loads the variation of the fracture toughness of the FGM is accounted. Thus the global critical load is determined as $P_{cr} = \min p_{crn}$ ($n = 1, 2, \ldots, N$).

Due to the mismatch of the thermal expansion coefficients of the FGC, a change of operating temperature leads to the appearance of thermal residual stresses in the considered structure. Besides, as noted in [2], microscopic and macroscopic residual stresses arise due to the mismatch of the thermal expansion coefficients between the ceramic and metal phases in the fabrication of FGMs. It is physically reasonable to assume that the non-homogeneity of the functionally graded material is a consequence of the form of the corresponding inhomogeneous stress distributions on the surfaces of the cracks, see [2,11]. Under this assumption, the properties of the FGM should vary slightly in the depth of the coating. These respective stresses that are applied to the crack faces are determined below.

2.3 Mechanical and physical properties for FGCs and inhomogeneity parameters

In functionally graded materials, the composition of the material gradually changes mainly in one direction. In particular, in our problem, the composition varies with the y-coordinate from the ceramic in the upper part of the FGC to the metal in the substrate. Consequently, the thermal and mechanical properties of an FGC also varies continuously with the thickness coordinate y. As in the previous works, e.g. [7,8], an exponential form of these properties is used:

$$\alpha_t(y) = \alpha_{t1}\exp(\varepsilon(y+h)),\ E(y) = E_1\exp(\omega(y+h)),\quad -h \le y \le 0. \tag{1}$$

Here, α_t is the coefficient of thermal expansion, and E is Young's modulus with non-homogeneity parameters ε and ω, respectively. E_1 and α_{t1} stand for the thermal and mechanical properties of the homogeneous substrate. Poisson's ratio is assumed to be constant, see [12], and is equal to the value of the homogeneous substrate. The values of the dimensionless inhomogeneity parameters εh and ωh (h is the thickness of the FGC) are obtained from Eq. (1) as follows

$$\varepsilon h = \ln(\alpha_{t2}/\alpha_{t1}),\quad \alpha_{t2} = \alpha_t(y)\big|_{y=0},\quad \alpha_{t1} = \alpha_t(y)\big|_{y=-h}, \tag{2}$$

$$\omega h = \ln(E_2/E_1),\quad E_2 = E(y)\big|_{y=0},\quad E_1 = E(y)\big|_{y=-h}. \tag{3}$$

Since the exponential law was chosen for the material parameters α_t and E, it is reasonable to use the exponential law also for the fracture toughness. But the comment in [4] that the fracture toughness can be overestimated in this case of the model have to be taken into account. Accordingly, it should be borne in mind that the fracture toughness values thus determined provide an upper bound for the possible values of the fracture toughness [4]. Thus the fracture toughness of a functionally graded material can be written as follows:

$$K_{Ic}(y) = K_{Ic1}\exp(\gamma(y+h)), \tag{4}$$

Materials Research Foundations **159** (2024) https://doi.org/10.21741/9781644902950

where γ is the inhomogeneous parameter of the fracture toughness. The nondimensional value γh is obtained as

$$\gamma h = \ln(K_{Ic2} / K_{Ic1}), \quad K_{Ic2} = K_{Ic}(y)_{|y=0}, \quad K_{Ic1} = K_{Ic}(y)_{|y=-h}. \tag{5}$$

In the local coordinate system (x_n, y_n) connected with the n-th crack, the function K_{Icn} is written as

$$K_{Icn}(x_n) = K_{Ic1}\exp(\gamma(h + y_n^0 - x_n \sin\beta_n)). \tag{6}$$

An example of a (ceramic/metal)/metal FGC/H material is (PSZ/steel)/steel with α_{t2} = 9 - 12.2 ($\cdot 10^{-6}$ K^{-1}), α_{t1} = 15($\cdot 10^{-6}$ K^{-1}), E_2 = 48-22, E_1 = 207 (GPa) for the temperature range 20°C - 1110°C (see [13]), and fracture toughness K_{Ic2} = 7-10, K_{Ic1} = 50 (MPa$\cdot$m$^{1/2}$). The corresponding inhomogeneity parameters are calculated by Eqs. (2), (3), (5) and are equal to εh = –0.5 - –0.2, ωh = –1.5 - –2.2, γh = –2.3. For this material combination, all of the listed material parameters, namely, thermal expansion coefficient, Young's modulus and fracture toughness, decrease towards the upper part of the FGC/H structure. The thermal conductivity of this material is also decreased in the direction to the ceramic top, from 16 to 2 ($Wm^{-1}K^{-1}$).

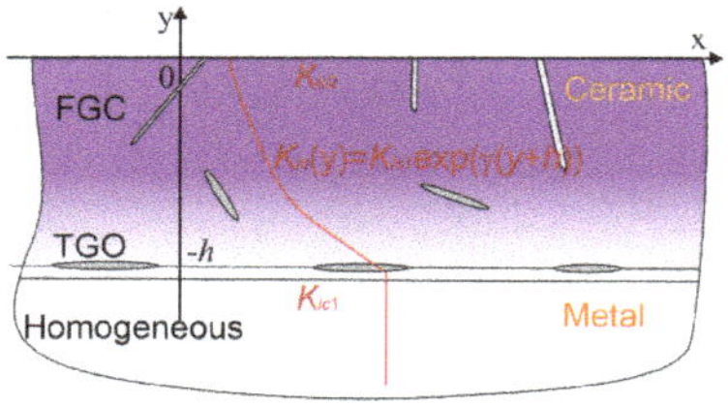

(a)

Figure 2. FGC/TGO/H with a system of cracks and changing fracture toughness.

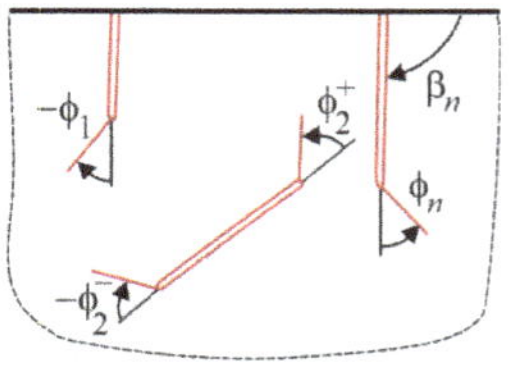

(b)

Figure 3. Edge and internal cracks with fracture angles.

2.4 Loadings on the crack faces

With changing the temperature, e.g. an FGM/H structure is cooled on ΔT, the residual stresses are arising due to mismatch in the coefficients of thermal expansion. In the presented model, the inhomogeneity of FGMs is taken into account through continuously varying residual stresses, and these stresses are the following [11]:

$$\sigma_{xx}^T(y) = [\alpha_t(y) - \alpha_{t1}]\Delta T E(y), \quad \sigma_{xx}^e(y) = [E(y)/E_1 - 1]\sigma_{xx}^0, \ \sigma_{xx}^0 = p, \tag{7}$$

α_{t1} and E_1 are, respectively, the thermal expansion coefficient and Young's modulus of a homogeneous substrate material and at the interface, $\alpha_t(y)$ and $E(y)$ are defined by Eq. (1).

The method of linear superposition is used to solve this problem, so that loads at infinity are reduced to the corresponding loads on the crack faces. Thus, the tensile load is reduced to the load p_n on the crack surfaces and written in complex form as

$$p_n = \sigma_n - i\tau_n = p(1-\exp(2i\beta_n))/2 \quad (n = 1, 2, \ldots, N). \tag{8}$$

In the common case of FGMs, the full load on the n-th crack consists of p_n, σ_n^T and σ_n^e, Eq. (7), where the index "n" denotes that the functions are written in the local coordinate system (x_n, y_n) connected with the n-th crack:

$$p_n + \sigma_n^e + \sigma_n^T = Q[p/Q\exp(\omega(h+y_n^0 - x_n\sin\beta_n)) + \exp(\varepsilon(h+y_n^0 - x_n\sin\beta_n)) - 1]$$

$$(n = 1, 2, \ldots, N), \tag{9}$$

$$Q = \alpha_{t1}\Delta T E_1 .$$

It is assumed that $p = Q$, otherwise an additional loading parameter p/Q should be considered.

3. Singular integral equations and solution

3.1 Singular integral equations

The boundary value problem of elasticity for a system of cracks is reduced to a system of singular integral equations [14]:

$$\int_{-a_n}^{a_n} \frac{g_n'(t)dt}{t-x} + \sum_{\substack{k=1\\k\neq n}}^{N} \int_{-a_k}^{a_k} [g_k'(t)R_{nk}(t,x) + \overline{g_k'(t)}S_{nk}(t,x)]dt = \pi p_n(x), \; |x| < a_n,$$

$$n = 1, 2, \ldots, N, \tag{10}$$

$$\int_{-a_n}^{a_n} g_n'(t)dt = 0, \quad \text{(for internal cracks)}$$

$$g_n'(x) = \frac{2\mu}{i(\kappa+1)}\frac{\partial}{\partial x}([u_n] + i[v_n]).$$

The number of equations N is equal to the number of cracks. The unknown functions $g_n'(x)$ contain the shear $[u_n]$ and normal $[v_n]$ displacement jumps on the n-th crack line, $\mu = E/2(1+v)$ is the shear modulus, E is Young's modulus, v is Poisson's ratio, $\kappa = 3 - 4v$ for the plane strain state, and $\kappa = (3 - v)/(1+ v)$ for the plane stress state. An overbar $\overline{(\ldots)}$ denotes the complex conjugate. The regular kernels R_{nk} and S_{nk} determine the geometry of the problem and can be found in [8] or [14]. In Eq. (10) the functions p_n are known functions determined by the load on the crack lines, Eq. (9).

3.2 Numerical solution

The singular integral equations (10) are solved numerically based on the Gauss-Chebyshev quadrature. Different versions of this method are used for the solution, and the effectiveness of the method has been proven in many studies [15]. In the present work, the version described in [14] is applied.

Eqs. (10) is rewritten in dimensionless form with the non-dimensionless coordinates $\xi = t / a_k$ and $\eta = x / a_n$, where $2a_k$ is the length of the *k*-th crack. The unknown function $g_n'(\eta)$ presents as

$$g_n'(\eta) = u_n(\eta) / \sqrt{1-\eta^2} \,,$$

Where, the function $u_n(\eta)$ is a bounded continuous function in the segment [-1,1] and $1/\sqrt{1-\eta^2}$ is the weight function, which is taking into account the square root singularities at the crack tips. For an edge crack, the function $g_n'(\eta)$ possesses a singularity less than $1/\sqrt{1+\eta}$ at the edge point η = – 1, this condition is accounted for as $u_n(-1) = 0$.

Using Gauss's quadrature formulae for regular and singular integrals, the singular integral equations are reduced to the following system of *N*x*M* (*N* – number of cracks, *M* – number of nodes) algebraic equations

$$\frac{1}{M}\sum_{m=1}^{M}\sum_{k=1}^{N}\left[u_k(\xi_m)R_{nk}(\xi_m,\eta_r)+\overline{u_k(\xi_m)}S_{nk}(\xi_m,\eta_r)\right]=\pi p_n(\eta_r) \tag{11}$$

$$\sum_{m=1}^{M}(-1)^m u_n(\xi_m)\tan\frac{2m-1}{4M}\pi = 0 \;\; \text{(for edge cracks)}$$

$$\sum_{m=1}^{M}u_n(\xi_m) = 0 \;\; \text{(for internal cracks)},$$

(n =1, 2, …, N; r = 1, 2, …, M-1),

$$\xi_m = \cos\frac{2m-1}{2M}\pi \;\; (m = 1, 2, \ldots, M), \;\; \eta_r = \cos\frac{\pi r}{M} \;\; (r = 1, 2, \ldots, M\text{-}1).$$

M is the total number of discrete points of the unknown functions $u_n(\eta)$ on the segment [- 1,1]. By applying the conjugate operation to the system (11), additional *N*x*M* equations are obtained, i.e. 2*N*x*M* equations should be solved, where *N* is the number of cracks.

The functions $u_n(\eta)$ are calculated by the interpolation formula:

$$u_n(\eta)=\frac{2}{M}\sum_{m=1}^{M}u_n(\xi_m)\sum_{r=0}^{M-1}T_r(\xi_m)T_r(\eta)-\frac{1}{M}\sum_{m=0}^{M}u_n(\xi_m) \tag{12}$$

T_r are Chebyshev polynomials of the first kind.

3.3 Stress intensity factors and critical loads

The stress intensity factors are obtained from the following formulas:

$$K_{nI}^{\pm} - iK_{nII}^{\pm} = \mp \lim_{\eta \to \pm 1} \sqrt{\pi a_n} \sqrt{1-\eta^2}\, g_n'(\eta) \tag{13}$$

$$K_{In}^{+} - iK_{IIn}^{+} = -\sqrt{\pi a_n} u_n(+1) = p_n \sqrt{\pi a_n} \frac{1}{M} \sum_{m=1}^{M} (-1)^m u_n(\xi_m) \cot \frac{2m-1}{4M}\pi, \tag{14}$$

$$K_{In}^{-} - iK_{IIn}^{-} = \sqrt{\pi a_n} u_n(-1) = p_n \sqrt{\pi a_n} \frac{1}{M} \sum_{m=1}^{M} (-1)^{M+m} u_n(\xi_m) \tan \frac{2m-1}{4M}\pi,$$

$n = 1, 2, \ldots, N$.

Here the signs "+" and "–" refer to the right and left crack tips, respectively.

The functions (12) written for $\eta = \pm 1$

$$u_n(1) = \frac{1}{M} \sum_{m=1}^{M} (-1)^{m+1} u_n(\xi_m) \cot \frac{2m-1}{4M}\pi$$

$$u_n(-1) = \frac{1}{M} \sum_{m=1}^{M} (-1)^{M+m} u_n(\xi_m) \tan \frac{2m-1}{4M}\pi$$

are used in the calculation of Eq. (14).

For predicting the crack growth and the determination of a direction of this growth, the criterion of maximum circumferential stresses [16] is used. According to this criterion, the crack deflection angle ϕ (or the so-called fracture angle, Fig. 3) and the critical stresses are calculated as

$$\phi_n = 2\arctan\left[\left(K_{In} - \sqrt{K_{In}^2 + 8K_{IIn}^2}\right) / 4K_{IIn}\right], \tag{15}$$

$$K_n^{eq} \equiv \cos^3(\phi_n/2)\left(K_{In} - 3K_{IIn}\tan(\phi_n/2)\right) = K_{Ic,tip} \text{ or } K_n^{eq} = K_{Ic,tip}. \tag{16}$$

Using a single crack subjected to a load p normal to the crack line as a reference crack with the stress intensity factor

$$K^0 = p\sqrt{\pi a}, \tag{17}$$

the corresponding critical load is obtained as

$$p_0 = K_{Ic1} / \sqrt{\pi a}, \tag{18}$$

where $a = \max_{n=1,\dots,N} a_n$.

In general, the SIFs are written as

$$K_{In} - iK_{IIn} = p\sqrt{\pi a}(k_{In} - ik_{IIn}). \tag{19}$$

By substituting (19) into condition (16), the critical loads are obtained

$$p_{crn} = \frac{K_{Ic}(y)}{\sqrt{\pi a_n}} \frac{1}{\cos^3(\phi_n/2)\left(k_{In} - 3k_{IIn}\tan(\phi_n/2)\right)}$$

or

$$\frac{p_{crn}}{p_0} = \frac{\exp(\gamma h(1 + y_n^0/h - (x_n/h)\sin\beta_n))}{\cos^3(\phi_n/2)\left(k_{In} - 3k_{IIn}\tan(\phi_n/2)\right)} \frac{\sqrt{a}}{\sqrt{a_n}}, \tag{20}$$

where p_0 is defined by Eq. (18) and K_{Ic} by Eqs. (4) and (6).

The fracture angle ϕ_n is shown in Fig. 3. First, the angle of the crack propagation (fracture angle) Eq. (15) is obtained using the results of the calculated stress intensity factors, Eqs. (13) and (14). Next, the local fracture stability is evaluated by Eq. (16). Then, the critical loads are obtained near the crack tips, Eq. (20). Finally, the weakest crack or crack tip is defined from the condition

$$P_{cr} = \min_n p_{crn}/p_0 \quad (n = 1, 2, \dots, N). \tag{21}$$

4. Results: Stress intensity factors and critical loads for a system of interacting cracks

As illustrative examples, consider the geometries shown in Fig. 4. Three edge cracks with inclination angles $\beta_n = \beta$ ($n = 1, 2, 3$) and three internal cracks with $\beta = 0$ in a week zone. The inclination angles for internal cracks are fixed and equal to zero, i.e. the cracks are parallel to the FGC boundary. Fig. 4a shows the geometry with different crack sizes ($a_n = 2a_1$), and Fig. 4b – for the same crack sizes. These crack patterns are observed in experiments, which were reported in the literature, e.g. see [1].

The dimensionless stress intensity factors (SIFs) are denoted as $k_{I,II} = K_{I,II}/K^0$, where K^0 is the SIF for a single crack, Eq. (17). The SIF for a single edge crack normal to the surface of the layer is equal to $K_I = 1.58\,p\,(\pi a)^{1/2}$ (a is the half-length of edge crack). This definition for SIFs is more convenient for the considered mixed system of internal and edge cracks than the commonly used one for the SIF for an edge crack: $K_I = 1.12p(2\pi a)^{1/2}$, where $2a$ is the full length of the edge crack.

The inhomogeneity parameters for the thermal expansion coefficient, Young's modulus and fracture toughness are the following: $\varepsilon h = -0.5$, $\omega h = -1.5$, $\gamma h = -2.3$. Other parameters (geometrical) are $h/a = 4$, $d/a = (2, 4, 6)$, $a = \max a_k$, $a_1 = 1\ mm$, $a_n = 2a_1$ (Fig. 4a) and $a_n = 0.5\ mm$ (Fig. 4b). The midpoint coordinates of the internal cracks are $(x + i\,y)/a = (-\,ih,\ d -$

ih, 2*d* – *ih*)/*a* (Figs. 4a and b). In Fig. 5, the non-dimensional distances *d*/*a* are denoted by *d*. In Figs. 6 and 7, the results for critical loads are obtained for $d/a = 2$.

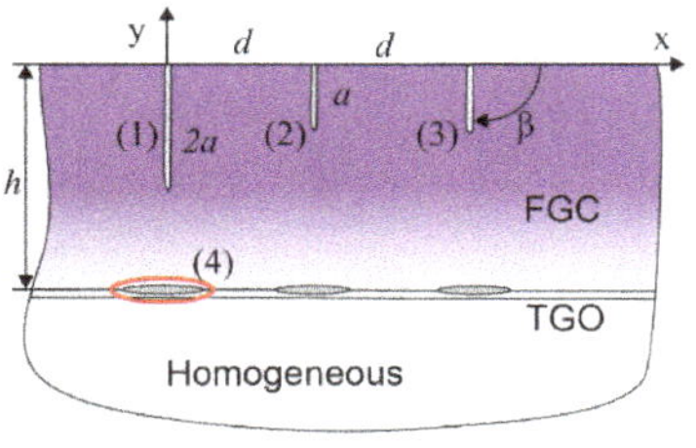

(a)

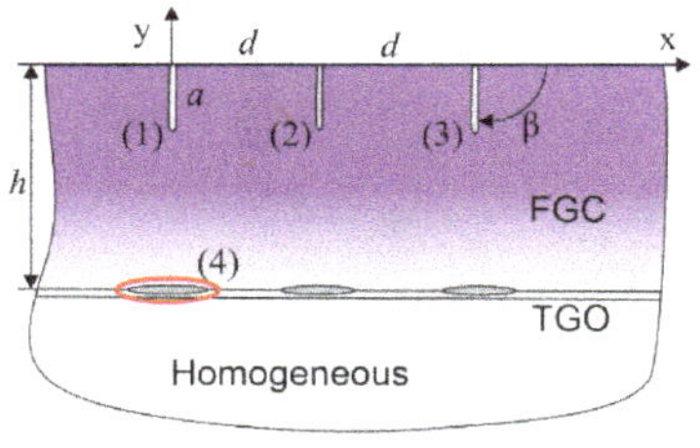

(b)

Figure 4. FGC/H structure with a system of three edge cracks and three internal cracks, (a) for $a_n = 2a_1$ *and (b) for the equal sizes of cracks.*

4.1. Stress Intensity Factors

Fig. 5 shows SIFs as functions of inclination angles β for edge cracks and for different distances *d*/*a* between the edge cracks for the geometry in Fig. 4a. Both k_I and k_{II} are non-zero, thus, the mixed–mode fracture conditions are realized. For most of the parameters β and *d*/*a*, the shielding effect is observed, because the values for k_I for edge interacting cracks do not exceed the value 1.58 for a single edge crack. A strong dependence of SIFs on the angle β is observed, especially for k_{II}. The effect of the distance *d*/*a* on the SIFs of the strongly interacting cracks 1 and 4 is negligible. The maximum effect of *d*/*a* is observed on SIFs k_I for edge cracks 2 and 3.

If the weakest crack is to be defined, and if this definition is based on the maximum SIF k_I, then edge crack 1 has the highest k_I value for $\beta = 90°$ and $d/a = 6$. That is, it looks like the large edge crack 1 would start to propagate first. In the next section, the critical loads are defined and the identification of the weakest cracks is discussed.

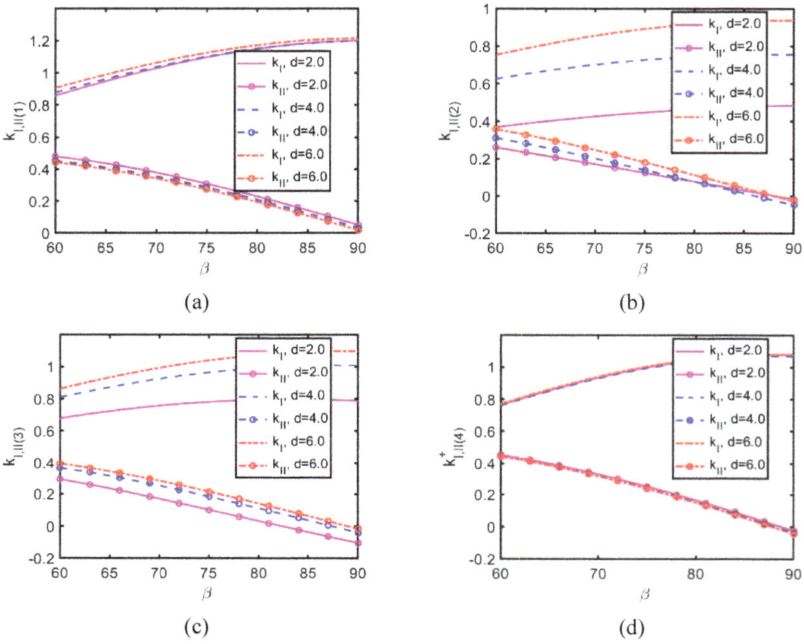

Figure 5. SIFs k_I and k_{II} as functions of edge crack angles β ($\beta_n = \beta$, n=1,2,3) and for different distances d between the edge cracks; (a) for edge crack 1 with half-length a_1; (b) and (c) for edge cracks 2 and 3 with half-length $a_2= a_3=0.5a_1$; (d) for right tip of internal crack 4 with half-length $a_4=0.5a_1$.

4.2. Critical loads

Figs. 6 and 7 represent the results for critical loads as functions of inclination angle β and for distance d/a = 2, Fig. 6 is for the geometry in Fig. 4a and Fig. 7 for the geometry in Fig. 4b (for cracks of equal length). As mentioned above, the calculations are performed for the inhomogeneity parameters $\varepsilon h = -0.5$, $\omega h = -1.5$, $\gamma h= -2.3$. These values correspond to material parameters that increase from the ceramic top to the metal substrate. Eq. (20) is used for dimensionless critical load, and Eq. (15) for fracture angles in Eq. (20).

Fig. 6a shows the critical loads for edge cracks from the system of cracks in Fig. 4a. The largest value for p_{cr}/p_0 is for crack 2 and the smallest one for crack 1. Thus, the weakest edge crack is the large crack 1.

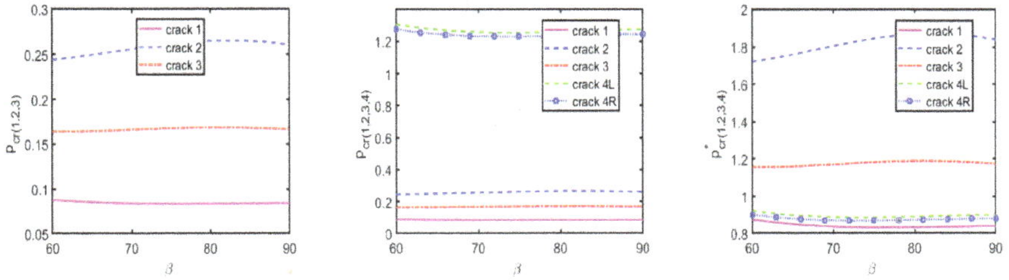

(a) (b) (c)

Figure 6. Critical loads as function of angles β for the geometry in Fig. 4a: (a) for edge cracks, (b) for edge cracks and internal cracks, (c) artificial results for edge cracks and internal cracks without taking into account the variation of fracture toughness for the FGC. The critical loads for other internal cracks show similar values as for crack 4 and are not shown here.

In Fig. 6b, the results for p_{cr}/p_0 for edge cracks and internal cracks are depicted for cracks in Fig. 4a. The critical loads for other internal cracks have similar values as for crack 4 and are not shown here. The largest value for p_{cr}/p_0 is for internal crack 4. The weakest crack has the smallest p_{cr}/p_0, and this is crack 1. Thus, the fracture starts from the large edge crack. The results in Fig. 6a and 6b take into account the variation of fracture toughness in accordance with the law in Eq. (4). Artificial results for p_{cr}/p_0 without taking into account the fracture toughness variation are shown in Fig. 6c. These results contrast with those in Fig. 6b, thus, in Fig. 6c, the lowest values for p_{cr}/p_0 are for crack 1 (the weakest crack) and for internals cracks, as was also defined in the comparison of the SIFs k_I in Section 4.1.

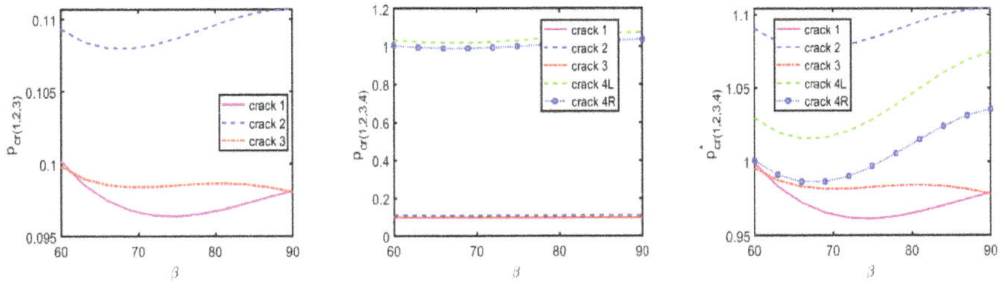

(a) (b) (c)

Figure 7. Critical loads as function of angles β for the geometry in Fig. 4b: (a) for edge cracks, (b) for edge cracks and internal cracks, (c) artificial results for edge cracks and internal cracks without taking into account the variation of fracture toughness for the FGC. The critical loads for other internal cracks have similar values as for crack 4 and are not shown here.

Fig. 7 shows p_{cr}/p_0 versus β for the geometry in Fig. 4b, i.e. for the cracks of the same length. As in the previous case Fig. 6a, the largest value of critical load is for edge crack 2 (Fig. 7a), but this time its value is about 50% smaller than those for crack 2 in Fig. 6a. At the same time, all values p_{cr}/p_0 for edge cracks 1, 2 and 3 are nearly the same.

In Fig. 7b, the results p_{cr}/p_0 for edge cracks and internal cracks are presented, and, as in the previous case in Fig. 6b, the lowest values of critical loads are for edge cracks, that is, these edge cracks are weakest cracks.

Artificial results for p_{cr}/p_0 without taking into account the fracture toughness variation are shown in Fig. 7c, where all values do not much differ from each other. The crack order with the smallest and lowest values for p_{cr}/p_0 is completely different from that shown in Fig. 7b and also different from the case of Fig. 6c.

It should be noted that the effect of the inclination angle β on critical loads is very small, Fig. 6 and 7.

5. Summary

A general theoretical formulation of the model for the thermal fracture analysis of functionally graded coatings on a homogeneous substrate (FGC/H) has been performed by means of integral equations. The stress intensity factors are calculated and critical stresses are obtained, using as fracture criterion the maximum hoop stresses criterion. Illustrative examples are presented to show the influence of the parameters of the problem, such as the crack sizes, the inclination angles and distances between the cracks as well as the inhomogeneity parameter of fracture toughness of FGCs, on the interaction of edge cracks with internal cracks in a weak zone as here the thermally grown oxide layer. The analysis of the parameters with respect to critical loads shows that for a system of edge and internal cracks in this weak zone, fracture starts from edge cracks in the top of the functionally graded coating, where the ceramic material predominates. It is important to consider the variation of the fracture toughness through the thickness of the FGCs. Optimal crack configurations can be determined at which the critical loads are maximal, to avoid dangerous cracks with minimal critical loads and to improve the fracture resistance of FGC/H structures.

Acknowledgement

The authors would like to acknowledge the financial support of the German Research Foundation under Grant SCHM 746/209-1.

References

[1] S. Rangaraj, K. Kokini, Multiple surface cracking and its effect on interface cracks in functionally graded thermal barrier coatings under thermal shock, Trans. ASME J. Appl. Mech. 70 (2003) 234-245. https://doi.org/10.1115/1.1533809

[2] K. Tohgo, M. Iizuka, H. Araki, Y. Shimamura, Influence of microstructure on fracture toughness distribution in ceramic–metal functionally graded materials, Eng. Fract. Mech. 75 (2008) 4529-4541. https://doi.org/10.1016/j.engfracmech.2008.05.005

[3] K. Tohgo, T. Suzuki, H. Araki, Evaluation of R-curve behavior of ceramic–metal functionally graded materials by stable crack growth, Eng. Fract. Mech. 72 (2005) 2359-2372. https://doi.org/10.1016/j.engfracmech.2005.03.006

[4] Z.-H. Jin, R.C. Batra, Some basic fracture mechanics concepts in functionally graded materials, J. Mech. Phys. Solids 44 (1996) 1221-1235. https://doi.org/10.1016/0022-

5096(96)00041-5

[5] Y.Z. Feng, Z.H. Jin, Thermal shock damage and residual strength behavior of a functionally graded plate with surface cracks of alternating lengths, J. Therm. Stresses 35 (2012) 30-47. https://doi.org/10.1080/01495739.2012.637457

[6] Y. Zhang, L. Guo, X. Wang, R. Shen, K. Huang, Thermal shock resistance of functionally graded materials with mixed-mode cracks, Int. J. Solids Struct. 164 (2019) 202-211. https://doi.org/10.1016/j.ijsolstr.2019.01.012

[7] V. Petrova, S. Schmauder, Modeling of thermo-mechanical fracture of FGMs with respect to multiple cracks interaction, Phys. Mesomech. 20 (2017) 241-249. https://doi.org/10.1134/S1029959917030018

[8] V. Petrova, S. Schmauder, A theoretical model for the study of thermal fracture of functionally graded thermal barrier coatings with a system of edge and internal cracks, Theor. Appl. Fract. Mech. 108 (2020) 102605. https://doi.org/10.1016/j.tafmec.2020.102605

[9] J.-H. Kim, G.H. Paulino, On Fracture Criteria for Mixed-Mode Crack Propagation in Functionally Graded Materials, Mech. Adv. Mater. Struct. 14 (2007) 227-244. https://doi.org/10.1080/15376490600790221

[10] X. Jin, L. Wu, L. Guo, H. Yu, Y. Sun, Experimental investigation of the mixed-mode crack propagation in ZrO2/NiCr functionally graded materials, Eng. Fract. Mech. 76 (2009) 1800-1810. https://doi.org/10.1016/j.engfracmech.2009.04.003

[11] A.M. Afsar, J.I. Song, J.I., Effect of FGM coating thickness on apparent fracture toughness of a thick-walled cylinder, Eng. Fract. Mech. 77 (2010) 2919-2926. https://doi.org/10.1016/j.engfracmech.2010.07.001

[12] J.W. Eischen, Fracture of nonhomogeneous materials, Int. J. Fracture 34 (1987) 3-22. https://doi.org/10.1007/BF00042121

[13] Y.C. Zhou, T. Hashida, Coupled effects of temperature gradient and oxidation on thermal stress in thermal barrier coating system, Int. J. Solids Struct. 38 (24-25) (2001) 4235-4264. https://doi.org/10.1016/S0020-7683(00)00309-7

[14] V.V. Panasyuk, M.P. Savruk, A.P. Datsyshin, Stress Distribution near Cracks in Plates and Shells (in Russian), Naukova Dumka, Kiev, 1976.

[15] F. Erdogan, G. Gupta, On the numerical solution of singular integral equations, Quart. Appl. Math. 29 (1972) 525-534. https://doi.org/10.1090/qam/408277

[16] F. Erdogan, G.C. Sih, On the crack extension in plates under plane loading and transverse shear, J. Basic. Eng. 85 (1963) 519-527. https://doi.org/10.1115/1.3656897

CHAPTER 12

A theoretical Model for the Study of Thermal Fracture of Functionally Graded Thermal Barrier Coatings with a System of Edge and Internal Cracks

Vera Petrova [1,2]*, Siegfried Schmauder [1]

[1] IMWF, University of Stuttgart, Pfaffenwaldring 32, D-70569 Stuttgart, Germany

[2] Voronezh State University, University Sq.1, Voronezh 394006, Russia

veraep@gmail.com *, Siegfried.Schmauder@imwf.uni-stuttgart.de

Abstract

The problem of fracture of functionally graded coatings (FGCs) on a homogeneous substrate (a semi-infinite medium) is investigated under the influence of thermal and/or mechanical loads (e.g. a heat flux, residual thermal stresses caused by cooling-heating, tension). These loads reflect the most important cases, which arise during the exploitation of FGC structures. The FGC contains pre-existing systems of cracks, such as edge, internal and interface cracks. The mathematical description of the model is based on singular integral equations. The properties of the FGC are continuous functions of the thickness coordinate. Furthermore, the non-homogeneity of the functionally graded material is revealed in the form of corresponding inhomogeneous traction distributions on the surfaces of cracks. This method is approximate and used with the assumption, that the gradation of material properties of the FGC with the depth of the layer is not abrupt. The influence of residual stresses caused by temperature changes on ΔT (e.g. cooling from operating temperatures) is investigated in detail. Different crack patterns (which are reported in experiments and available in the literature) are studied by carrying out numerical experiments with respect to stress intensity factors and fracture angles (a deviation of cracks from the initial direction of propagation). The proposed model in combination with a detailed parametric analysis can help to optimize FGCs in order to improve the fracture resistance of FGC/homogeneous structures.

Keywords

Thermo-Mechanical Load; Functionally Graded Coating; Stress Intensity Factors; Fracture Angles; Singular Integral Equations; Semi-Analytical Solution

1. Introduction

Thermal barrier coatings (TBCs) have wide application in various engineering systems that operate at elevated temperatures, for example, in the aerospace and airplane industries, as well as in power engineering. TBCs are used to prevent melting of metal parts of structures, therefore, they are made of materials with low thermal conductivity. Ceramics possess these properties, therefore they are used for TBCs. However, due to differences in the thermal expansion coefficients of ceramics and metals, high residual stresses arise at interfaces between the ceramic coating and metal substrates, which lead to cracking and debonding along the interface and, thus, can cause the deterioration of the entire structure. In order to increase the effectiveness and durability of engineering components, new materials and new concepts for designing the materials are challenging tasks for researchers, see [1]. In this regard, the concept of functional gradient materials (FGMs) has arisen [2]. The properties of FGMs vary continuously mainly in one direction, which is achieved by changing the composition of the material and its structure, for example, a gradually varying concentration of porosity. In spite of numerous investigations in this field, the fundamental aspects for modeling of fracture in functionally graded thermal barrier coatings (FGCs) under elevated temperatures are still challenging tasks for researches [1].

In the authors' previous work [3], a general theoretical formulation of the model was presented for the thermal fracture analysis of functionally graded coatings on a homogeneous substrate (FGC/H) under thermo-mechanical loading. Models for functionally gradient materials and some special models for cracks, such as partially thermally permeable cracks and cracks with contact, which can be used in the considered problem, were discussed, as well as numerical and analytical methods for the solution of singular integral equations for the problem were considered. In [4], a series of numerical calculations was performed to study the interaction of a system of edge cracks in FGC/H structures under thermomechanical loading. The influence of the geometry of the problem and the nonhomogeneity of the FGC on the fracture angles of cracks (the deviation of a crack from the initial direction of propagation) was investigated for some material parameters.

The present work is devoted to the study of the interaction of multiple cracks in FGCs with respect to the main characteristics of fracture, such as stress intensity factors and fracture angles. The mechanisms of crack formation in TBCs under the influence of high-temperature were discussed in many papers, e.g. see [5-7]. These mechanisms for conventional TBCs are schematically shown in Fig. 1. The experimental study of functionally graded thermal barrier coatings (FGCs) has shown a tendency to form multiple surface cracks with increased the coating gradation level, as well as a greater resistance to horizontal cracking was observed with increasing coating gradation, see [5].

New requirements for engineering components that must withstand very high temperatures and changes in temperatures, and at the same time withstand mechanical loads demand new investigations in this field. TBCs are critical components in gas-turbine engines, and because of operating temperatures are typically higher than the melting point of the underlying metal parts, any TBC failure endangers the engine [1]. At these high temperatures, a thermally grown oxide (TGO) layer forms between the ceramic coatings and the underlying alloys as a result of oxidation processes. Consequently, it is important to consider TBCs as a complex system consisting of a FGC, an underlying metal and some other layers between them. These functionally graded coatings on homogeneous substrate (FGC/H) systems and/or systems with an intermediate TGO layer (FGC/TGO/H) are reflected in the geometry of the problem in the presented work (Fig. 2a).

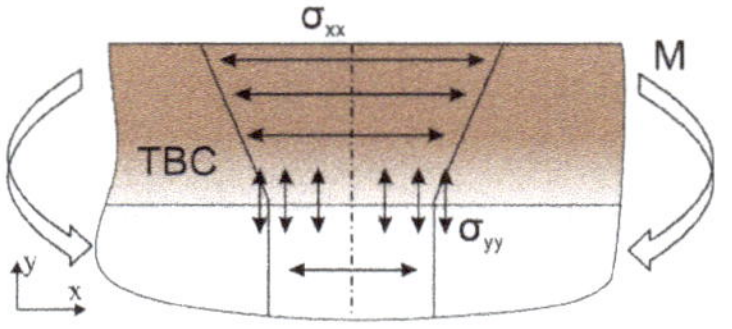

(a)

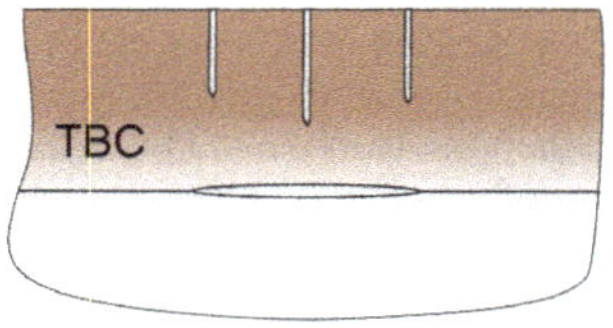

(b)

Figure 1. (a) Stress-state in a thermal barrier coating (TBC) after heating-cooling cycle and (b) resulting surface and interface cracks.

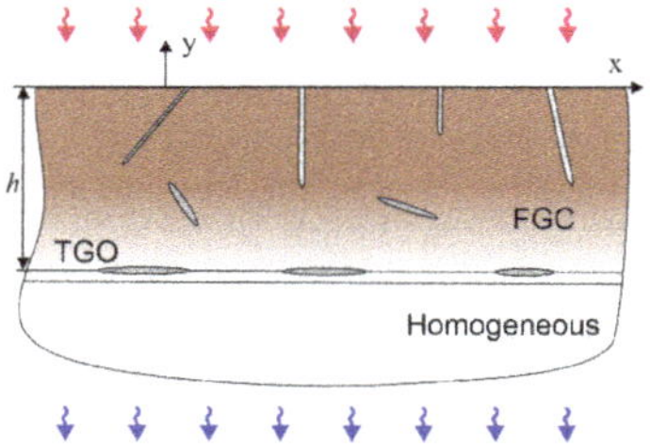

(a)

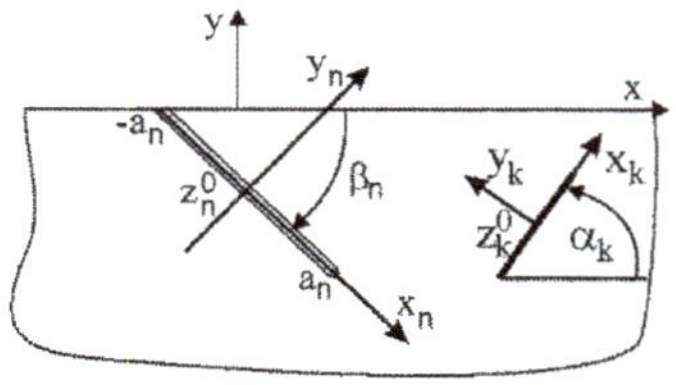

(b)

Figure 2. Geometry of the problem: (a) FGC/TGO/H with a system of cracks, and (b) coordinate systems connected with cracks.

2. Formulation of the problem and solution

2.1 General description of the problem

The general case of the geometry of the problem is shown in Fig. 2a. The upper layer of thickness h is made from an FGM, and the semi-infinite substrate consists of a homogeneous material. Between the FGM layer and the homogeneous substrate, there is a weak thin layer, which models the thermal growth oxide (TGO). The FGC contains arbitrary located cracks of length $2a_k$ ($k = 1,\ldots,N$), which can be internal and/or edge cracks, TGO models by cracks located at this thin layer. The coordinate systems are chosen as follows: the global coordinates (x, y) with the x-axis located on the surface of the FGC/H (or FGC/TGO/H) structure and the local coordinates (x_k, y_k) connected with cracks and with centers at the midpoint of the k-th crack. The crack positions are determined explicitly by the midpoint coordinates (x_k^0, y_k^0) or in the complex form $z_k^0 = x_k^0 + iy_k^0$ (i is the imaginary unite), and the inclination angles α_k to the x-axis (or β_k for edge cracks, $\beta_k = -\alpha_k$) (see Fig. 2b).

The following loads are considered: the tension parallel to this surface (in the x-direction), and cooling by ΔT, $\Delta T > 0$ (e.g. cooling from operating or processing temperatures). In the case of cooling of the FGC/H structure tensile residual stresses are observed as shown in the experimental investigations [6]. Since the problem is considered in the linear elastic formulation, the results for each load can be superimposed. The uncoupled, quasi-static thermo-elasticity theory is applied. In this theory, the temperature distribution is independent of the mechanical field. In the case of the FGC/H structure under a heat flux, the problem is solved in two steps: first, the thermal problem for the FGC/H structure with cracks, and then the thermo-elastic problem for the same geometry.

The following assumptions are used for this problem:

The thermal and mechanical properties of an FGC are continuous functions of the thickness coordinate y.

The non-homogeneity of the functionally graded material is revealed in the form of the corresponding inhomogeneous stress distributions on the surfaces of the cracks, see [8-10]. In the frame of this assumption, the properties of the FGM should vary slightly with the depth of the layer.

The edges of the cracks are free of tractions. The FGC and the substrate are perfectly bonded, i.e. ideal thermal and mechanical conditions are fulfilled outside of interface cracks, namely, the tractions and displacements are equal at the interface.

The problem is formulated employing the method of complex variables [11], as well as the superposition principle is used. Due to the superposition principle, the problem is decomposed into sub-problems, each of which contains one crack. Besides, the loads at infinity are reduced to the corresponding loads on the crack faces. The solution of each sub-problem can be obtained in explicit form (and can be analyzed separately) or in an integral form, and this integral form is used for constructing the equations of the complete problem. The methods for constructing singular integral equations for systems of cracks in a homogeneous material can be found in [12, 13], and for the antiplane problem for a bimaterial with cracks, see [14].

The thermal and mechanical properties of an FGC are continuous functions of the thickness coordinate y. As in the previous works, e.g. [3, 4], the exponential form of these properties is used:

$$f(y) = f_1 \exp(\zeta_n (y+h)), \; -h \le y \le 0, \tag{1}$$

$$f = (k, \alpha_t, E), \; f_1 = (k_1, \alpha_{t1}, E_1), \; \zeta_n = (\zeta_1, \zeta_2, \zeta_3) = (\delta, \varepsilon, \omega) \tag{2}$$

In Eq. (2), f refers to the thermal and mechanical properties of an FGM, namely, k is the thermal conductivity, α_t is the coefficient of thermal expansion, and E is Young's modulus with non-homogeneity parameters δ, ε and ω, respectively. The notation ζ_n (n = 1, 2, 3) is introduced for these inhomogeneity parameters (δ, ε, and ω). f_1 stands for the thermal and mechanical properties of a homogeneous substrate. Poisson's ratio is assumed to be constant and is equal to the value of the homogeneous substrate. Previous studies show that the effect of Poisson's ratio on the stress intensity factors is negligibly small, see [15]. The values of the dimensionless graded parameters $\zeta_n h$ (h is the thickness of the FGC) are obtained from Eq. (1) as follows

$$\zeta_n h = \ln(f_2 / f_1), \; f_2 = f(y)_{|y=0}, \; f_1 = f(y)_{|y=-h} \tag{3}$$

An example of a ceramic/metal FGC/H material is (ZrO_2/Ti-6Al-4V)/Ti-6Al-4V with f_2 / f_1 =(2/6.7, 10/8.6, 200/114) and with the corresponding inhomogeneity parameters (δh, εh, ωh) = (−1.2, 0.15, 0.56) (for material properties see [16]). For this material combination the coefficient of thermal expansion and Young's modulus increase towards the upper part of the FGC/H structure, while the coefficient of thermal conductivity decreases. Some examples for other functional models for FGCs and the rules for determination of corresponding inhomogeneity parameters can be found in [3, 4].

Table 1 presents examples of FGC/H systems consisting of ceramics, such as partially stabilized zirconia (PSZ) and mullite as TBCs, and Ni-superalloy and steel as substrates. Material parameters are given for different temperatures in the range from 20°C to 1100°C. These parameters are listed in the paper [17], in which the thermal stress fields in an undamaged TBC system were studied for a cylindrical shell. In the present work, some of these material combinations of materials are used for numerical results.

Table 1. Material properties of some FGMs and inhomogeneity coefficients

	Thermal expansion CTE ($*10^{-6}$ K^{-1})	Inhomogeneity coefficient of CTE	Young modulus E (GPa)	Inhomogeneity coefficient of E	Thermal conduct. ($Wm^{-1}K^{-1}$)
FGM/H (PSZ/Ni superalloy)/ Ni superalloy					
PSZ Ni sup.	α_{t2} = 9 - 12.2 α_{t1} = 14.8 - 18	α_{t2}/α_{t1} <1 εh = −0.5÷ −0.4	E_2=48-22 E_1=220-120	E_2/E_1 < 1 ωh = −1.5÷ −1.7	k_2 = 2-1.7 k_1= 88-69
FGM/H (mullite/Ni superalloy)/ Ni superalloy					
Mullite Ni sup.	α_{t2} = 4.5 α_{t1} = 14.8 - 18	α_{t2}/α_{t1} <1 εh = −1.2÷ −1.4	E_2=30 E_1=220-120	E_2/E_1 < 1 ωh = −2.÷ −1.4	k_2 = 5.2-0.4 k_1= 88-69
FGM/H (PSZ/steel)/ steel					
PSZ Steel	α_{t2} = 9 - 12.2 α_{t1} = 15	α_{t2}/α_{t1} <1 εh = −0.5÷ −0.2	E_2=48-22 E_1=207	E_2/E_1 < 1 ωh = −1.5÷ −2.2	k_2 = 2-1.7 k_1= 16
FGM/H (mullite/steel)/ steel					
Mullite Steel	α_{t2} = 4.5 α_{t1} = 15	α_{t2}/α_{t1} <1 εh = −1.2	E_2=30 E_1=207	E_2/E_1 < 1 ωh = −1.9	k_2 = 5.2-0.4 k_1= 16

2.2 Thermal and mechanical loadings

If an FGM/H structure is cooled, then residual stresses are arising due to mismatch in the coefficients of thermal expansion [8, 9]. In the presented model, the inhomogeneity of FGMs is taken into account through continuously varying residual stresses, and these stresses are the following [8]:

$$\sigma_{xx}^{T}(y)=[\alpha_t(y)-\alpha_{t1}]\Delta TE(y),\ \sigma_{xx}^{e}(y)=[E(y)/E_1-1]\sigma_{xx}^{0},\ \sigma_{xx}^{0}=p \tag{4}$$

α_{t1} and E_1 are, respectively, the thermal expansion coefficient and Young's modulus of a homogeneous substrate material and at the interface, i.e. in the region $y\le -h$; $\alpha_t(y)$ and $E(y)$ are defined by formulas (1) and (2).

The method of linear superposition is used to solve this problem, so that loads at infinity are reduced to the corresponding loads on the crack faces. Thus, the tensile load is reduced to the load p_n on the crack surfaces and written in complex form as

$$p_n=\sigma_n-i\tau_n=p(1-\exp(2i\beta_n))/2 \quad (n=1,2,..,N). \tag{5}$$

In the common case of FGMs, the full load on the n-th crack consists of p_n, σ_n^T and σ_n^e, Eq. (4), where the index "n" denotes that the functions are written in the local coordinate systems (x_n,y_n) connected with cracks.

There are materials with a similar Young's modulus; they can be considered as elastically homogeneous materials. In this case, $E_1 = E(y)$ and, consequently, $\sigma_{xx}^{e}=0$ in Eq. (4), and the graded parameter of the Young's modulus is $\omega = 0$, see examples of elastically homogeneous materials in [3, 4].

2.3 Singular integral equations for the problem and solution

As it was mentioned in Section 2.1, assuming that the inhomogeneity of the FGM is revealed in the form of the corresponding inhomogeneous stress distributions on the surfaces of cracks, the solution of the boundary value problem of elasticity is reduced to the solution of the system of singular integral equations:

$$\int_{-a_n}^{a_n}\frac{g_n'(t)dt}{t-x}+\sum_{\substack{k=1\\k\neq n}}^{N}\int_{-a_k}^{a_k}[g_k'(t)R_{nk}(t,x)+\overline{g_k'(t)}S_{nk}(t,x)]dt=\pi p_n(x),\ |x|<a_n,$$

$$n=1,2,\ldots,N, \tag{6}$$

$$g_n'(x)=\frac{2\mu}{i(\kappa+1)}\frac{\partial}{\partial x}([u_n]+i[v_n]).$$

The number of equations N is equal to the number of cracks. In Eq. (6), unknown functions $g_n'(x)$ are derivatives of displacement jumps on the crack lines; $[u_n]$ and $[v_n]$ are the shear and normal displacement jumps, respectively, on the n-th crack line, $\mu = E/2(1+v)$ is the shear modulus, E is Young's modulus, v is Poisson's ratio, $\kappa = 3 - 4v$ for the plane strain state, and $\kappa = (3 - v)/(1+ v)$ for the plane stress state. An overbar $\overline{(\ldots)}$ denotes the complex conjugate. The regular kernels R_{nk} and S_{nk} determine the geometry of the problem and can be found in Appendix A, Eqs. (A.1)-(A.5). On the right side of equations (6) the functions p_n are known functions determined by the load on the crack lines. For internal cracks, the condition of displacement continuity at the crack tips has to be taken into account:

$$\int_{-a_n}^{a_n} \frac{g_n'(t)dt}{t-x} = 0. \tag{7}$$

The solution of the equations is obtained by the method of mechanical quadratures, see [12, 13, 18]. Applying quadrature formulas based on Chebyshev polynomials, the integral equations (6) are reduced to systems of algebraic equations. This solution scheme can be found in Appendix B, Eqs. (B.1)-(B.7).

After the solution of the thermo-elastic problem the main characteristics of the fracture mechanics are calculated, in particular, the stress intensity factors (SIFs) at the crack tips. In the case of elastically homogeneous materials the classical definition of the SIFs can be used. For cracks in FGMs, the crack tip singular field has the same form as in homogeneous media, and the concept of the SIFs can be applied directly to these cracks. Besides, the interface crack between the FGM and the homogeneous material with smooth transition between these materials is also a classical crack with square-root singularities at the crack tips. The SIFs at the crack tips are written as

$$K_{nI}^{\pm} - iK_{nII}^{\pm} = \mp \lim_{x_n \to \pm a_n} \sqrt{\pi}\sqrt{(a_n^2 - x_n^2)/a_n}\, g_n'(x_n), \quad (n = 1, 2, \ldots, N), \tag{8}$$

where the upper part of the "±" or "∓" signs concerns to the right tip and the lower to the left tip of a crack.

2.4 Analysis by means of fracture criteria

In the considered thermo-elastic problems for FGC/H (or FGC/TGO/H) the cracks, in general, are under mixed mode loading conditions (i.e. both SIFs mode I and mode II are non-zero). Under these conditions, the cracks deviate from their initial propagation direction, Fig. 3.

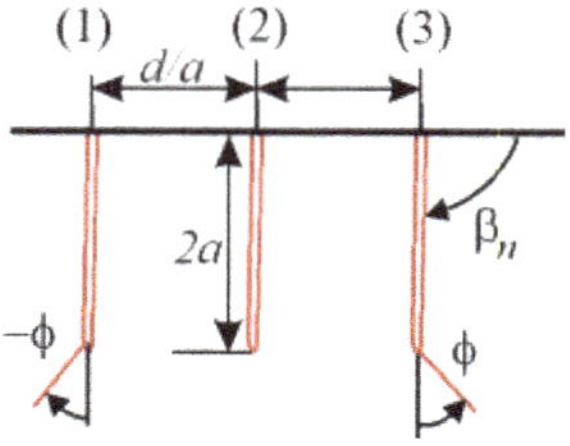

Figure 3. Edge cracks with fracture angles ϕ.

For predicting the crack growth and the determination of a direction of this growth, different fracture criteria are applied. In previous work [19], the results were compared for three criteria: the maximum circumferential stress criterion, the minimum strain energy density function and the maximum energy release rate, and these results are very close to each other. In the present work, the maximum circumferential stress criterion [20] is used. Applying a fracture criterion for

FGMs, it is important to take into account that the value of fracture toughness has to be determined near the crack tip, i.e. $K_{Ic,tip}$.

According to the criterion of maximum normal stresses [20], the crack deflection angle (or the fracture angle) and the critical stresses are calculated as

$$\phi = 2\arctan\left[\left(K_I - \sqrt{K_I^2 + 8K_{II}^2}\right) / 4K_{II}\right], \cos^3(\phi/2)\left(K_I - 3K_{II}\tan(\phi/2)\right) = K_{Ic,tip} / \sqrt{\pi}, \quad (9)$$

where K_I and K_{II} are the SIFs, and $K_{Ic,tip}$ is the fracture toughness of the material near the crack tip. The fracture angle ϕ is shown in Fig. 3. First, the angle of the crack propagation (fracture angle) is obtained using the results of calculated stress intensity factors. Then, the local fracture stability is evaluated. This criterion is the simplest one to employ. As a result (using this criterion), the fracture angles are derived as functions of the geometry of the problem and of the non-homogeneity parameters of FGCs with additional parameters due to thermal loads. After this, the critical stresses are obtained near the cracks.

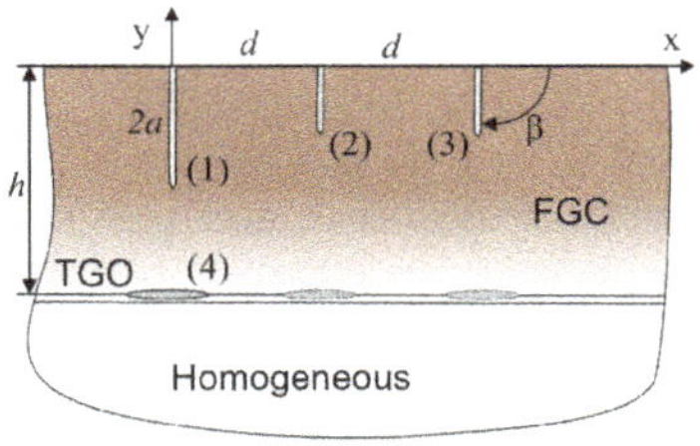

Figure 4. FGC/H structure with a system of three edge cracks and an internal crack in three different positions.

3. Results

The presented model allows to analyze the influence of geometric parameters (crack sizes, midpoint coordinates of cracks and their inclination angles , as well as FGC thickness), and parameters of material inhomogeneity (Eqs. (1)-(3)) on the main fracture characteristics of the problem, e.g. stress intensity factors (SIFs) and fracture angles.

A series of numerical calculations was performed to study the interaction of edge cracks with internal cracks in a weak zone in FGC/H structures. The SIFs are normalized, i.e. $k_{I,II} = K_{I,II} / K^0$, where K^0 is the SIF for a single crack, $K^0 = p\sqrt{\pi a}$. It should be noted, that the SIF for a single edge crack, normal to the surface, is $K_I = 1.12p\sqrt{2\pi a}$, or in more convenient values for interacting edge and internal cracks, $K_I = 1.58p\sqrt{\pi a}$, where p is the tensile load applied to the FGC/H structure normal to the edge crack. The normalized SIFs mode I for a single edge crack is $k_I = 1.58$ and for an internal crack $k_I = 1$.

Illustrative examples for the geometry shown in Fig. 4 are presented in Figs. 5-8 for SIFs and in Figs. 9-12 for fracture angles as functions of inclination angles β of edge cracks. The inclination angle for internal crack is fixed and equal to 0°, i.e. the crack is parallel to the boundary of the

FGC. The inhomogeneity parameters of the thermal expansion and Young's modulus are $(\varepsilon h, \omega h) = (-0.5, -1.5)$ in Figs. 5-7 and Figs. 9-11, and $(\varepsilon h, \omega h) = (-1.4, -1.4)$ in Figs. 8 and 12, and other parameters are $h/a = 4$, $d/a = (2, 4, 6)$, $a = \max a_k$, $a = 1$mm, $a_2 = a_3 = a_4 = 0.5a_1$. Here h/a is the non-dimensional thickness of the FGC, and d/a is the non-dimensional distance between the edge cracks, Fig. 4. The midpoint coordinates of the internal crack are $(x + i\,y)/a = (-ih, d - ih, 2d - ih)/a$ (see Fig. 4), and these three cases will be called geometry one, geometry two and geometry three. In Figs. 5-12, the non-dimensional distances d/a are denoted by d.

Negative values for εh and ωh correspond to lower values of the thermal expansion coefficient and Young's modulus, respectively, in the upper part of the FGC (examples of materials are given in Table 1).

When the values for SIFs for the left ($k^-_{I,II}$) and right ($k^+_{I,II}$) tips for the internal crack do not differ much, then only k^+_I is shown in Figs. 5g, h, 7g, h and 8g, h. For the edge cracks $k_{I,II(2)} \approx k_{I,II(3)}$, so only $k_{I,II(2)}$ is depicted in Fig. 6c. Here, the subscripts contain the crack numbers.

Figs. 5-8 show that k_I and k_{II} are both non-zero, that is, the mixed-mode stress-strain conditions are revealed. Besides, the shielding effect is observed for most values of the inclination angles β and distances d/a, because the values for k_I for edge interacting cracks (cracks 1, 2 and 3) are not exceed 1.58, i.e. they are less than for a single edge crack.

A strong dependence of the SIFs on the inclination angles β is observed. The angle β affects the SIFs of edge cracks 1, 2, and 3, as well as SIFs for the internal crack, despite the fact that β_4 is fixed and equal to zero.

The influence of the distance d/a on the SIFs mode I and mode II is observed for most crack configurations. The SIF mode I for edge cracks decreases with decreasing distance d/a, i.e. a shielding effect is revealed for edge cracks, Figs. 5-8. However, for interacting edge crack 1 and internal crack 4, the distance d/a does not influence on the SIF values, see Fig. 5 a, b, g, h for $(\varepsilon h, \omega h) = (-0.5, -1.5)$ and Fig. 8 a, b, g, h for $(\varepsilon h, \omega h) = (-1.4, -1.4)$. In the configuration of interacting cracks 1 and 4, crack 1 is two times larger than crack 4.

Materials Research Foundations **159** (2024) https://doi.org/10.21741/9781644902950

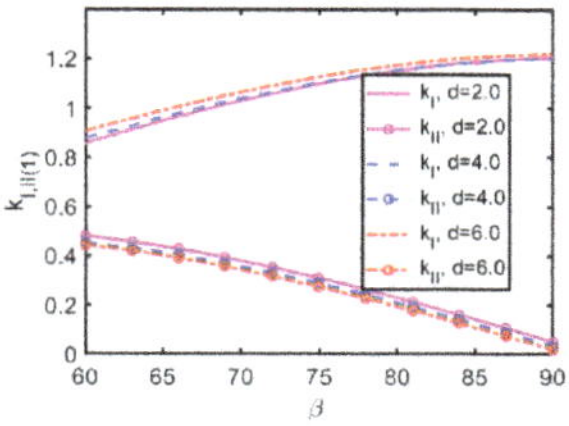

(a) Crack 1

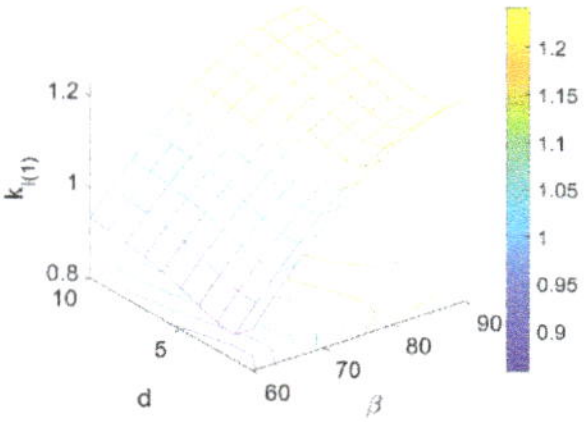

(b) Crack 1

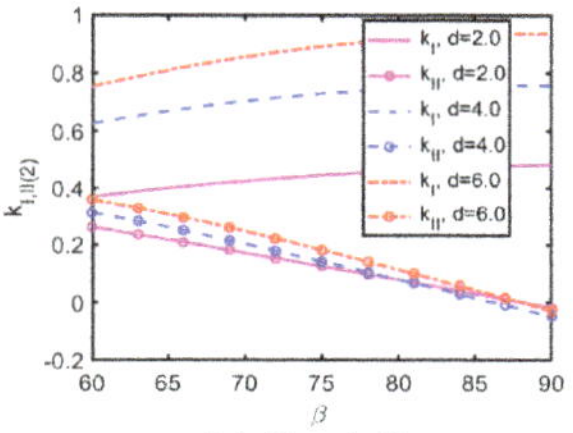

(c) Crack 2

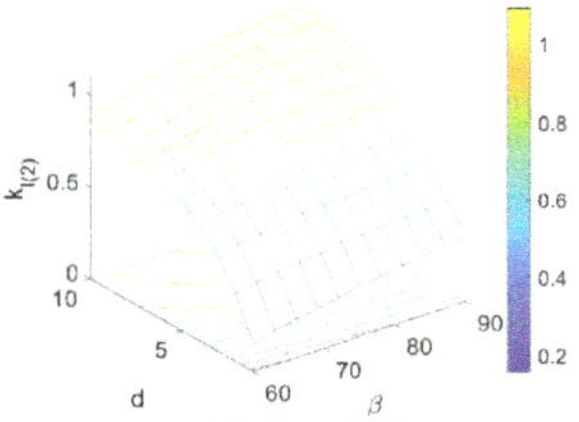

(d) Crack 2

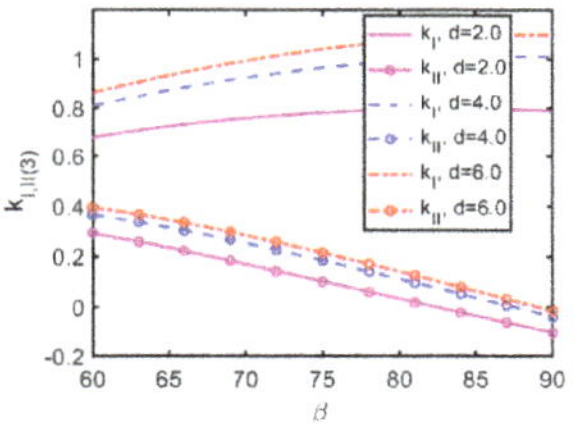

(e) Crack 3

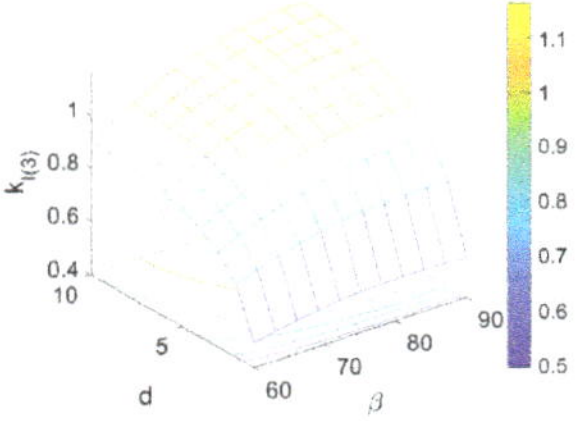

(f) Crack 3

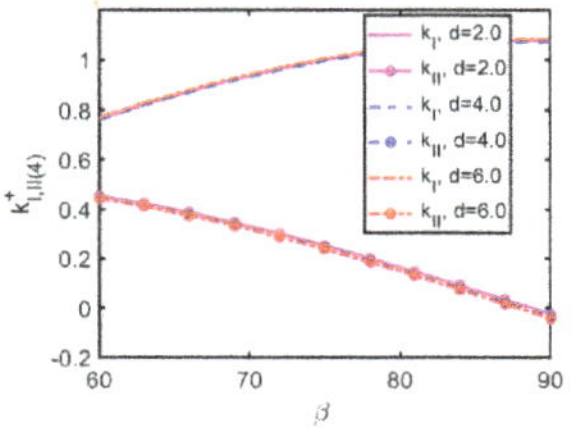

(g) Crack 4, right tip

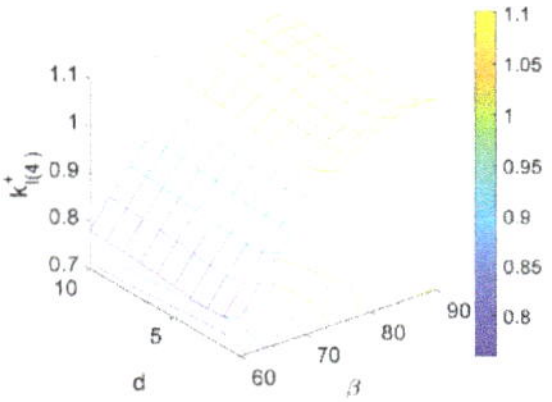

(h) Crack 4, right tip

Figure 5. SIFs k_I and k_{II} and three-dimensional plots for k_I as functions of edge crack angles β ($\beta_n = \beta$, $n=1,2,3$, $\beta_4 = 0$) and for different distances d between the edge cracks; (a), (b) for edge crack 1 with half-length a_1; (c), (d) for edge crack 2 and (e),(f) for edge crack 3 with half-length $a_2 = a_3 = 0.5a_1$; (g), (h) for internal crack 4 with half-length $a_4 = 0.5a_1$; the midpoint coordinate of internal crack is (0,–4); $\varepsilon h = -0.5$; $\omega h = -1.5$.

The influence of large edge crack 1 on internal crack 4 is more pronounce, than the influence of cracks 2 or 3 on the SIFs of crack 4. Figs. 5-7 show that the values of $k_{I(4)}$ for geometry one (Fig. 5) is larger than $k_{I(4)}$ for geometries 2 and 3 (Figs. 6 and 7). For closely located edge cracks, the larger edge crack 1 is more dangerous with respect to the propagation of the internal crack at the weak zone, i.e. a large edge crack initiates the propagation of an internal crack (this can lead to FGC debonding). Besides, the dominant effect of the large edge crack on the neighboring edge crack 2 is also observed.

Fig. 8 shows the results for inhomogeneity parameters $\varepsilon h = -1.4$ and $\omega h = -1.4$, which correspond to the combination of materials (mullite/ steel)/steel at high temperatures, see Table 1. In this case, the k_I values are in the same order as the k_{II} values and the mode of mixity is greater. The k_I values much less than k_I for the previous case for $\varepsilon h = -0.5$ and $\omega h = -1.5$ in Figs. 5-7.

Figs. 9-11 show the results for fracture angles ϕ as functions of β and d/a, and Fig. 13 schematically represents the crack patterns due to the interaction of the cracks. Even for edge cracks located normally to the surface ($\beta_{1,2,3} = 90^o$), the fracture angles are not equal to zero. For a single edge crack with the inclination angle $\beta = 60^o$, the fracture angle is $\phi = 31^o$ and for $\beta = 90^o - \phi = 0^o$. The magnitude for ϕ for $\beta = 60^o$ is larger than 31°; the magnitudes for ϕ for inhomogeneity parameters $(\varepsilon h, \omega h) = (-1.4, -1.4)$ (Fig. 12) is larger than ϕ for $(\varepsilon h, \omega h) = (-0.5, -1.5)$ (Fig. 9).

It should be emphasized that a small deviation of inclination angles β from 90^0 for the system of edge cracks leads to a deviation of the direction of crack propagation, as well as to a deviation of the crack propagation of the internal crack in the weak zone, see Figs. 9-12 and Fig. 13. The change of the fracture angle of edge cracks can cause the spallation of FGCs.

At high temperatures, the values of material parameters change, see Table 1, and the behavior of the interacting cracks also changes, as can be seen from a comparison of Figs. 5 and 8 for SIFs and Figs. 9 and 12 for fracture angles.

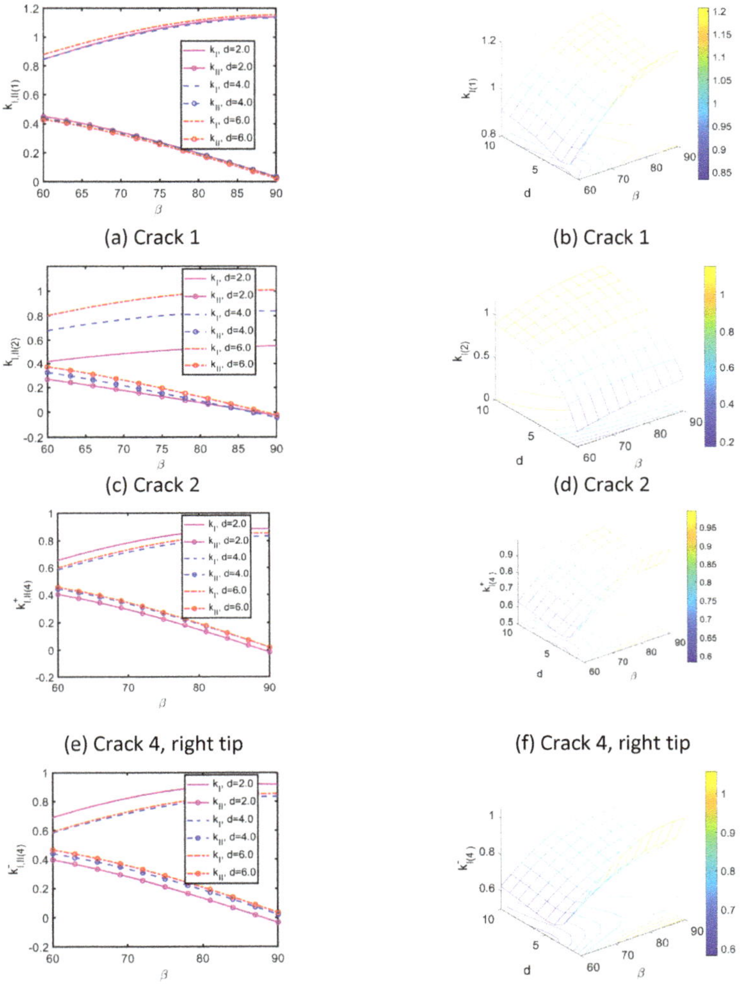

(a) Crack 1 (b) Crack 1

(c) Crack 2 (d) Crack 2

(e) Crack 4, right tip (f) Crack 4, right tip

(g) Crack 4, left tip (h) Crack 4, left tip

Figure 6. SIFs k_I and k_{II} and three-dimensional plots for k_I as functions of edge crack angles β ($\beta_n = \beta$, n=1,2,3, $\beta_4 = 0$) and for different distances d between the edge cracks; (a), (b) for edge crack 1 with half-length a_1; (c), (d) for edge crack 2 and (e), (f) for edge crack 3 with half-length $a_2 = a_3 = 0.5\, a_1$; (g), (h) for internal crack 4 with half-length $a_4 = 0.5a_1$; the midpoint coordinate of internal crack is (d, –4); $\varepsilon h = -0.5$; $\omega h = -1.5$.

Materials Research Foundations **159** (2024) https://doi.org/10.21741/9781644902950

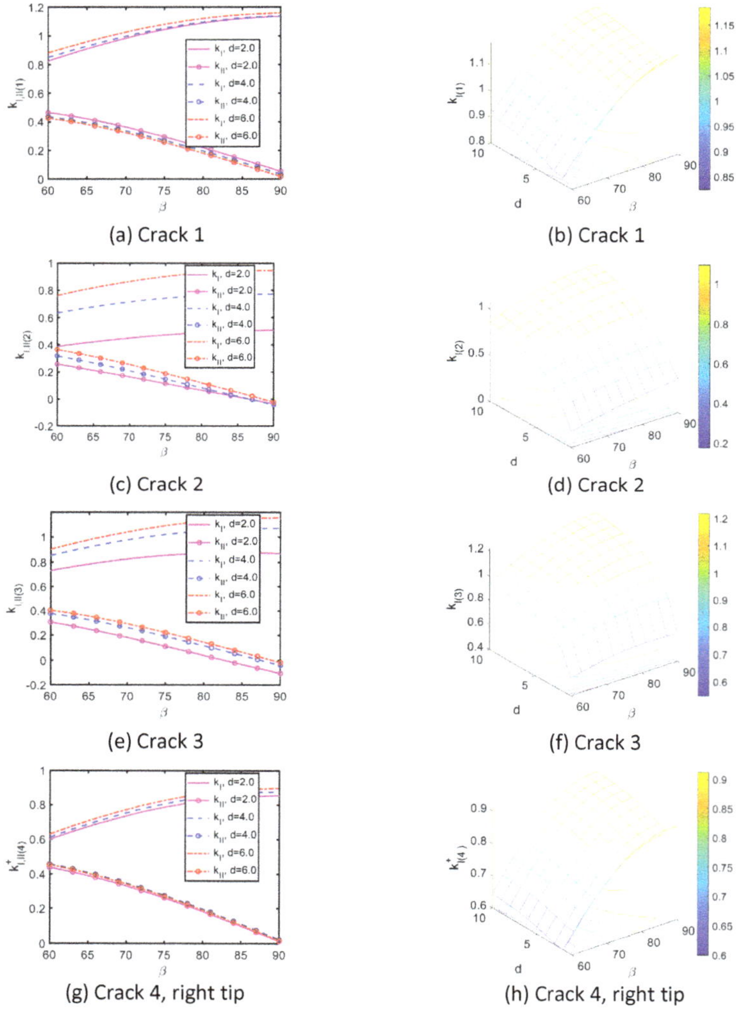

Figure 7. SIFs k_I and k_{II} and three-dimensional plots for k_I as functions of edge crack angles β ($\beta_n = \beta$, $n=1,2,3$, $\beta_4 = 0$) and for different distances d between the edge cracks; (a), (b) for edge crack 1 with half-length a_1; (c), (d) for edge crack 2 and (e), (f) for edge crack 3 with half-length $a_2= a_3=0.5\ a_1$; (g), (h) for internal crack 4 with half-length $a_4=0.5a_1$; the midpoint coordinate of internal crack is (2d, –4); $\varepsilon h = -0.5$; $\omega h = -1.5$.

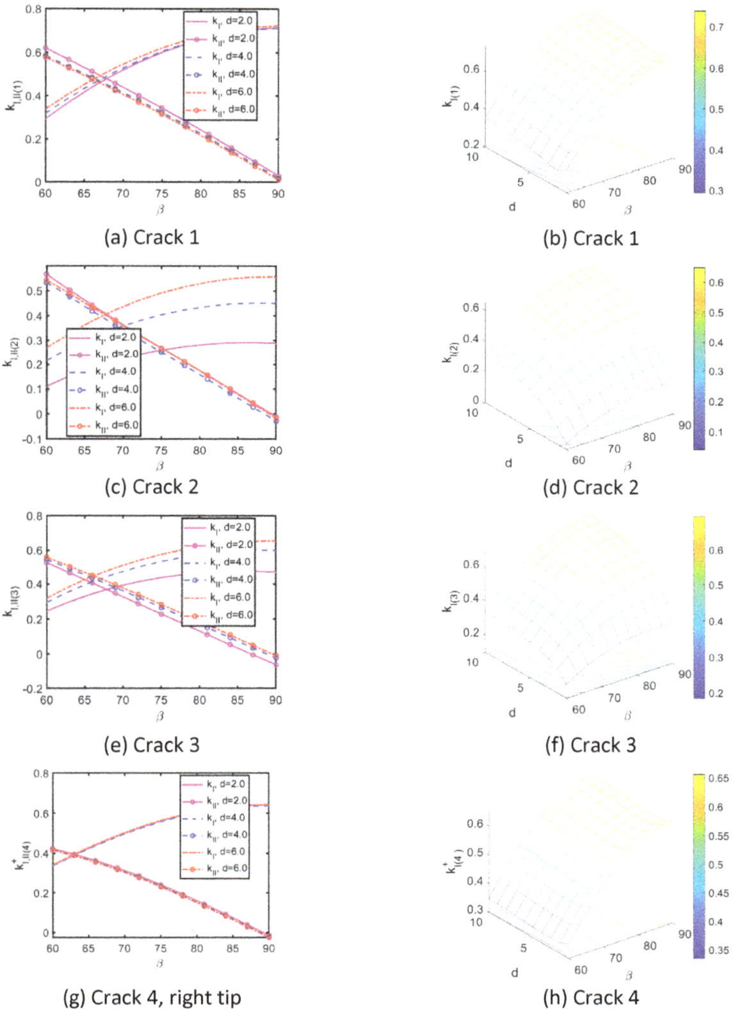

Figure 8. SIFs k_I and k_{II} and three-dimensional plots for k_I as functions of edge crack angles β ($\beta_n = \beta$, n=1,2,3, $\beta_4 = 0$) and for different distances d between the edge cracks; (a), (b) for edge crack 1 with half-length a_1; (c), (d) for edge crack 2 and (e), (f) for edge crack 3 with half-length $a_2 = a_3 = 0.5\, a_1$; (g), (h) for internal crack 4 with half-length $a_4 = 0.5a_1$; the midpoint coordinate of internal crack is (0, –4); $\varepsilon h = -1.4$; $\omega h = -1.4$.

Materials Research Foundations **159** (2024) https://doi.org/10.21741/9781644902950

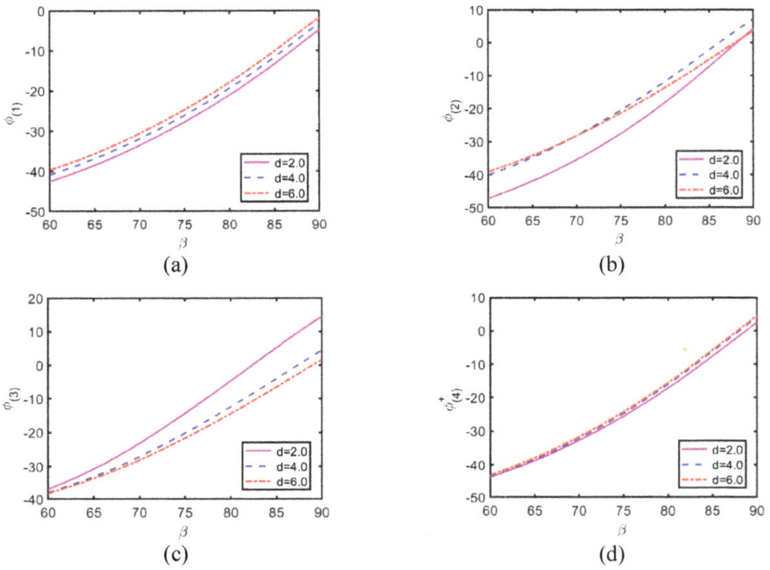

Figure 9. Fracture angles ϕ as functions of β for different distances d between the edge cracks; (a) for edge crack 1; (b) and (c) for edge cracks 2 and 3, (d) for internal crack 4 at right tip. The midpoint coordinate of internal crack is (0, –4); εh = –0.5; ωh = –1.5.

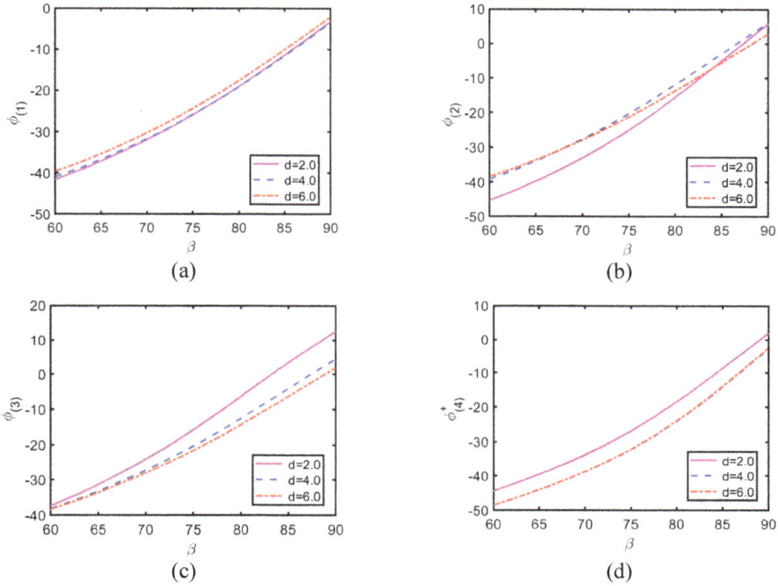

Figure 10. Fracture angles ϕ as functions of β for different distances d between the edge cracks; (a) for edge crack 1; (b) and (c) for edge cracks 2 and 3, d for internal crack 4 at right tip. The midpoint coordinate of internal crack is (d, –4); $\varepsilon h = -0.5$; $\omega h = -1.5$.

Materials Research Foundations **159** (2024) https://doi.org/10.21741/9781644902950

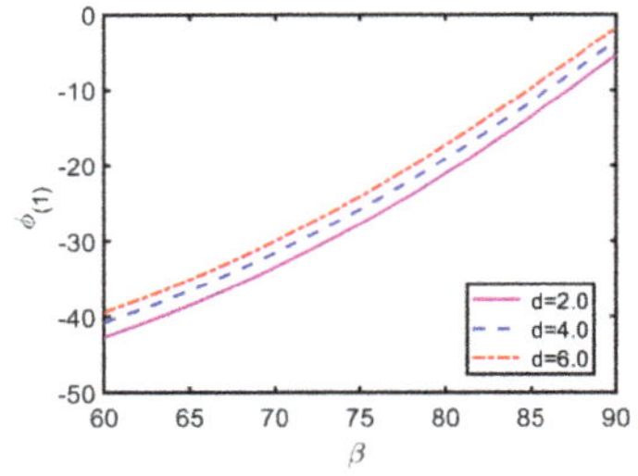

(a)

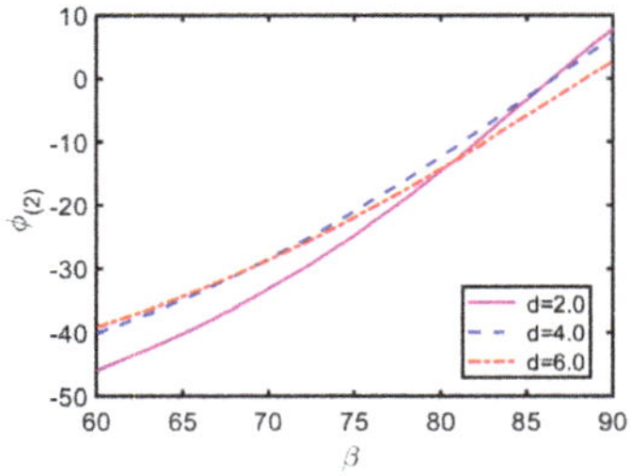

(b)

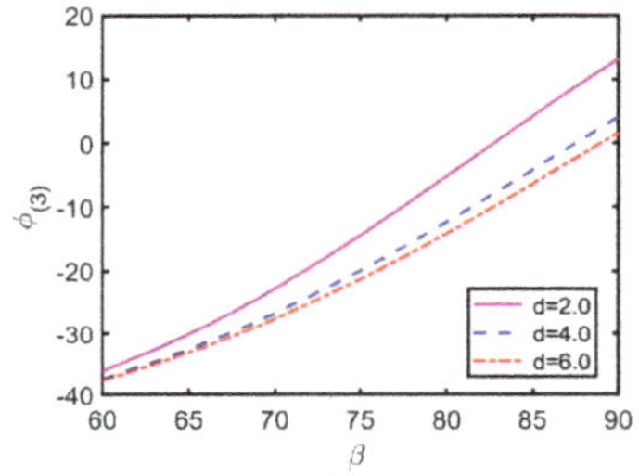

(c)

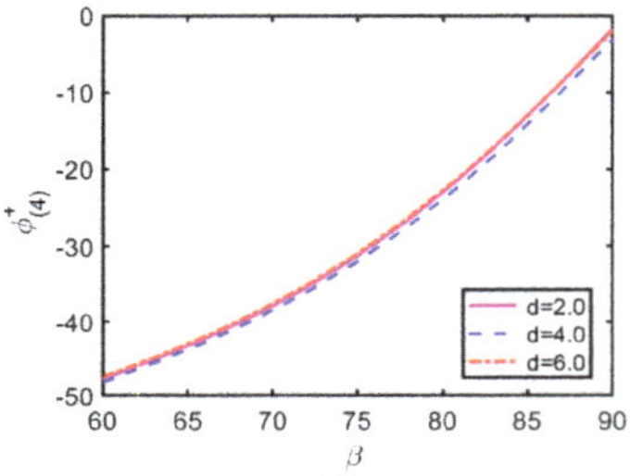

(d)

Figure 11. Fracture angles ϕ as functions of β for different distances d between the edge cracks; (a) for edge crack 1; (b) and (c) for edge cracks 2 and 3, (d) for internal crack 4 at right tip. The midpoint coordinate of the internal crack is (2d, –4); $\varepsilon h = -0.5$; $\omega h = -1.5$.

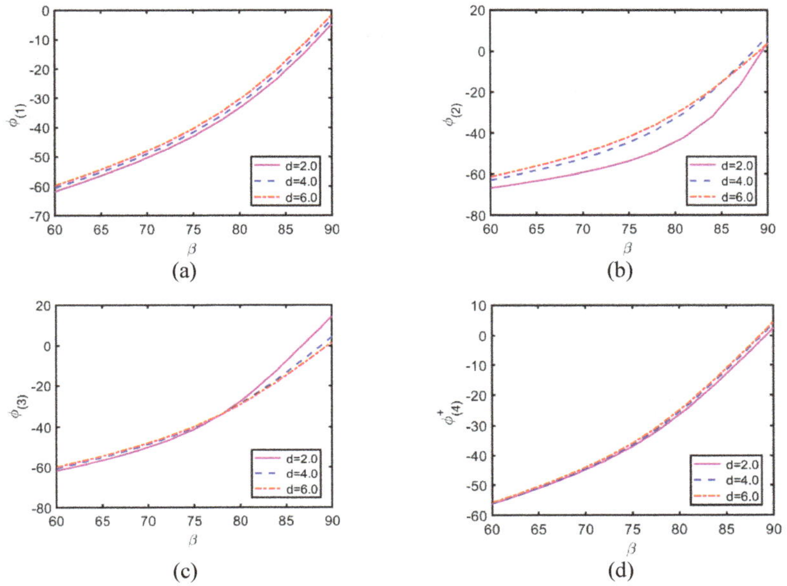

Figure 12. Fracture angles ϕ as functions of β for different distances d between the edge cracks; (a) for edge crack 1; (b) and (c) for edge cracks 2 and 3, (d) for internal crack 4 at right tip. The midpoint coordinate of internal crack is (0, –4); εh = –1.4; ωh = –1.4.

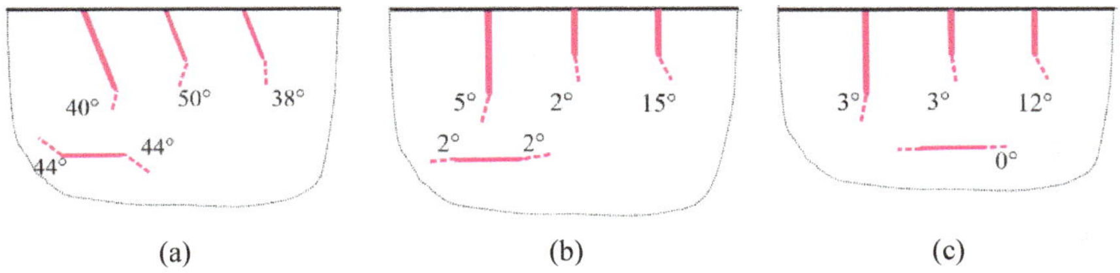

Figure 13. Fracture angles ϕ for an internal crack and three edge cracks (a) with inclination angle β = 60°, (b) and (c) β = 90°; ϕ = 31° for a single edge crack with inclination angle β = 60°; εh = –0.5, ωh = –1.5.

Conclusions

A general theoretical formulation of the model for the thermal fracture analysis of functionally graded coatings on a homogeneous substrate (FGCs/H) with a weak layer TGO between them has been derived by means of integral equations. Illustrative examples are presented to show the influence of the parameters of the problem on the interaction of a system of edge cracks with an internal crack in the weak layer. On the basis of the obtained results the following conclusions can be made:

1. The shielding effect is observed with respect to SIFs mode I for most of the considered geometrical and material parameters, and this effect is enhanced with decreasing the distances d/a between the edge cracks. However, for the strongly interacting cracks (the large edge crack and the internal crack) the distance *d/a* does not affect the SIF values and, accordingly, does not change the shielding level.

2. A strong influence of the inclination angle β of the edge cracks on the SIFs mode I and II was observed both for edge cracks and for the internal one, despite the fact that the inclination angle of the internal crack is fixed and equal to zero.

3. The influence of the distance *d/a* on the SIFs mode I and mode II is more pronounced for the edge crack 2 located between two other edge cracks, and practically does not affect the SIFs for strongly interacting cracks, i.e. for crack 1 and the internal crack.

The influence of the distance on the SIF is more noticeable for the edge crack 2 located between two other edge cracks, and practically does not affect the SIF for strongly interacting cracks, that is, for crack 1 and the internal crack.

4. For closely spaced edge cracks, the larger edge crack is more dangerous with respect to the propagation of an internal crack in the weak layer, i.e. a large edge crack initiates the propagation of an internal crack and also causes its deviation (this can lead to FGC/H debonding).

5. The results for fracture angles as functions of inclination angles β: a slight deviation of the inclination angles β from 90^0 for the edge cracks leads to a deviation of the direction of their propagation, as well as to a deviation of the propagation of the internal crack in the weak layer. Changing the fracture angle of the edge cracks can lead to spallation of FGCs.

6. At high temperatures, the values of the material parameters change, and the behavior of the interacting cracks changes, Figs. 8 and 12. The values of the SIFs mode I decrease, while the values of the SIFs mode II increase slightly. The increase in SIFs mode II leads to an increase in the fracture angles, which becomes relevant with respect to both spallation of FGCs as well as its debonding.

In previous papers of the authors, [4] (for FGC/H) and [21] (for edge cracks in an homogeneous half plane), the influence of the geometry of the problem, i.e. the crack sizes, the inclination angles and distances between the cracks, on the fracture angles of pure edge cracks was studied. The mutual interaction of edge cracks was considered on the basis of systems of two and three arbitrary inclined edge cracks. In [4] the FGM was supposed to be elastically homogeneous, that is the constituent materials have similar Young's moduli. It was shown that the directions of crack propagation mainly depend on the geometry of the problem (as in [21]), while the influence of inhomogeneity of the thermal expansion coefficients of FGM is negligible.

On the basis of these results optimal crack configurations can be determined at which the stress intensity factors at the crack tips possess minimum values or at which the critical loads are maximal and, accordingly, the fracture resistance of FGC/H structures can be improved.

Acknowledgement

The authors V. Petrova and S. Schmauder would like to acknowledge the financial support of the German Research Foundation under Grant Schm 746/209-1.

References

[1] D. Clarke, M. Oechsner, N. Padture, Thermal-barrier coatings for more efficient gas-turbine engines, MRS Bulletin 37(2012) 891-941. https://doi.org/10.1557/mrs.2012.232

[2] Y. Miyamoto, W.A. Kaysser, B.H. Rabin, A. Kawasaki, R.G. Ford, Functionally Graded Materials: Design, Processing and Applications, Kluwer Academic, Dordrecht, 1999. https://doi.org/10.1007/978-1-4615-5301-4

[3] V. Petrova, S. Schmauder, Fracture of functionally graded thermal barrier coating on a homogeneous substrate: models, methods, analysis, J. Phys.: Conf. Ser. 973 (2018) 012017. https://doi.org/10.1088/1742-6596/973/1/012017

[4] V. Petrova, S. Schmauder, Modeling of thermo-mechanical fracture of FGMs with respect to multiple cracks interaction, Phys. Mesomech. 20 (2017) 241-249. https://doi.org/10.1134/S1029959917030018

[5] S. Rangaraj, K. Kokini, Multiple surface cracking and its effect on interface cracks in functionally graded thermal barrier coatings under thermal shock, Trans. ASME J. Appl. Mech. 70 (2003), 234-245. https://doi.org/10.1115/1.1533809

[6] A. Gilbert, K. Kokini, S. Sankarasubramanian, Thermal fracture of zirconia-mullite composite thermal barrier coatings under thermal shock: An experimental study, Surf. Coat. Technol. 202(10) (2008), 2152-2161. https://doi.org/10.1016/j.surfcoat.2007.09.001

[7] A. Gilbert, K. Kokini, S. Sankarasubramanian, Thermal fracture of zirconia-mullite composite thermal barrier coatings under thermal shock: A numerical study, Surf. Coat. Technol. 203 (1-2) (2008), 91-98. https://doi.org/10.1016/j.surfcoat.2008.08.003

[8] A.M. Afsar, H. Sekine, Crack spacing effect on the brittle fracture characteristics of semi-infinite functionally graded materials with periodic edge cracks. Int. J. Fract. 102 (2000) L61- L66.

[9] H. Sekine, A.M. Afsar, Composition profile for improving the brittle fracture characteristics in semi-infinite functionally graded materials, JSME Int. Journal, Ser. A, 42 (4) (1999) 592-600. https://doi.org/10.1299/jsmea.42.592

[10] V. Petrova, T. Sadowski, Theoretical modeling and analysis of thermal fracture of semi-infinite functionally graded materials with edge cracks, Meccanica 49 (2014) 2603-2615. https://doi.org/10.1007/s11012-014-9941-x

[11] N.I. Muskhelishvili, Some Basic Problems of Mathematical Theory of Elasticity, Noordhoff, Groningen, The Netherlands, 1953.

[12] V.V. Panasyuk, M. P. Savruk, A. P. Datsyshin, Stress Distribution near Cracks in Plates and Shells (in Russian), Naukova Dumka, Kiev, 1976.

[13] O. Datsyshyn, V. Panasyuk, Singular Integral Equations for Some Contact Problems of Elasticity Theory for Bodies with Cracks, Structural Integrity Assessment of Engineering

Components under Cyclic Contact. Structural Integrity, Springer, Cham, 9 (2020) 65-138. https://doi.org/10.1007/978-3-030-23069-2_3

[14] V. Petrova, S. Schmauder, M. Ordyan, A. Shashkin, Revisit of antiplane shear problems for an interface crack. Does the stress intensity factor for the interface Mode III crack depend on the bimaterial modulus? Eng. Fract. Mech. 216 (2019) 106524. https://doi.org/10.1016/j.engfracmech.2019.106524

[15] J.W. Eischen, Fracture of nonhomogeneous materials, Int. J. Fracture 34 (1987) 3-22. https://doi.org/10.1007/BF00042121

[16] J.F. Shackelford, W. Alexander, CRC Materials Science and Engineering Handbook, CRC Press, Boca Raton, 2001. https://doi.org/10.1201/9781420038408

[17] Y.C. Zhou, T. Hashida, Coupled effects of temperature gradient and oxidation on thermal stress in thermal barrier coating system, Int. J. Solids Struct. 38 (24-25) (2001) 4235-4264. https://doi.org/10.1016/S0020-7683(00)00309-7

[18] F. Erdogan, G. Gupta, On the numerical solution of singular integral equations, Quart. Appl. Math. 29 (1972) 525-534. https://doi.org/10.1090/qam/408277

[19] V. Petrova, S. Schmauder, FGM/homogeneous bimaterials with systems of cracks under thermo-mechanical loading: Analysis by fracture criteria, Eng. Fract. Mech. 130 (2014) 12-20. https://doi.org/10.1016/j.engfracmech.2014.01.014

[20] F. Erdogan, G.C. Sih, On the crack extension in plates under plane loading and transverse shear, J. Basic. Eng. 85 (1963) 519-527. https://doi.org/10.1115/1.3656897

[21] V. Petrova, S. Schmauder, S. Shashkin, Modeling of edge cracks interaction, Frattura ed Integrità Strutturale (Fracture and Structural Integrity) 36 (2016) 8-26. https://doi.org/10.3221/IGF-ESIS.36.02

Appendix A

In Eq. (6), the regular kernels $R_{nk}(t,x)$ and $S_{nk}(t,x)$ contain the geometry of the problem and are written as

$$R_{nk}(t,x) = (1-\delta_{nk})K_{nk}(t,x) + \frac{e^{i\alpha_k}}{2}\left\{\frac{1}{X_n - \overline{T}_k} + \frac{e^{-2i\alpha_n}}{\overline{X}_n - T_k} + \right.$$

$$\left. +(\overline{T}_k - T_k)\left[\frac{1+e^{-2i\alpha_n}}{(\overline{X}_n - T_k)^2} - \frac{2e^{-2i\alpha_n}(X_n - T_k)}{(\overline{X}_n - T_k)^3}\right]\right\}, \tag{A.1}$$

$$S_{nk}(t,x) = (1-\delta_{nk})L_{nk}(t,x) + \frac{e^{-i\alpha_k}}{2}\left[\frac{T_k - \overline{T}_k}{(X_n - \overline{T}_k)^2} + \frac{1}{\overline{X}_n - T_k} - e^{-2i\alpha_n}\frac{X_n - T_k}{(\overline{X}_n - T_k)^2}\right], \tag{A.2}$$

$$T_k = te^{i\alpha_k} + z_k^0 \quad X_n = xe^{i\alpha_n} + z_n^0, \tag{A.3}$$

 Materials Research Forum LLC
https://doi.org/10.21741/9781644902950

and

$$\delta_{nk} = \begin{cases} 0 & \text{for } n \neq k \\ 1 & \text{for } n = k \end{cases}$$

with kernels $K_{nk}(t,x)$ and $L_{nk}(t,x)$

$$K_{nk}(t,x) = \frac{e^{i\alpha_k}}{2}\left(\frac{1}{T_k - X_n} + \frac{e^{-2i\alpha_n}}{\overline{T}_k - \overline{X}_n}\right) \quad ,\text{(A.4)}$$

$$L_{nk}(t,x) = \frac{e^{-i\alpha_k}}{2}\left(\frac{1}{\overline{T}_k - \overline{X}_n} + \frac{T_k - X_n}{(\overline{T}_k - \overline{X}_n)^2}e^{-2i\alpha_n}\right), \quad \text{(A.5)}$$

which are the same as for the system of cracks in an infinite plane. Additional terms (in addition to K_{nk} and L_{nk}) in Eqs. (A.1) and (A.2) take into account the influence of the edge of the half plane. α_n is the inclination angle of n-th crack to the x-axis, and $\alpha_n = -\beta_n$; $z^0{}_n$ is the coordinate of the center of crack in the global coordinate system (x,y), see Fig. 2b.

Appendix B

The solution of singular integral equations (6) and (7) is obtained by a numerical method which is based on Gauss-Chebyshev quadrature. Eqs. (6) and (7) are rewritten in dimensionless form with the non-dimensionless coordinates $\xi = t / a_k$ and $\eta = x / a_n$, where $2a_k$ is the length of the k-th crack. The unknown function $g_n'(\eta)$ consists of a function $u_n(\eta)$ (a bounded continuous function in the segment [-1,1]) and the weight function $1/\sqrt{1-\eta^2}$, that is,

$$g_n'(\eta) = u_n(\eta) / \sqrt{1-\eta^2} \quad \text{(B.1)}$$

For edge cracks, the function $g_n'(\eta)$ possess a singularity less than $1/\sqrt{1+\eta}$ at the edge point $\eta = -1$, and this condition is accounted as [12, 13]

$$u_n(-1) = 0. \quad \text{(B.2)}$$

In spite of that the exact singularity at the edge points is not taken into account, the numerical results have shown good accuracy for calculating the SIF at the internal crack tip ($\eta = 1$) [12]. If the stress-strain state in the vicinity of the point $\eta = -1$ of the edge crack is examined then the exact order of singularity in this tip should be taken into account.

Using Gauss's quadrature formulae for the regular and singular integrals, the singular integral equations are reduced to the following system of NxM (N – number of cracks, M – number of nodes) algebraic equations

Materials Research Foundations **159** (2024) https://doi.org/10.21741/9781644902950

$$\frac{1}{M}\sum_{m=1}^{M}\sum_{k=1}^{N}\left[u_k(\xi_m)R_{nk}(\xi_m,\eta_r)+\overline{u_k(\xi_m)}S_{nk}(\xi_m,\eta_r)\right]=\pi p_n(\eta_r), \quad \text{(B.3)}$$

$$\sum_{m=1}^{M}(-1)^m u_n(\xi_m)\tan\frac{2m-1}{4M}\pi=0 \qquad (n=1,2,\ \ldots,N;\ r=1,2,\ldots,M\text{-}1), \quad \text{(B.4)}$$

with

$$\xi_m=\cos\frac{2m-1}{2M}\pi \ \ (\mathrm{m}=1,2,\ldots,\mathrm{M});\ \ \eta_r=\cos\frac{\pi r}{M} \qquad (r=1,2,\ \ldots,M\text{-}1).$$

M is the total number of discrete points of the unknown functions $u_n(\eta)$ within the interval (-1,1). Applying the conjugate operation to the system (B.3) and (B.4), additional NxM equations are obtained, i.e. 2NxM equations should be solved, where N is the number of cracks.

If internal cracks are considered, then, instead of Eq. (B.4), the following equation should be used

$$\sum_{m=1}^{M}u_n(\xi_m)=0.$$

After solving the algebraic system (B.3) and (B.4), the functions $u_n(\eta)$ are calculated by the interpolation formula:

$$u_n(\eta)=\frac{2}{M}\sum_{m=1}^{M}u_n(\xi_m)\sum_{r=0}^{M-1}T_r(\xi_m)T_r(\eta)-\frac{1}{M}\sum_{m=0}^{M}u_n(\xi_m) \quad \text{(B.5)}$$

The functions T_r are Chebyshev polynomials of the first kind.

The stress intensity factors are obtained from the following formulas [12, 13]:

$$K_{nI}^{\pm}-iK_{nII}^{\pm}=\mp\lim_{\eta\to\pm1}\sqrt{\pi a_n}\sqrt{1-\eta^2}\,g_n'(\eta),$$

$$K_{In}^{+}-iK_{IIn}^{+}=-\sqrt{\pi a_n}u_n(+1)$$

$$=p_n\sqrt{\pi a_n}\frac{1}{M}\sum_{m=1}^{M}(-1)^m u_n(\xi_m)\cot\frac{2m-1}{4M}\pi, \quad \text{(B.6)}$$

$$K_{In}^{-}-iK_{IIn}^{-}=\sqrt{\pi a_n}u_n(-1)$$

$$=p_n\sqrt{\pi a_n}\frac{1}{M}\sum_{m=1}^{M}(-1)^{M+m}u_n(\xi_m)\tan\frac{2m-1}{4M}\pi \quad (n=1,2,\ \ldots,N). \quad \text{(B.7)}$$

The functions (B.5) are used in this calculation.

CHAPTER 13

Thermal Fracture of Functionally Graded Thermal Barrier Coatings with Pre-Existing Edge Cracks and Multiple Internal Cracks Imitating a Curved Interface

Vera Petrova [1,2]*, Siegfried Schmauder [1]

[1] IMWF, University of Stuttgart, Pfaffenwaldring 32, D-70569 Stuttgart, Germany

[2] Voronezh State University, University Sq.1, Voronezh 394006, Russia

veraep@gmail.com *, Siegfried.Schmauder@imwf.uni-stuttgart.de

Abstract

This work is devoted to the problem of thermal fracture of a functionally graded coating on a homogeneous substrate (FGC/H) with an emphasis on the analysis of a special system of cracks that simulates a curved interface. The FGC/H structure contains the pre-existing crack system in the FGC, both edge cracks (which are often seen in FGC/H structures) and internal cracks. The stress intensity factors are calculated (generally, both Mode I and Mode II are non-zero). Then, using the appropriate fracture criterion for mixed-mode fracture conditions, the crack propagation direction (so-called fracture angles) and critical loads, when this propagation is initiated, are determined. The application of fracture criteria requires knowledge of the fracture toughness near the crack tips. Thus, it is assumed that the fracture toughness of an FGC, as well as other material properties, continuously vary through the thickness of the coating. For multiple cracks, it is also important to know the weakest crack that starts to propagate first, and the initial direction of this growth. Therefore, main attention is paid to the evaluation of the fracture angles for the cracks for different parameters of the FGC/H structure. Both cases of a homogeneous semi-infinite medium with a system of cracks imitating a curved interface, and FGC/H structures with identical crack systems are studied.

Keywords

Thermal Fracture; System of Cracks; Functionally Graded Coatings; Fracture Toughness; Fracture Angles

1. Introduction

Thermal barrier coatings (TBCs), in particular functionally graded coatings (FGCs), are used in different engineering structures, and the most important one is the application in components where protection against elevated temperatures is required. An example of this application is gas turbine engine parts in which TBCs protect metal parts from overheating and melting [1]. Functionally graded materials (FGMs) are special types of composite materials, the properties of which gradually change along a spatial coordinate. This is achieved by creating variations in material compositions and/or structurers. The application of FGC as thermal barrier coatings helps to reduce thermal residual stresses and increases the fracture resistance of the coating.

However, under very high temperatures and complex mechanical loadings, complicated diffusion, oxidation and mechanical interaction processes occur in FGCs structures. As a result of oxidative processes, a thermal growing oxide (TGO) layer is formed [1 - 3]. As reported in the literature [4 - 6], the oxide layer can serve as a source of fracture (cracking). Besides, cracks can occur as a result of initial defects or microcracks during manufacturing. Previous investigations [5 - 9] have revealed that thermal fracture of FGCs is significantly affected by a complex crack interaction mechanism, e.g. interacting cracks can enhance or suppress the propagation of each other. Therefore, the study of fracture of FGCs is important for a better understanding of the fracture resistance of graded coatings.

The interface or oxide layer between coatings and a substrate is rarely perfectly flat. This is usually a wavy, rough interface [4, 10, 11]. The problem of the roughness of the interface between coatings and a substrate has received a lot of attention in the literature. Different aspects of this problem were discussed in numerous publications, e.g. see [4, 9 - 12]. Experimental studies of thermal barrier coatings [4] have reported that a rough interface leads to a higher interface toughness, and the contribution of friction at the interface between the TBC and bond coat may be the cause of this effect. An analytical theoretical study of a wavy non-damaged interface between a thin-film and a substrate is presented in [12]. The film thickness and interface roughness are in the nano-meter range. The perturbation technique was used within the first-order approximation, where the ratio between the maximum deviation of the interface from the flat state and a perturbation wavelength is considered as a small parameter. The numerical analysis [12] has shown that the interface stress reduces the stress concentration factor, when the residual interface stresses are neglected. This effect decreases when the size of interface asperities (the width and depth), stiffness of the covering film and curvature radius of cavities increase. In [13], analytical solutions are presented for a periodic set of edge dislocations and point forces interacting with a planar traction-free surface of a semi-infinite elastic solid. The fundamental solutions obtained in [13] can be used for applying the boundary integral equation method to an analysis of defects such as cracks and inhomogeneities, periodically distributed at the nanometer distance from the boundary. In [14], a rate-dependent cohesive zone model was presented with a novel traction-separation-law with the capability to simulate heat conduction through the interface. The applications of cohesive elements to an FE simulation to describe the cracking of a coating, and the delamination of this cracked coating from the substrate material at high temperatures were demonstrated. Interface crack problems were also theoretically studied in [15] for Mode III interface cracks, and in [16] for an interface crack and multiple internal cracks in infinite FGM/H bimaterials under thermo-mechanical loading. Flat interfaces were considered in [15, 16]. In [11], an experiment and finite element analysis modeling was used to study the influence of a temperature gradient across an YSZ topcoat on the thermal cyclic lifetime of this TBC. A study of the effect of interface roughness on the strain energy release rate and surface cracking behavior in air plasma sprayed thermal barrier coating system was presented in [10]. The extended FEM and periodic boundary conditions were used. Predictions for the stress field and the driving force of multiple surface cracks in the film/substrate system were presented. It was shown that the interface roughness significantly affects the strain energy release rate, the interfacial stress distribution, and the crack formation patterns.

The present work is devoted to the thermal fracture problem for a functionally graded coating on a homogeneous substrate (FGM/H). The model presented in authors' previous papers [17-20] is used in this study in an extension to a special crack system. The present FGC/H structure contains a pre-existing system of cracks in the FGC. In addition to edge cracks, there are internal/interface cracks imitating a partially wavy interface. The coordinates of the centers of these internal cracks are located on a straight line parallel to the surface, but the inclination angles of these cracks vary symmetrically from 0° to a certain angle, see Fig. 1. The Mode I and Mode II stress intensity factors are calculated. Then, using the appropriate fracture criterion for mixed-mode fracture conditions, the crack propagation direction and the critical loads at which this propagation begins are determined. For multiple cracks, it is also important to know the weakest crack that starts to propagate first, and the initial direction of this growth, which may indicate the interface roughness in the case of interface cracks. Therefore, the main attention is paid to the evaluation of the fracture angles for the cracks for different parameters of the FGC/H structure. In addition to the FGC/H structures, a homogeneous semi-infinite medium with an identical crack system is also studied as a special case.

This paper is organized as follows. Section 2 presents the description of the problem, namely, the geometry of the problem, the properties of functionally graded coatings and the corresponding load on the crack faces. Section 3 describes the solution of the problem, which includes the main singular integral equations, determination of stress intensity factors, fracture angles and critical loads. Section 4 presents the results for two cases: for a particular case for a homogeneous medium and for an FGC/TGO/H structure. The parametric results are analyzed with respect to stress intensity factors, fracture angles and critical loads for a special system of cracks, that is, edge cracks and internal-interface cracks, imitating a partly curved weakened interface. The conclusions are presented in section 5.

2. Description of the problem

2.1 Geometry of the problem and loading

A structure consisting of a functionally graded layer of thickness h on a semi-infinite homogeneous substrate is shown in Fig. 1a. For curved-rough interfaces, the thickness is determined by the distance from the top surface to the center line of the curved profile. For use in thermal barrier coatings (TBCs), the layer consists of a ceramic on the top layer and mixtures of ceramic and metal gradually transforming into metal on the substrate. TBCs operate at high temperatures at which oxidation processes can occur, resulting in the formation of an oxide layer between the coating and the underlying alloy. That is, the structure under consideration is a functionally graded coating (FGC) (with a material gradient perpendicular to the interface) on a homogeneous substrate with the thermally grown oxide (TGO) layer between them. Special arrangement of internal cracks imitates a weak interface.

In the functionally graded coating a system of N cracks of length $2a_k$ ($k = 1, 2, \ldots, N$) is located, which can be edge and/or internal cracks. In the present work, a special system of cracks is considered: three edge cracks normal to the surface and internal-interface cracks with inclination angles $\alpha_{2n} = -\alpha_{2n+1}$ ($n = 2, 3, 4$) in a curved weak zone, Fig. 1a and b. The numbering of cracks in the system under consideration is shown in Fig. 1c. The global coordinate system (x, y) is set at the top of the FGC surface, and the local coordinates (x_k, y_k) are referring to each crack with the x-axis on the crack lines as shown in Fig. 2a. Using these coordinate systems, the positions of the cracks are determined by their midpoint coordinates (x_k^0, y_k^0) and the inclination angles α_k to the x-axis, Fig. 2a. Further, the complex variables method will be used in the formulation of the equations of the problem; therefore, it is convenient to represent the midpoint coordinates of the cracks in complex form as $z_k^0 = x_k^0 + iy_k^0$ (i is the imaginary unit).

The FGC/TGO/H structure is cooled by ΔT, $\Delta T > 0$ (this can be cooling from operating temperatures). This sudden cooling causes thermal residual stresses in the structure and, in addition, a tensile load p is applied to the structure parallel to the surface.

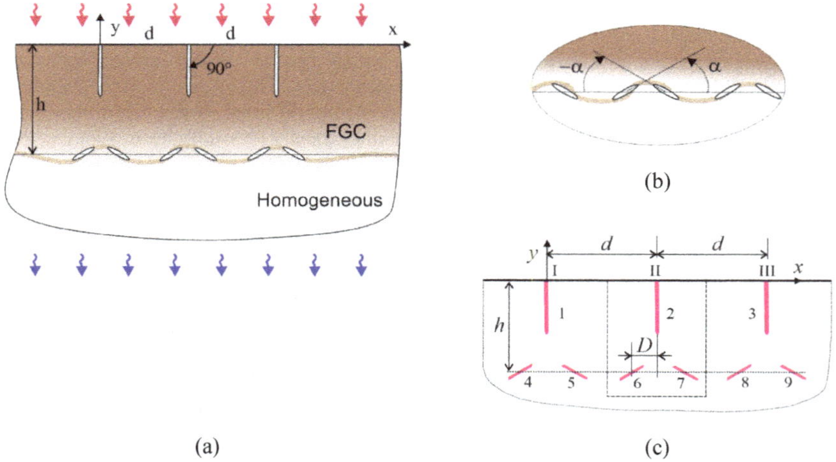

Figure 1. Geometry of the problem: (a) FGC/TGO/H with a system of cracks, (b) curved weak layer with cracks, and (c) the crack system with crack numbering.

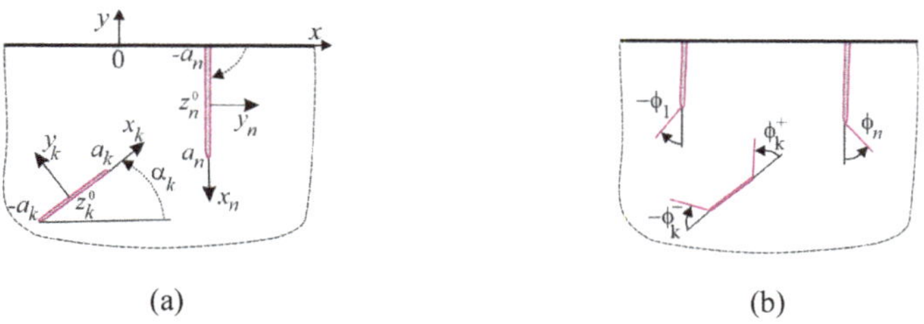

Figure 2. Geometry of the problem: (a) coordinate systems connected with cracks, (b) edge and internal cracks with fracture angles.

2.2 Model for mechanical and physical properties for an FGC

In functionally graded materials, the composition of the material gradually changes mainly in one direction. In particular, in our problem, the composition varies with the y-coordinate from the ceramic in the upper part of the FGC to the metal in the substrate. Consequently, the thermal and mechanical properties of an FGC also varies continuously with the thickness coordinate y. In this work, an exponential form of these properties is used:

$$\alpha_t(y) = \alpha_{t1}\exp(\varepsilon(y+h)),\ E(y) = E_1\exp(\omega(y+h)),\ -h \le y \le 0. \tag{1}$$

Here, α_t is the coefficient of thermal expansion, and E is Young's modulus with non-homogeneity parameters ε and ω, respectively. E_1 and α_{t1} stand for the thermal and mechanical properties of the homogeneous substrate. Poisson's ratio is assumed to be constant, and is equal to the value of the homogeneous substrate. The values of the dimensionless inhomogeneity parameters εh and ωh (h is the thickness of the FGC) are obtained from Eq. (1) as follows

$$\varepsilon h = \ln(\alpha_{t2}/\alpha_{t1}),\ \ \alpha_{t2} = \alpha_t(y)\big|_{y=0},\ \ \alpha_{t1} = \alpha_t(y)\big|_{y=-h}, \tag{2}$$

$$\omega h = \ln(E_2/E_1),\ \ E_2 = E(y)\big|_{y=0},\ \ E_1 = E(y)\big|_{y=-h}. \tag{3}$$

Since the exponential law was chosen for the material parameters α_t and E, it is reasonable to use the exponential law also for the fracture toughness, e.g. see [21]. Thus, the fracture toughness of a functionally graded material can be written as follows:

$$K_{Ic}(y) = K_{Ic1}\exp(\gamma(y+h)), \tag{4}$$

where γ is the inhomogeneous parameter of the fracture toughness. The nondimensional value γh is obtained as

$$\gamma h = \ln(K_{Ic2}/K_{Ic1}), K_{Ic2} = K_{Ic}(y)_{|y=0},\ K_{Ic1} = K_{Ic}(y)_{|y=-h}. \tag{5}$$

In the local coordinate system (x_n, y_n) connected with the n-th crack, the function K_{Icn} is written as

$$K_{Icn}(x_n) = K_{Ic1}\exp(\gamma(h + y_n^0 - x_n\sin\beta_n)). \tag{6}$$

Similar expressions are also written for α_t and E in the local coordinates (x_n, y_n).

2.3 Loadings on the crack faces

With changing the temperature, e.g. an FGM/H structure is cooled on ΔT, the residual stresses are arising due to mismatch in the coefficients of thermal expansion [22, 23]. In the presented model, the inhomogeneity of FGMs is taken into account through continuously varying residual stresses, and these stresses are the following [21]:

$$\sigma_{xx}^T(y) = [\alpha_t(y) - \alpha_{t1}]\Delta T E(y),\ \ \sigma_{xx}^e(y) = [E(y)/E_1 - 1]\sigma_{xx}^0,\ \sigma_{xx}^0 = p, \tag{7}$$

α_{t1} and E_1 are, respectively, the thermal expansion coefficient and Young's modulus of a homogeneous substrate material and at the interface, $\alpha_t(y)$ and $E(y)$ are defined by Eq. (1).

The method of superposition is used to solve this problem, so that loads at infinity are reduced to the corresponding loads on the crack faces. Thus, the tensile load is reduced to the load p_n on the crack surfaces and written in complex form as

$$p_n = \sigma_n - i\tau_n = p(1-\exp(2i\beta_n))/2 = pf(\beta_n) \quad (n = 1, 2, \ldots, N) \tag{8}$$

In the common case of FGMs, the full load on the n-th crack consists of p_n, σ_n^T and σ_n^e, Eq. (7), where the index "n" denotes that the functions are written in the local coordinate system (x_n, y_n) connected with the n-th crack:

$$p_n + \sigma_n^e + \sigma_n^T = \exp(\omega(h + y_n^0 - x_n \sin\beta_n))\left[pf(\beta_n) + Q\exp(\varepsilon(h + y_n^0 - x_n \sin\beta_n)) - Q\right]$$

$$(n = 1, 2, \ldots, N) \tag{9}$$

$$Q = \alpha_{t1}\Delta T E_1 .$$

It is assumed that $p = Q$, or an additional loading parameter p/Q should be considered.

3. Solution of the problem

3.1 Singular integral equations

The boundary value problem of elasticity for a system of N cracks is reduced to a system of N singular integral equations [24] with respect to the unknown functions $g_n'(x)$ contain the shear $[u_n]$ and normal $[v_n]$ displacement jumps on the n-th crack line. These equations are written in Appendix A, Eq. (A.1)-(A.6).

The singular integral equations (A.1) are solved numerically based on the Gauss-Chebyshev quadrature. Different versions of this method were used for the solution, the effectiveness of which has been proven in many studies [25]. In the present work, the version described in [24] is used, and the solution scheme for this method is given in Appendix B.

3.2 Stress intensity factors

The stress intensity factors at tips of the n-th crack are obtained from the following formulas:

$$K_{nI}^{\pm} - iK_{nII}^{\pm} = \mp \lim_{\eta \to \pm 1} \sqrt{\pi a_n}\sqrt{1-\eta^2}\, g_n'(\eta) \tag{10}$$

Here the signs "+" and "–" refer to the right and left crack tips, respectively. a_n is the half-length of the crack.

In general, the SIFs can be written as

$$K_{In} - iK_{IIn} = p\sqrt{\pi a_n}(k_{In} - ik_{IIn}) . \tag{11}$$

3.3 Stress intensity factors, fracture angles and critical loads

For predicting the crack growth and the determination of the direction of this growing crack, the criterion of maximum circumferential stresses [26] is used. According to this criterion, the crack deflection angle ϕ (or the so-called fracture angle, Fig.2b), the critical stress intensity factor, and then the critical stresses are calculated from the following relations:

$$\phi_n = 2\arctan\left[\left(K_{In} - \sqrt{K_{In}^2 + 8K_{IIn}^2}\right) / 4K_{IIn}\right], \tag{12}$$

$$K_n^{eq} \equiv \cos^3(\phi_n/2)\left(K_{In} - 3K_{IIn}\tan(\phi_n/2)\right) = K_{Ic,tip} \text{ or } K_n^{eq} = K_{Ic,tip}. \tag{13}$$

Using a single crack subjected to a load p normal to the crack line as a reference crack with the stress intensity factor

$$K^0 = p\sqrt{\pi a}, \tag{14}$$

the corresponding critical load is obtained as

$$p_0 = K_{Ic1} \big/ \sqrt{\pi a}, \tag{15}$$

where $a = \max\limits_{n=1,\ldots,N} a_n$.

By substituting (11) into condition (13), the critical loads are obtained as

$$p_{cr\,n} = \frac{K_{Ic}(y)}{\sqrt{\pi a_n}} \frac{1}{\cos^3(\phi_n/2)\left(k_{In} - 3k_{IIn}\tan(\phi_n/2)\right)}$$

or

$$\frac{p_{cr\,n}}{p_0} = \frac{\exp(\gamma(h + y_n^0 - x_n \sin\beta_n))}{\cos^3(\phi_n/2)\left(k_{In} - 3k_{IIn}\tan(\phi_n/2)\right)} \frac{\sqrt{a}}{\sqrt{a_n}}, \tag{16}$$

where p_0 is defined by Eq. (15) and K_{Ic} by Eqs. (4) and (6).

The fracture angle ϕ_n is shown in Fig. 2b. First, the angle of the crack propagation (fracture angle) Eq. (12) is obtained using the results of the calculated stress intensity factors, Eq. (10). Next, the local fracture stability is evaluated by Eq. (13). Then, the critical loads are obtained near the crack tips, Eq. (16). Finally, the weakest crack or crack tip is defined from the condition

$$P_{cr} = \min_n p_{cr\,n} / p_0 \; (n = 1, 2, \ldots, N) \tag{17}$$

4. Results: stress intensity factors, fracture angles and critical loads for a system of interacting cracks

Fig. 1a shows the geometry, which is used for numerical example. Three edge cracks are normal to the surface, and the internal cracks have inclination angles $\alpha_{2n} = -\alpha_{2n+1}$ ($n = 2, 3, 4$) in the curved weak zone. The crack patterns, consisting of edge and internal cracks, have been observed in experiments reported in the literature, e.g. see [5, 8]. Table 1 introduces the coordinates of the crack centers. According to this definition of coordinates, when the distance d between the edge cracks changes, the distance between the pairs of cracks 4 and 5, 6 and 7, 8 and 9 also changes. The simplest repeating "crack pattern" in our crack system consists of one edge crack and two internal cracks. There are three such sub-systems: pattern I includes cracks 1, 4 and 5, II – cracks 2, 6 and 7, and III – cracks 3, 8 and 9, see Fig. 1c. Pattern II is an inner one with two neighboring patterns I and III. The behavior of pattern II may reflect the behavior of these cracks in a periodic crack system with a repeating pattern II. The results for these subsystems will be done in the following sections.

The results are presented for the following parameters: the width of the coating is $h = 2.5$ *mm*, the distance between the edge cracks is $d = 2, 4, 6$ (*mm*), and $D = 0.25$ *mm*; the dimensionless parameters are used in the numerical calculations: h/a, d/a and D/a, where $a = \max a_n$ ($n = 1, 2, \ldots, 9$). Results are also obtained for other values for cracks and distances, which are not shown here, but are mentioned in some cases in the discussion. In the following figures, the dimensionless distance d/a is denoted by d. The sizes of cracks are: edge cracks $a_n = 1$ *mm* ($n = 1, 2, 3$), internal cracks $a_n = 0.22$ *mm* ($n = 4, 5, \ldots, 9$); a_n denotes the half-length of the n-th crack. In the calculations the non-dimensional size is used a_n/a, $a = \max a_n$. The following designations are also used: a_{edge} and a_{int} for edge and internal cracks, respectively.

The stress intensity factors (SIFs) are used in dimensionless form: $k_{I,II} = K_{I,II}/K^0$, where $K^0 = p(\pi a)^{1/2}$ is the SIF for a single crack, Eq. (14). A single edge crack of half-length a, normal to the surface of the layer, has SIF equal to $K_I = 1.58p(\pi a)^{1/2}$. This definition for SIFs is more convenient for the considered mixed system of internal and edge cracks than the commonly used one for the SIF for an edge crack: $K_I = 1.12p(2\pi a)^{1/2}$, where $2a$ is the full length of the edge crack.

Further, two cases are considered: a particular case of a homogeneous semi-infinite medium and a case of a FGC/H structure. In both cases, the values of stress intensity factors and fracture angles are approximately the same for the following cracks: *for crack 4 and 9, for crack 5 and 8, for crack 6 and 7*. The critical loads are also approximately the same for the listed cracks. Therefore, the results are presented not for all internal cracks, but mostly for cracks 4, 5 and 6.

Table 1. The coordinates of the centers of cracks

Edge cracks	1		2		3	
x_k^0 y_k^0	0 $-a_1$		d $-a_2$		$2d$ $-a_3$	
Internal cracks	4	5	6	7	8	9
x_k^0 y_k^0	$-D$ $-h$	D $-h$	$d-D$ $-h$	$d+D$ $-h$	$2d-D$ $-h$	$2d+D$ $-h$

4.1 Homogeneous semi-infinite medium

4.1.1 Stress intensity factors (SIFs)

Fig. 3 presents the dimentionless stress intensity factors (SIFs) $k_{I,II}$ for edge cracks depending on α, $0° \le \alpha \le 60°$, and for different distances d/a. The SIFs for internal cracks are shown in Fig. 4 as functions of α, and $k^-_{I,II}$ refers to the left crack tips (Figs. 4 a, c, e), and $k^+_{I,II}$ - to the right crack tips (Figs. 4 b, d, f). Both k_I and k_{II} are non-zero, thus, the mixed-mode conditions are realized.

The influence of α (the inclination angle of internal cracks) on the SIFs for edge cracks is negligible, Fig. 3. The distance d/a between the cracks affects SIFs: the larger the distance, the greater the k_I, and the value k_I tends to $k_I = 1.58$, while k_{II} tends to zero, to the values for a single crack. Besides, k_I less than 1.58 for all parameters, that is, the shielding effect is observed. For internal cracks, both α and d/a affect $k_{I,II}$, see Fig. 4. The weakest crack can be defined by the largest value for k_I . As seen in Figs. 3 and 4, outer edge cracks 1 and 3 and outer internal crack 4 for $\alpha = 60°$ (and also crack 9) have largest values for k_I (greater than 1). That is, from these cracks, propagation can start.

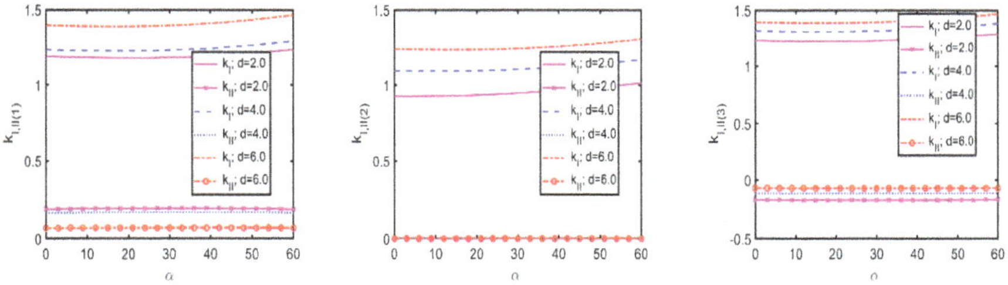

(a) (b) (c)

Figure 3. SIFs k_I and k_{II} for edge cracks as functions of internal crack angles α and for different distances d between the edge cracks; (a) for crack 1, (b) for middle crack 2, (c) for crack 3. Homogeneous material.

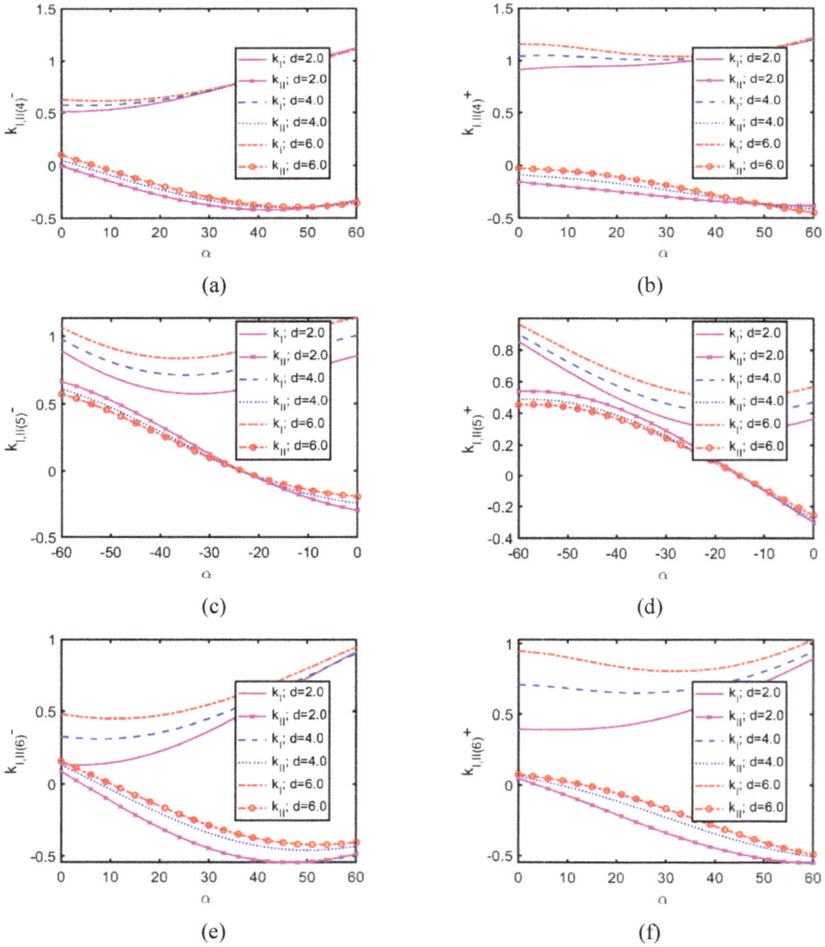

Figure 4. SIFs k_I and k_{II} for internal cracks as functions of angles α and for different distances d between the cracks: (a) and (b) for crack 4 at left and right crack tips, respectively, (c) and (d) for crack 5 at left and right crack tips, (e) and (f) for crack 5 at left and right crack tips. Homogeneous material.

4.1.2 Fracture angles

The fracture angles (angles of the crack propagation direction) depending on α and for different d are shown in Fig. 5 for edge cracks 1, 2, 3, and in Fig. 6 for internal cracks 4, 5, 6 *(remind that for cracks 4 and 9, 5 and 8, 6 and 7 the values of ϕ are approximately the same).*

Behavior of edge cracks in the system of edge + internal cracks relative to the crack deviation path. If all cracks in the system in Fig. 1 are small cracks (e.g. $a_n = 0.1/a$, $n = 1, 2, \ldots, 9$, these figures are not shown here) the edge cracks have zero or close to zero fracture angles. Thus, if they begin to propagate, the propagation path will be straight. However, as the edge cracks become larger, the fracture angles for outer edge cracks (crack 1 and 3) become larger, Fig. 5. The angle α has little effect on ϕ for edge cracks. This influence is slightly more pronounced for large cracks, both edge and internal, as would be expected due to the stronger interaction of large cracks. For example, for $a_{edge} = 1/a$ and $a_{int} = 0.22/a$, the maximum range of change ϕ is about 1° or 7% for $d = 2$, Fig. 5. The influence of the distance d between cracks on ϕ is stronger: the smaller d, the larger ϕ, see Fig. 5 and Table 3. The middle edge crack 2 has $\phi = 0°$ for all considered parameters of the problem.

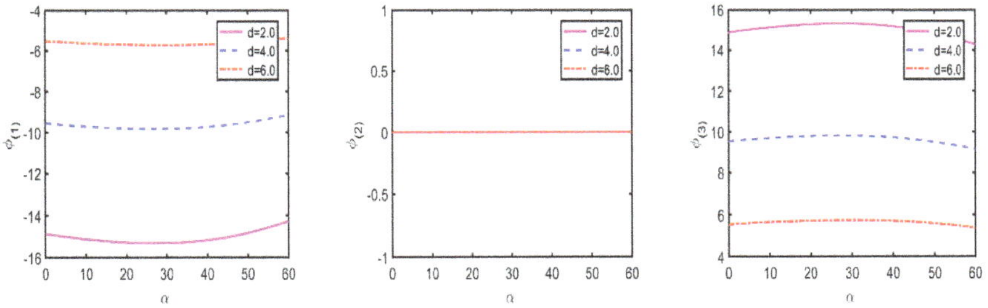

(a) (b) (c)

Figure 5. Fracture angles for edge cracks as functions of angles α and for different distances d between the edge cracks: (a) for crack 1, (b) for middle crack 2, (c) for crack 3. Homogeneous material.

Table 2. Fracture angles for edge cracks

Crack #	1	2	3
d=2 homogeneous	–15°	0°	15°
FGC	–15°	–0.15°	15°
d=4 homogeneous	–9.5°	0°	9.5°
FGC	–9.9°	–0.04°	9.9°
d=6 homogeneous	–5.5°	0°	5.5°
FGC	–5.9°	0°	5.9°

Materials Research Foundations **159** (2024) https://doi.org/10.21741/9781644902950

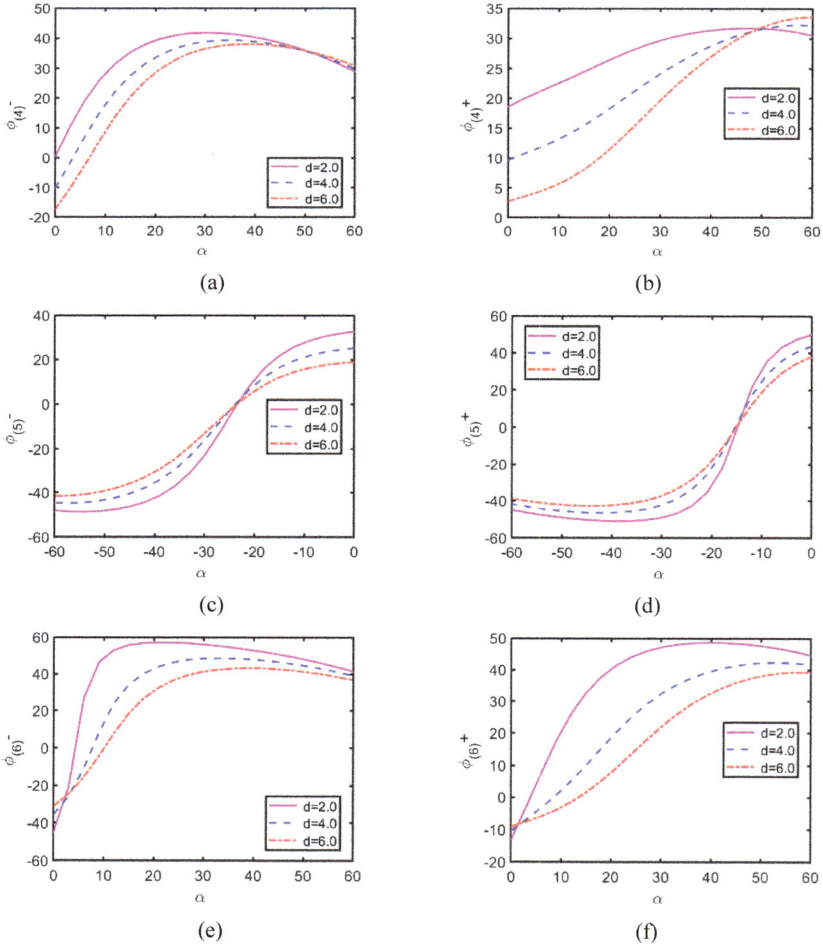

Figure 6. Fracture angles for edge cracks as functions of angles α and for different distances d between the cracks: (a) and (b) for crack 4 at left and right crack tips, respectively, (c) and (d) for crack 5 at left and right crack tips, (e) and (f) for crack 5 at left and right crack tips. Homogeneous material.

Behavior of internal cracks in the system of edge + internal cracks relative to the crack deviation path. Fig. 6 shows the fracture angles ϕ for internal cracks at left and right crack tips, and Fig. 9 a, b schematically illustrates the behavior of internal cracks for some parameters of the problem.

Note that the fracture angle ϕ is highly sensitive to the inclination angle α, see Fig. 6. For collinear cracks (α = 0) with crack sizes a_{edge} = 1/a and a_{int} = 0.22/a, and distance d = 2, the ϕ value is changed from 0° to 50°. What is more, the maximum value ϕ has inner cracks in the row of these collinear cracks, i.e. cracks 5 - 8, while the outer cracks 4 and 9 have smaller ϕ, 0° for the outer tips and 18° for the inner ones, see Fig. 6 and Fig. 9a. For small internal cracks a_{int} =0.1/a (with the size for edge cracks a_{edge} = 1) the results for outer cracks are opposite: 0° for the inner tips and 18° for the outer ones.

The distance d also affects ϕ. The influence of d on ϕ is greatly pronounced for large cracks (and, accordingly, with stronger interaction), see Fig. 6. For small cracks, a = 0.1/a, the effect of d on the fracture parameters is negligible.

Behavior of cracks in the system of edge + collinear internal cracks, weak cracks and deviation path. For collinear internal cracks, α = 0°, the SIF k_I at outer crack tips (crack 4 left tip and crack 9 right tip) have smaller values in comparison to inner crack tips for internal cracks and also for edge cracks. It means that propagation rarely starts from these tips. More dangerous are outer edge cracks 1 and 3, right tip for crack 4 and left for 5, right tip for crack 8 and left tip for crack 9, so these internal cracks will grow towards each other.

A deviation of cracks from the straight path indicates that cracks can cause some roughness. For example, an initially flat interface tends to become rougher due to this crack deviation path. For the collinear cracks in our system, the magnitude of fracture angles changes from zero to approximately 50°, as can been seen in Fig. 6 and in the scheme in Fig. 9a. The instability of collinear cracks relative to the deviations of the crack path has been observed in many experimental and theoretical studies, e.g. see [28-30]. It was noticed, when two collinear Mode I cracks were growing towards each other, they do not merge tip to tip, but instead repel each other [28]. The origin of this effect has been discussed by several authors. In [29], it was theoretically proved that the straight crack path is unstable, i.e. that tip to tip coalescence will not take place. This was done by considering a periodic array of approximately collinear but slightly curved cracks. In our problem, collinear cracks are loaded parallel to the cracks; in the classical case, the cracks are loaded perpendicular to the cracks (Mode I), and in both these cases, the interaction of cracks causes strong perturbations of the crack path, as can be seen from the change in the fracture angles, Fig. 9.

4.2 FGC/H structures

The inhomogeneity parameters are shown in Table 3 for two FGMs and for temperatures 20°C and 1000°C, they are cited as in [3]. For an illustrative example, the parameters of inhomogeneity are taken as follows: $\varepsilon h = -0.5$, $\omega h = -1.5$, $\gamma h = -2.3$. These values correspond to (PSZ/Ni)/Ni with material parameters that increase from the ceramic top to the metal substrate. The geometrical parameters are the same as in the previous section. The fracture angles ϕ as functions of α and for different distances d/a are presented in Fig. 7 for edge cracks, and in Fig. 8 – for internal cracks. The critical loads are shown in Figs. 10 and 11. Eq. (16) is used for dimensionless critical load, and Eq. (12) for fracture angles.

Table 3. The inhomogeneity parameters εh, ωh and γh for the thermal expansion coefficient (α_t), Young's modulus (E) and fracture toughness (K_{Ic}), respectively

FGM	T (°C)	εh (α_t)	ωh (E)	γh (K_{Ic})
(PSZ/Ni)/Ni	20°C	–0.5	–1.5	–2.3
	1000°C	–0.4	–1.7	–
(PSZ/Steel)/Steel	20°C	–0.2	–2.2	–2.3
	1000°C	–0.4	–0.2	–

4.2.1 Fracture angles

As in the homogeneous case, the inclination angle α does not strongly affect ϕ of edge cracks, while the influence of the distance *d/a* on ϕ is noticeable, see Fig. 7 and Table 2. For middle crack 2, the fracture angle ϕ is approximately equal to zero. For internal cracks, both α and the distance *d/a* strongly affect the fracture angles ϕ, Fig. 8. However, the maximum ϕ values in this case are less than in the previous homogeneous one, these maxima are equal to 25° and 50° for FGC and homogeneous materials, respectively. Fig. 9 c, d schematically illustrates the behavior of edge and internal cracks for inclination angles $\alpha = 0°$ and 30° ($d/a = 2$). It can be seen that the deviation of the edge cracks is approximately the same as for the homogeneous case, Fig. 9 a, c. However, the internal cracks have a smaller fracture angle, so the crack path for internal cracks in FGM is smoother than in a homogeneous medium for both collinear and oblique cracks, see Fig. 9.

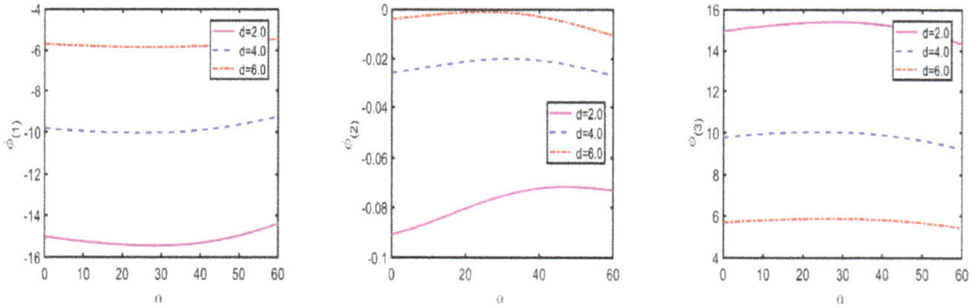

(a) (b) (c)

Figure 7. Fracture angles ϕ for edge cracks as functions of angles α and for different distances d between the cracks: (a) for crack 1, (b) for middle crack 2, (c) for crack 3. FGC/H structure.

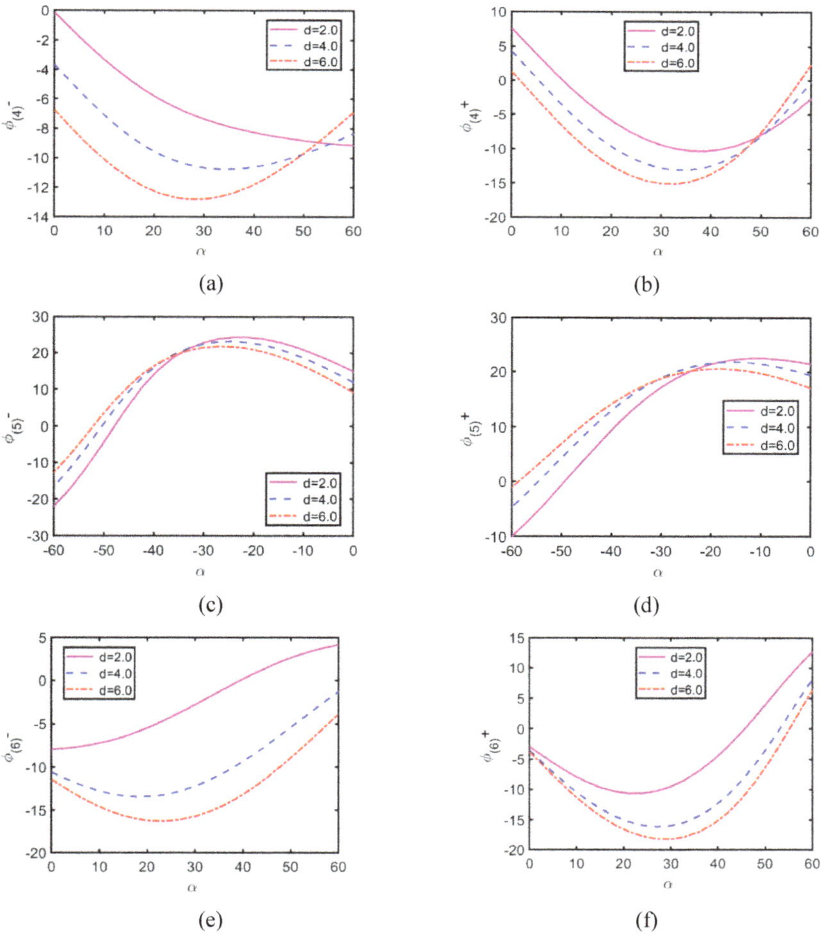

Figure 8. Fracture angles ϕ for internal cracks as functions of angles α and for different distances d between the cracks; (a) and (b) for crack 4 at left and right crack tips, respectively; (c) and (d) for crack 5 at left and right crack tips, (e) and (f) for crack 5 at left and right crack tips. FGC/H structure.

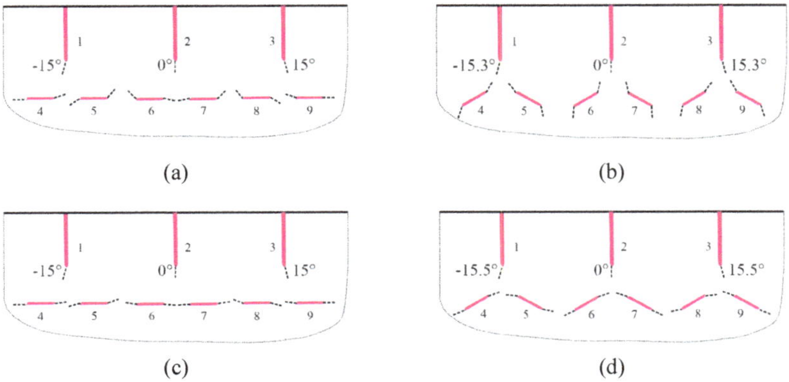

Figure 9. Scheme for fracture angles ϕ for internal cracks with inclination angles $\alpha = 0°$ and $30°$ and three edge cracks: (a) and (b) – for homogeneous medium; (c) and (d) – for an FGC/H structure. $a_{edge}/a = 1$, $a_{int}/a = 0.22$, $d/a = 2$.

4.2.2 Critical loads

The critical loads as functions of α are presented in Fig. 10 for internal cracks at left (Figs. 10 a, c, e) and right (Figs. 10 b, d, f) crack tips for $d/a = 2$, 4, 6. The other figures refer to the following cases: Fig. 11a for edge cracks, Fig. 11b for internal cracks, and Fig. 11c for edge cracks and internal crack 4, all are for $d/a = 2$. The results for critical loads take into account the variation of fracture toughness in accordance with the law in Eq. (4).

The critical loads are influenced by both the angle α and the distance d/a, Figs. 10 and 11. The weakest crack has the smallest p_{cr}/p_0. For edge cracks in Fig. 11a, the largest value for p_{cr}/p_0 is for crack 2 and the smallest one for cracks 1 and 3. Thus, the fracture can start from the edge cracks 1 and 3. In Fig. 11b, the critical loads p_{cr}/p_0 are shown for all internal cracks.

It can been seen that the values p_{cr}/p_0 for crack pairs 4 and 9, 5 and 8, 6 and 7 are equal, as mentioned earlier. The smallest p_{cr}/p_0 values are for crack 4 at right tip (and accordingly for crack 9 at left tip) for $0° < \alpha < 30°$, and for crack 4 at left tip (and accordingly for crack 9 at right tip) for $30° < \alpha < 50°$. Fig. 11c shows p_{cr}/p_0 for internal crack 4 and for edge cracks, as can be seen, the weakest cracks are edge crack with the lowest values of critical loads.

Materials Research Foundations **159** (2024) https://doi.org/10.21741/9781644902950

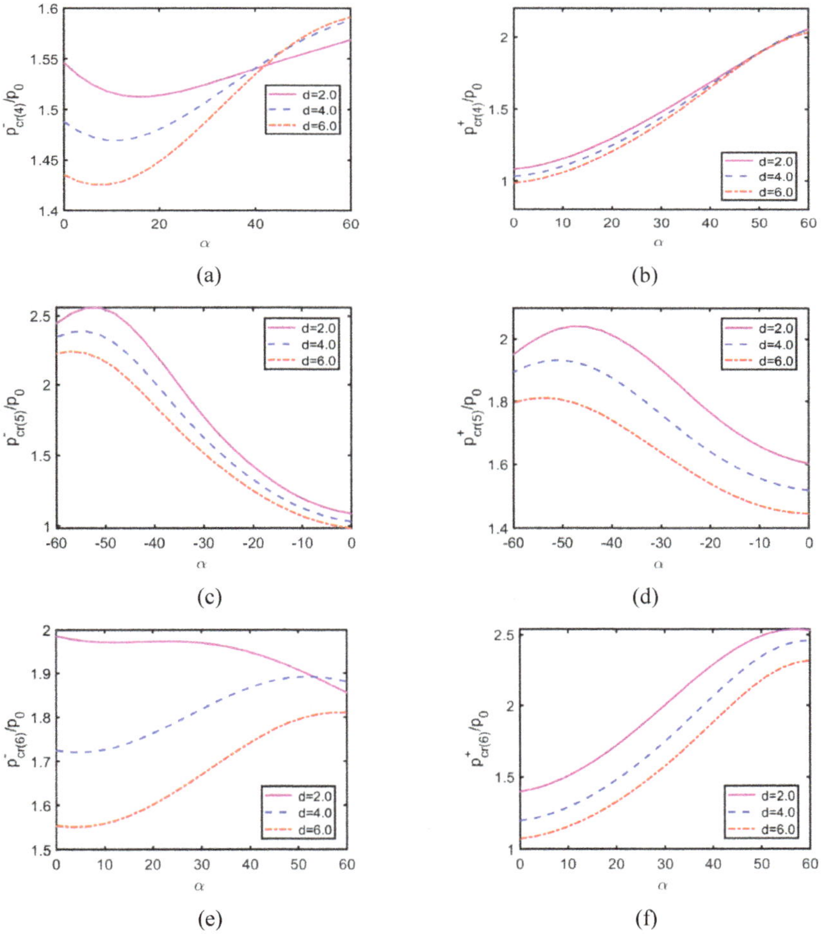

Figure 10. Critical loads as functions of angles α and for different distances d between the edge cracks; (a) and (b) for crack 4 at left and right crack tips, respectively, (c) and (d) for crack 5 at left and right crack tips, (e) and (f) for crack 5 at left and right crack tips; $a_{edge}/a = 1$, $a_{int}/a = 0.22$. FGC/H structure.

Materials Research Foundations **159** (2024) https://doi.org/10.21741/9781644902950

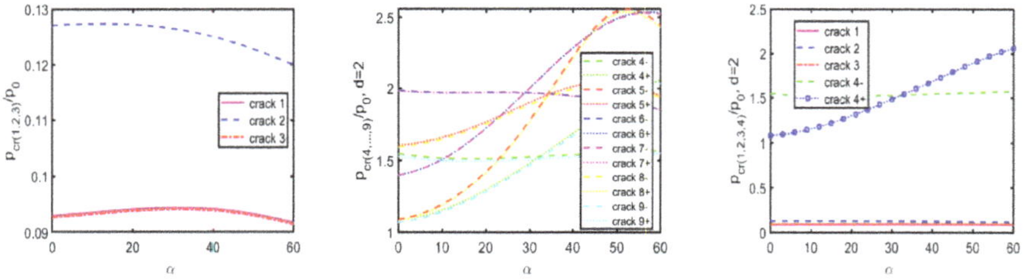

(a) (b) (c)

Figure 11. Critical loads as function of angles α: (a) for edge cracks, (b) for internal cracks, (c) for edge cracks and internal crack 4; $a_{edge}/a = 1$, $a_{int}/a = 0.22$, $d/a = 2$. FGC/H structure.

4.2.3 Behavior of substructure II

The behavior of the crack pattern II (Fig. 1c) may reflect the behavior of a periodic crack system with pattern II as a period. Comparing the behavior of pattern II with I and III, one can see that this subsystem is more resistant to a possible crack propagation. In both cases, of a homogeneous material and of a FG material, these cracks are less dangerous compared to cracks in patterns I and III, Figs. 4, 10 and 11. A shielding effect is observed for crack 2. At the same time, the crack deviations (fracture angles) of internal cracks (imitating a partly curved interface) are greater than in patterns I and III for some parameters of the problem, see Figs. 6, 8, and 9. The fracture angles for edge cracks are zero in pattern II.

5. Summary

Summarizing the obtained results, the following conclusions can been drawn.

The weakest cracks in the considered system are the outer edge cracks (1 and 3) and the outer internal cracks (4 and 9) in a homogeneous medium and the outer edge cracks (1 and 3) in the considered case of FGC. In the homogeneous case, these weak (critical) cracks are determined by the stress intensity factors, and in the case of the FGC - by the critical loads. The fracture criterion of maximal hoop stresses was used, and the structural variation of fracture toughness for the FGM was taken into account. If the material parameters of the FGC tend to homogeneous, critical cracks will be the same as in the homogeneous case.

The fracture angles at the crack tips determine the initial crack paths. These values are strongly influenced by geometric parameters (i.e. crack sizes, distances between the cracks as well as inclination angles). Besides, the inhomogeneity of the FGC also affects the fracture angles. In the considered FGC/H structure, the deviation of cracks from their straight path will be less than in a homogeneous semi-infinite medium with an identical crack system.

Considering the system of internal collinear cracks, which can imitate a flat interface, it should be noted that the cracks deviate from their straight path, thus these cracks lead to some roughness of the interface. In the case of a homogeneous medium, this roughness is more pronounced than in the case of FGCs, because of the larger fracture angles of these collinear cracks (Fig. 9a and Fig. 9c).

A system of inclined cracks, imitating a partly curved interface, tends to increase the roughness of the interface in the case of a homogeneous medium (Fig. 9b). However, in the case of the FGC/H system, there is a tendency towards smoothing out the interface roughness (Fig. 9d).

The detailed study of the representative crack patterns, which are observed in thermal barrier coatings, can help find ways to improve the fracture resistance of FGC/H structures.

Acknowledgement

The authors would like to acknowledge the financial support of the German Research Foundation under Grant SCHM 746/209-1.

Compliance with Ethical Standards

This study was funded by German Research Foundation under Grant SCHM 746/209-1.

Ethical approval

This article does not contain any studies with human participants or animals performed by any of the authors.

References

[1] D. Clarke, M. Oechsner, N. Padture, Thermal-barrier coatings for more efficient gas-turbine engines, MRS Bull. 37 (2012) 891-941. https://doi.org/10.1557/mrs.2012.232

[2] T.S. Hille, S. Turteltaub, A.S.J. Suiker, Oxide growth and damage evolution in thermal barrier coatings, Eng. Fract. Mech. 78 (2011) 2139-2152. https://doi.org/10.1016/j.engfracmech.2011.04.003

[3] Y.C. Zhou, T. Hashida, Coupled effects of temperature gradient and oxidation on thermal stress in thermal barrier coating system, Int. J. Solids Struct. 38 (24-25) (2001) 4235-4264. https://doi.org/10.1016/S0020-7683(00)00309-7

[4] Y. Kagawa, M. Tanaka, M. Hasegawa, Interface delamination analysis of dissimilar materials: Application to thermal barrier coatings, in: Schmauder, S. et al. (Eds.) Handbook of Mechanics of Materials, Springer, Singapore, 2019, pp. 1373-1412. https://doi.org/10.1007/978-981-10-6884-3_83

[5] S. Rangaraj, K. Kokini, Multiple surface cracking and its effect on interface cracks in functionally graded thermal barrier coatings under thermal shock, Trans. ASME J. Appl. Mech. 70 (2003) 234-245. https://doi.org/10.1115/1.1533809

[6] T.S. Hille, A.S.J. Suiker, S. Turteltaub, Microcrack nucleation in thermal barrier coating systems, Eng. Fract. Mech. 76 (2009) 813-825. https://doi.org/10.1016/j.engfracmech.2008.12.010

[7] B. Zhou, K. Kokini, Effect of surface pre-crack morphology on the fracture of thermal barrier coatings under thermal shock, Acta Mater. 52 (2004) 4189-4197. https://doi.org/10.1016/j.actamat.2004.05.035

[8] A. Gilbert, K. Kokini, S. Sankarasubramanian, Thermal fracture of zirconia-mullite

composite thermal barrier coatings under thermal shock: An experimental study, Surf. Coat. Technol. 202(10) (2008) 2152-216. https://doi.org/10.1016/j.surfcoat.2007.09.001

[9] A. Gilbert, K. Kokini, S. Sankarasubramanian, Thermal fracture of zirconia-mullite composite thermal barrier coatings under thermal shock: A numerical study, Surf. Coat. Technol. 203 (2008) 91-98. https://doi.org/10.1016/j.surfcoat.2008.08.003

[10] W.X. Zhang, X.L. Fan, T.J. Wang, The surface cracking behavior in air plasma sprayed thermal barrier coating system incorporating interface roughness effect, Appl. Surf. Sci. 258 (2011) 811-817. https://doi.org/10.1016/j.apsusc.2011.08.103

[11] H. Dong, G.-J. Yang, H.-N. Cai, H. Ding, C.-X. Li, C.-J. Li, The influence of temperature gradient across YSZ on thermal cyclic lifetime of plasma-sprayed thermal barrier coatings. Ceram. Int. 41 (9) (2015)11046-11056. https://doi.org/10.1016/j.ceramint.2015.05.049

[12] S. Kostyrko, M. Grekov, H. Altenbach, Stress concentration analysis of nanosized thin-film coating with rough interface. Continuum Mech. Thermodyn. 31 (2019) 1863-1871. https://doi.org/10.1007/s00161-019-00780-4

[13] M.A. Grekov, T.S. Sergeeva, Y.G. Pronina, O.S. Sedova, A periodic set of edge dislocations in an elastic semi-infinite solid with a planar boundary incorporating surface effects. Eng. Fract. Mech. 186 (2017) 423-435. https://doi.org/10.1016/j.engfracmech.2017.11.005

[14] J. Nordmann, K. Naumenko, H. Altenbach, Cohesive zone models—theory, numerics and usage in high-temperature applications to describe cracking and delamination. In: Naumenko, K., Krüger, M. (eds.) Advances in Mechanics of High-Temperature Materials, Advanced Structured Materials, vol. 117, pp. 131-168. Springer, Cham. 2020. https://doi.org/10.1007/978-3-030-23869-8_7

[15] V. Petrova, S. Schmauder, M. Ordyan, A. Shashkin, Revisit of antiplane shear problems for an interface crack: Does the stress intensity factor for the interface Mode III crack depend on the bimaterial modulus? Eng. Fract. Mech. 216 (2019) 106524. https://doi.org/10.1016/j.engfracmech.2019.106524

[16] V. Petrova, S. Schmauder, FGM/homogeneous bimaterials with systems of cracks under thermo-mechanical loading: Analysis by fracture criteria. Eng. Fract. Mech. 130 (2014) 12-20. https://doi.org/10.1016/j.engfracmech.2014.01.014

[17] V. Petrova, S. Schmauder, Modeling of thermo-mechanical fracture of FGMs with respect to multiple cracks interaction, Phys. Mesomech. 20 (2017) 241-249. https://doi.org/10.1134/S1029959917030018

[18] V. Petrova, S. Schmauder, A theoretical model for the study of thermal fracture of functionally graded thermal barrier coatings, Procedia Struct. Integr. 23 (2019) 407-412. https://doi.org/10.1016/j.prostr.2020.01.121

[19] V. Petrova, S. Schmauder, A theoretical model for the study of thermal fracture of functionally graded thermal barrier coatings with a system of edge and internal cracks, Theor. Appl. Fract. Mech. 108 (2020) 102605. https://doi.org/10.1016/j.tafmec.2020.102605

[20] V. Petrova, S. Schmauder, Analysis of interacting cracks in functionally graded thermal barrier coatings. Procedia Struct. Integr. 28 (2020) 608-618. https://doi.org/10.1016/j.prostr.2020.10.071

[21] Z.-H. Jin, R.C. Batra, Some basic fracture mechanics concepts in functionally graded materials, J. Mech. Phys. Solids 44 (1996) 1221-1235. https://doi.org/10.1016/0022-5096(96)00041-5

[22] A.M. Afsar, H. Sekine, Crack spacing effect on the brittle fracture characteristics of semi-infinite functionally graded materials with periodic edge cracks, Int. J. Fract. 102 (2000) L61-L66.

[23] K. Tohgo, M. Iizuka, H. Araki, Y. Shimamura, Influence of microstructure on fracture toughness distribution in ceramic–metal functionally graded materials, Eng. Fract. Mech. 75 (2008) 4529-4541. https://doi.org/10.1016/j.engfracmech.2008.05.005

[24] V.V. Panasyuk, M. P. Savruk, A. P. Datsyshin, Stress Distribution near Cracks in Plates and Shells (in Russian), Naukova Dumka, Kiev, 1976.

[25] F. Erdogan, G. Gupta, On the numerical solution of singular integral equations, Q. Appl. Math. 29 (1972) 525-534. https://doi.org/10.1090/qam/408277

[26] F. Erdogan, G.C. Sih, On the crack extension in plates under plane loading and transverse shear, J. Basic. Eng. 85 (1963) 519-527. https://doi.org/10.1115/1.3656897

[27] Y.C. Zhou, T. Hashida, Coupled effects of temperature gradient and oxidation on thermal stress in thermal barrier coating system, Int. J. Solids Struct. 38 (24-25) (2001) 4235-4264. https://doi.org/10.1016/S0020-7683(00)00309-7

[28] P.-P. Cortet, G. Huillard, L. Vanel, S. Ciliberto, Attractive and repulsive cracks in a heterogeneous material, J. Stat. Mech.: Theory and Experiment 10 (2008) P10022. https://doi.org/10.1088/1742-5468/2008/10/P10022

[29] S. Melin, Why do cracks avoid each other? Int. J. Fract. 23 (1983) 37-45. https://doi.org/10.1007/BF00020156

[30] S. Melin, Which is the most unfavourable crack orientation? Int. J. Fract. 51 (1991) 255-263. https://doi.org/10.1007/BF00045811

Appendix A.

Singular integral equations

The boundary value problem of elasticity for a system of cracks is reduced to a system of singular integral equations [24]:

$$\int_{-a_n}^{a_n} \frac{g_n'(t)dt}{t-x} + \sum_{\substack{k=1\\k\neq n}}^{N} \int_{-a_k}^{a_k} [g_k'(t)R_{nk}(t,x) + \overline{g_k'(t)}S_{nk}(t,x)]dt = \pi p_n(x)\,,\ |\,x\,|< a_n\,,$$

n = 1, 2, …, N, (A.1)

Materials Research Foundations **159** (2024) https://doi.org/10.21741/9781644902950

$$\int_{-a_n}^{a_n} g_n'(t)dt = 0 \quad \text{(for internal cracks)},$$

$$g_n'(x) = \frac{2\mu}{i(\kappa+1)}\frac{\partial}{\partial x}([u_n]+i[v_n]).$$

The number of equations N is equal to the number of cracks. The unknown functions $g_n'(x)$ contain the shear $[u_n]$ and normal $[v_n]$ displacement jumps on the n-th crack line, $\mu = E/2(1+\nu)$ is the shear modulus, E is Young's modulus, ν is Poisson's ratio, $\kappa = 3 - 4\nu$ for the plane strain state, and $\kappa = (3 - \nu)/(1+ \nu)$ for the plane stress state. An overbar $\overline{(...)}$ denotes the complex conjugate. In Eq. (A.1), the functions p_n are known functions determined by the load on the crack lines, Eq. (9). The regular kernels $R_{nk}(t,x)$ and $S_{nk}(t,x)$ contain the geometry of the problem and are written as

$$R_{nk}(t,x) = (1-\delta_{nk})K_{nk}(t,x) + \frac{e^{i\alpha_k}}{2}\left\{\frac{1}{X_n - \overline{T}_k} + \frac{e^{-2i\alpha_n}}{\overline{X}_n - T_k} + \right.$$

$$\left. +(\overline{T}_k - T_k)\left[\frac{1+e^{-2i\alpha_n}}{(\overline{X}_n - T_k)^2} - \frac{2e^{-2i\alpha_n}(X_n - T_k)}{(\overline{X}_n - T_k)^3}\right]\right\}, \tag{A.1}$$

$$S_{nk}(t,x) = (1-\delta_{nk})L_{nk}(t,x) + \frac{e^{-i\alpha_k}}{2}\left[\frac{T_k - \overline{T}_k}{(X_n - \overline{T}_k)^2} + \frac{1}{\overline{X}_n - T_k} - e^{-2i\alpha_n}\frac{X_n - T_k}{(\overline{X}_n - T_k)^2}\right], \tag{A.2}$$

$$T_k = te^{i\alpha_k} + z_k^0 \quad X_n = xe^{i\alpha_n} + z_n^0, \tag{A.3}$$

and

$$\delta_{nk} = \begin{cases} 0 & \text{for } n \neq k \\ 1 & \text{for } n = k \end{cases}$$

with kernels $K_{nk}(t,x)$ and $L_{nk}(t,x)$

$$K_{nk}(t,x) = \frac{e^{i\alpha_k}}{2}\left(\frac{1}{T_k - X_n} + \frac{e^{-2i\alpha_n}}{\overline{T}_k - \overline{X}_n}\right) \quad ,\text{(A.4)}$$

$$L_{nk}(t,x) = \frac{e^{-i\alpha_k}}{2}\left(\frac{1}{\overline{T}_k - \overline{X}_n} + \frac{T_k - X_n}{(\overline{T}_k - \overline{X}_n)^2}e^{-2i\alpha_n}\right), \tag{A.5}$$

Materials Research Foundations **159** (2024) https://doi.org/10.21741/9781644902950

which are the same as for the system of cracks in an infinite plane. Additional terms (in addition to K_{nk} and L_{nk}) in Eqs. (A.2) and (A.3) take into account the influence of the edge of the half plane. α_n is the inclination angle of n-th crack to the x-axis; z^0_n is the coordinate of the center of crack in the global coordinate system (x,y), see Fig. 2a.

Appendix B.

Numerical solution

The singular integral equations (A.1) are solved numerically based on the Gauss-Chebyshev quadrature. Different versions of this method are used for the solution, and the effectiveness of the method has been proven in many studies [25]. In the present work, the version described in [24] is applied.

Eqs. (A.1) are rewritten in dimensionless form with the non-dimensionless coordinates $\xi = t / a_k$ and $\eta = x / a_n$, where $2a_k$ is the length of the *k*-th crack. The unknown function $g'_n(\eta)$ presents as

$$g'_n(\eta) = u_n(\eta) / \sqrt{1-\eta^2}\ , \tag{B.1}$$

where the function $u_n(\eta)$ is a bounded continuous function in the segment [-1,1] and $1/\sqrt{1-\eta^2}$ is the weight function, which is taking into account the square root singularities at the crack tips. For an edge crack, the function $g'_n(\eta)$ possesses a singularity less than $1/\sqrt{1+\eta}$ at the edge point $\eta = -1$, this condition is accounted for as $u_n(-1) = 0$.

Using Gauss's quadrature formulae for regular and singular integrals, the singular integral equations are reduced to the following system of *N*x*M* (*N* – number of cracks, *M* – number of nodes) algebraic equations

$$\frac{1}{M}\sum_{m=1}^{M}\sum_{k=1}^{N}\left[u_k(\xi_m)R_{nk}(\xi_m,\eta_r)+\overline{u_k(\xi_m)}S_{nk}(\xi_m,\eta_r)\right]=\pi p_n(\eta_r) \tag{B.2}$$

$$\sum_{m=1}^{M}(-1)^m u_n(\xi_m)\tan\frac{2m-1}{4M}\pi = 0 \quad \text{(for edge cracks) or}$$

$$\sum_{m=1}^{M} u_n(\xi_m) = 0 \quad \text{(for internal cracks)}, \quad (n = 1, 2, \ldots, N;\ r = 1, 2, \ldots, M\text{-}1)$$

$$\xi_m = \cos\frac{2m-1}{2M}\pi \ \ (m = 1, 2, \ldots, M),\ \ \eta_r = \cos\frac{\pi r}{M} \ \ (r = 1, 2, \ldots, M\text{-}1).$$

M is the total number of discrete points of the unknown functions $u_n(\eta)$ on the segment [-1,1]. By applying the conjugate operation to the system (B.1), additional *N*x*M* equations are obtained, i.e. 2*N*x*M* equations should be solved, where *N* is the number of cracks.

The functions $u_n(\eta)$ are calculated by the interpolation formula:

$$u_n(\eta) = \frac{2}{M}\sum_{m=1}^{M} u_n(\xi_m)\sum_{r=0}^{M-1} T_r(\xi_m)T_r(\eta) - \frac{1}{M}\sum_{m=0}^{M} u_n(\xi_m) \qquad \text{(B.3)}$$

Tr are Chebyshev polynomials of the first kind.

Taking into account Eqs.(10) and (B.1), the stress intensity factors are obtained as:

$$K_{In}^{+} - iK_{IIn}^{+} = -\sqrt{\pi a_n}u_n(+1) = p_n\sqrt{\pi a_n}\frac{1}{M}\sum_{m=1}^{M} (-1)^m u_n(\xi_m)\cot\frac{2m-1}{4M}\pi, \qquad \text{(B.4)}$$

$$K_{In}^{-} - iK_{IIn}^{-} = \sqrt{\pi a_n}u_n(-1) = p_n\sqrt{\pi a_n}\frac{1}{M}\sum_{m=1}^{M} (-1)^{M+m} u_n(\xi_m)\tan\frac{2m-1}{4M}\pi,$$

$n = 1, 2, \ldots, N$

Here the signs "+" and "–" refer to the right and left crack tips, respectively.

The functions (B.23) written for $\eta = \pm 1$

$$u_n(1) = \frac{1}{M}\sum_{m=1}^{M} (-1)^{m+1} u_n(\xi_m)\cot\frac{2m-1}{4M}\pi$$

$$u_n(-1) = \frac{1}{M}\sum_{m=1}^{M} (-1)^{M+m} u_n(\xi_m)\tan\frac{2m-1}{4M}\pi$$

are used in the calculation of SIFs, Eq. (B.4).

CHAPTER 14

Thermal Fracture Resistance of Functionally Graded Thermal Barrier Coatings with Systems of Multiple Cracks. Application of Rule of Mixtures

Vera Petrova [1,2]*, Siegfried Schmauder [1]

[1] IMWF, University of Stuttgart, Pfaffenwaldring 32, D-70569 Stuttgart, Germany

[2] Voronezh State University, University Sq.1, Voronezh 394006, Russia

veraep@gmail.com *, Siegfried.Schmauder@imwf.uni-stuttgart.de

Abstract

The problem of thermal fracture of functionally graded thermal barrier coatings (FGCs) on a homogeneous substrate under the influence of thermo-mechanical loadings is studied in the presented work. It is assumed that FGCs contain a pre-existing system of multiple cracks, edge and/or internal. The method of analytical functions of complex variables and singular integral equations, as well as a mechanical quadrature method for solving singular integral equations were used. The functionally graded material properties (including the fracture toughness) are modelled by the formulas based on the rule of mixtures (RoM) with a power-law coefficient λ as the grading parameter. The following fracture characteristics are calculated: stress intensity factors, critical stresses and fracture angles. A series of computational experiments is carried out for FGC/H structures with typical systems of multiple cracks in the FGC at different grading parameters. As illustrative examples, systems of edge cracks and edge cracks with internal cracks are investigated. The results show that the grading parameter λ has a dominant effect on the fracture characteristics. The presented model allows to correlate the material parameters, geometrical parameters and the main fracture characteristics, which can be the basis for the solution for an optimization problem, e.g., to find the grading parameter λ and the crack arrangements that increase the fracture toughness.

Keywords

Functionally Graded Coating, Fracture Resistance, Thermo-Mechanical Load

1. Introduction

Functionally graded materials (FGMs) are a special type of composites, the properties of which are continuously varying mainly in one direction, which is achieved by changing the composition of the material and its structure. FGMs have wide engineering applications, in particular, in thermal barrier coatings (TBCs), where ceramics are used on the coating top and then continuously vary up to metal in the substrate. Functionally graded coatings (FGCs) with a gradual compositional variation from heat resistant ceramics to fracture resistant metals have been proposed in order to reduce thermal residual stresses causing delamination and debonding

Materials Research Foundations **159** (2024) https://doi.org/10.21741/9781644902950

of interfaces, enhancing coating toughness, and improving the long-term performance of TBCs, see [1]. However, cracks can occur because of initial defects or microcracks that appear during manufacturing or operation. Therefore, the study of fracture of FGCs is important for a better understanding of the fracture resistance of graded coatings.

The study of FGMs involves many challenging mechanical problems, including the prediction and measurement of the effective properties, thermal stress distribution, and fracture of FGMs. Modelling and evaluation of the effective properties of FGMs was considered in many works, see [2,3]. The evaluation of fracture toughness of FGMs is also a very important problem for studying of FGCs fracture [4,5,6,7]. In [8,9], the analysis of the fracture parameters with respect to critical loads for a system of edge and internal cracks in FGCs showed the importance of considering the variation of fracture toughness through the thickness of FGCs, and accordingly the importance of using critical stresses as the main fracture characteristic for FGCs.

In previous papers by the authors [8,9,10,11,12], different aspects of thermal fracture of FGC/H structures were investigated using an exponential model for material properties evaluation. In the present work, the problem for thermal fracture of FGC/H structures is considered with application of the functions based on the rule of mixtures (RoM) with corresponding gradation parameters. The RoM, originally applied to conventional composites, has been successfully used in FGMs ([3,4]) and has more possibilities for optimizing FGMs in terms of improving fracture toughness. In the current paper, the general formulation of the problem is the same as in previous works, e.g. in Petrova and Schmauder, 2020b, so we only briefly repeat this formulation for completeness, see Sections 2.1 and 3. Section 2.2 presents the model for material properties. The results and their discussion are given in Section 4, and in Section 5 are conclusions.

2. Formulation of the problem

2.1 Geometry of the problem and loading

The general geometry of the problem is shown in Fig. 1a, i.e. an FGC of thickness h, an underlying homogeneous substrate (H) and a layer of thermally grown oxide (TGO) that can form between them (FGC/TGO/H structures). For TBCs the top layer is made from ceramics, materials with low thermal conductivity, and the homogeneous substrate is made of metal. A FGC/TGO/H structure is cooled by ΔT, $\Delta T > 0$ (e.g. due to cooling from operating temperatures) and a tensile load p is applied parallel to the surface. A pre-existing system of cracks (length $2a_k$) is located in the FGC. The global coordinate system (x, y) and the local coordinates (x_k, y_k), which are referred to each crack with the x_k-axis on the crack lines, are shown in Fig. 1b. The position of cracks is determined explicitly by their midpoint coordinates (x_k^0, y_k^0) (or in the complex form $z_k^0 = x_k^0 + iy_k^0$, i is the imaginary unit) and the inclination angles α_k to the x-axis, or β_k for edge cracks, $\beta_k = -\alpha_k$, Fig. 1b.

Materials Research Forum LLC
https://doi.org/10.21741/9781644902950

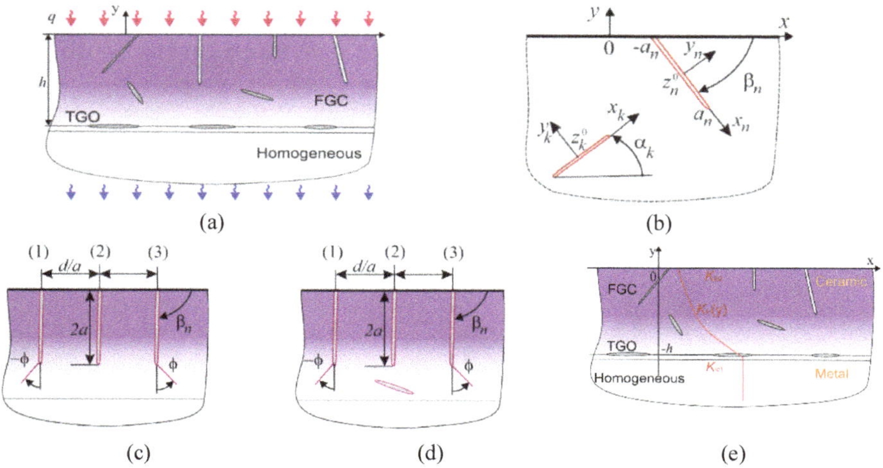

Figure 1. (a) FGC/TGO/H with a system of cracks; (b) global (x, y) and local (xk, yk) coordinate systems; (c) three edge cracks, (d) three edge cracks and an internal crack; (e) variation of fracture toughness in an FGC.

As in the previous authors works (e.g. see [8,9,11,12]), the following assumptions are used: the uncoupled, quasi-static thermo-elasticity theory is applied, that is, the temperature distribution is independent of the mechanical field; the thermal and mechanical properties of an FGC are continuous functions of the thickness coordinate *y*; the non-homogeneity of the functionally graded material is revealed in the form of the corresponding inhomogeneous stress distributions on the surfaces of cracks [6,13].

2.2 Material properties for FGCs and residual stresses

The presented method is applicable to different material combinations. However, due to the application of FGCs for TBCs, only (ceramic/metal)/metal coatings are considered in this study. Thus, the material composition varies with the *y*-coordinate from ceramic at the top of the FGC to metal in the substrate, Fig. 1e. Consequently, the properties of the FGC also vary continuously with the thickness coordinate *y* and can be mathematically described by a continuous function. In previous works [9,11], an exponential form of Young's modulus and thermal expansion coefficient was used, and in [8] – a linear model. Poisson's ratio was assumed to be constant and equal to the value of the homogeneous substrate. Along with the change in these mechanical properties, the change in fracture toughness was also taken into account, see Fig. 1e.

In the present study, the thermal and mechanical properties as well as the fracture toughness (K_{Ic}) of the FGC are modeled by functions based on the RoM. The RoM with its various modifications has long been used for conventional composites. In contrast to conventional composites, the volume fraction of one material in another in FGMs varies, so the effective properties of FGMs depend on this volume fraction. The FGC consists of two constituents, ceramic on the top and metal as a homogeneous substrate, with their volume fractions V_c and V_m, respectively, determined by a power law function

$$V_m(y)=\left[\frac{y}{h}\right]^{\lambda}=\left[\frac{1}{h}\left(h+\mathrm{Im}\left(z_n^0\right)-x_n\sin\left(\beta_n\right)\right)\right]^{\lambda},\quad V_c(y)=1-V_m(y) \tag{1}$$

with a power coefficient λ as the grading parameter for the FGC. This parameter can be set to different values to realize different volume fractions (as desired) and profile shapes. The coefficient of thermal expansion and Young's modulus of the FGC are assumed to take the following forms [3]:

$$\alpha_t(y)=\frac{\alpha_{tm}V_m(y)E_m/(1-\nu)+\alpha_{tc}V_c(y)E_c/(1-\nu)}{V_m(y)E_m/(1-\nu)+V_c(y)E_c/(1-\nu)},$$

$$E(y)=E_c\left[\frac{E_c+(E_m-E_c)V_m(y)^{(2/3)}}{E_c+(E_m-E_c)\left(V_m(y)^{(2/3)}-V_m(y)\right)}\right], \tag{2}$$

here v is Poisson's ratio.

The fracture toughness (K_{Ic}) for FGCs also changes with the coordinate y and can be determined using one of the models, see [3,4]. In the present work, the second function in Eq. (2) is used for K_{Ic}, where E should be replaced by K_{Ic}.

The method of superposition is used to solve this problem, so the loads at infinity are reduced to the corresponding loads on the crack faces (p_n). Besides, with changing the temperature, e.g. the structure is cooled on ΔT, the residual stresses (σ_n^T) are arising due to mismatch in the coefficients of thermal expansion. Thus, the full load on the n-th crack consists of p_n, σ_n^T and σ_n^e, where the index "n" denotes that the functions are written in the local coordinate system (x_n,y_n) connected with the n-th crack and σ_n^e is due to $E(y)$ variation [13]:

$$p_n^*=p_n+\sigma_n^e+\sigma_n^T=pf(\beta_n)+pf(\beta_n)[E(y)-1]+Q[\alpha_t(y)-1]E(y)$$

$$(\mathrm{n}=1,2,\ldots,\mathrm{N}), \tag{3}$$

$$pf(\beta_n)=p_n=\sigma_n-i\tau_n=p(1-\exp(2i\beta_n))/2,\quad Q=\alpha_{t1}\Delta TE_1$$

It is assumed that $p = Q$, otherwise the additional loading parameter p/Q should be considered. E_1 and α_{t1} are the material parameters of the substrate, $\alpha_t(y)$ and $E(y)$ are Eq. (2).

3. Solution and determination of main fracture characteristics

The method of analytical functions of complex variables is used for formulation of the singular integral equations (see [14]). The equations are solved numerically using the method of mechanical quadrature [14,15]. With this method, the singular integral equations are reduced to a system of algebraic equations. The unknown derivatives of the displacements jumps on the crack lines are determined; then the stress intensity factors near the crack tips are calculated. The details of this solution can be found in [9,11,12].

The cracks are under mixed-mode conditions due to their interaction and applied thermo-mechanical loading. That is, both stress intensity factors, Mode I and Mode II, are generally non-zero. In this case, the cracks will deviate from their initial paths. To predict the crack growth and determine the direction of crack growth, the criterion of maximum circumferential stresses [16] is applied. According to this criterion, the crack deflection angle ϕ (or the so-called fracture angle, Fig. 1c), the critical stress intensity factors, and then the critical stresses are calculated from the following relations:

$$\phi_n = 2\arctan\left[\left(K_{In} - \sqrt{K_{In}^2 + 8K_{IIn}^2}\right) / 4K_{IIn}\right],$$

$$K_n^{eq} \equiv \cos^3(\phi_n/2)\left(K_{In} - 3K_{IIn}\tan(\phi_n/2)\right) = K_{Ic,tip} \text{ or } K_n^{eq} = K_{Ic,tip}. \tag{5}$$

From Eq. (5), the critical loads are obtained as

$$\frac{p_{cr\,n}}{p_0} = \frac{K_{Ic,n\,tip}}{\cos^3(\phi_n/2)\left(k_{In} - 3k_{IIn}\tan(\phi_n/2)\right)}\frac{\sqrt{a}}{\sqrt{a_n}}, \tag{6}$$

where $p_0 = K_{Ic1}/\sqrt{\pi a}$ is the critical load for a reference single crack subjected to a load p normal to the crack line with the stress intensity factor $K^0 = p\sqrt{\pi a}$ and $a = \max a_n$ (n =1, 2, …, N), here N is the number of cracks. K_{Ic} is defined by an expression similar to Eq. (2) for $E(y)$. In Eqs. (4), (5), and (6), dimensional and non-dimensional stress intensity factors (SIFs) are related as follows:

$$K_{In} - iK_{IIn} = p\sqrt{\pi a_n}(k_{In} - ik_{IIn}). \tag{7}$$

The weakest crack is defined from the condition

$$P_{cr} = \min_n p_{crn} / p_0 \quad (\text{n} = 1, 2, \ldots, \text{N}).$$

4. Results and discussion

Typical crack patterns resulting from thermo-mechanical loads are multiple surface cracks at the ceramic top of FGCs [17]. Many studies are devoted to different edge crack systems, such as periodic edge cracks [13], or three or more edge cracks [10]. In a number of works [13,18], it is reported that the fracture resistance of a TBC can be increased by introducing dense vertical cracks in ceramic coatings; this mechanism is due to the shielding effect in these crack systems. A system of a finite number of edge cracks is a good candidate for investigating the effects of crack interaction, in particular crack shielding, i.e. the possibility of increasing fracture resistance. At the same time, additional defects, such as internal cracks can disrupt this shielding effect. As illustrating examples, two crack geometries are studied: a system of three edge cracks (Fig. 1c) and edge cracks with one internal crack (Fig. 1d). The results were obtained using functions based on the RoM (Eq. 2) to evaluate the FGM properties. In the following, the results for crack 1 and for middle crack 2 are presented. The behavior of fracture characteristics for

Materials Research Forum LLC
https://doi.org/10.21741/9781644902950

crack 3 is similar to that of crack 1 and is not shown in the figures. The following parameters are used in the calculations: d/a = 2.0-10.0 (in the figures, d/a is denoted by d), h/a = 4.0, $2a$ – crack size, $2a$ = 2; ΔT = 300 °C. Material parameters are listed in Table 1.

Table 1. Material parameters

Material property	Top coat (ceramic) PSZ	Bottom coat (metal) Steel
Young's modulus [GPa]	48.0	207.0
Fracture toughness [MPa m$^{1/2}$]	7.0	50.0
Thermal exp. coeff. [10^{-6} K^{-1}]	9.0	15.0

4.1 Three edge cracks

Fig. 2 shows the normalized critical loads p_{cr}/p_0 as functions of inclination angle $\beta = \beta_{1,2,3}$ for three FGM models, exponential, linear and RoM [8]. Fig. 2a refers to crack 1 and Fig. 2b - for the middle crack 2 8see geometry in Fig 1c). The weakest crack is crack 1 (and crack 3) since $p_{cr,2}/p_0 > p_{cr,1}/p_0$, and a shielding effect is observed for crack 2. The lowest p_{cr}/p_0 value is for the RoM model. Different values of the critical load for different material models, but for the same crack, show a difference in the fracture toughness values determined by these models, which in turn shows a different concentration of ceramics (metal) near the crack tip.

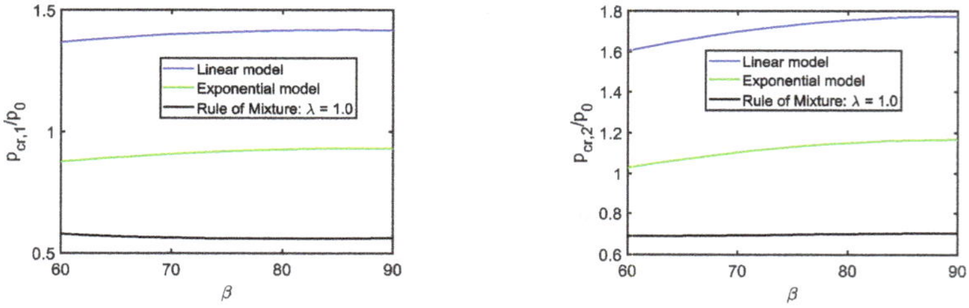

(a) (b)

Figure 2. Critical loads as a function of β for three FG models, (a) for crack 1and (b) for the middle crack 2.

Figs. 3 a and b present normalized SIFs k_I and critical loads p_{cr}/p_0 as functions of inclination angle β and distances d for the RoM for FGM (λ=3). The shapes of 3D plots for $k_{I(1)}$ and $k_{I(2)}$ are similar, as well as for $p_{cr,1}/p_0$ and $p_{cr,2}/p_0$. But the values of $k_{I(1)}$ and $k_{I(2)}$ are different and of $p_{cr,1}/p_0$ and $p_{cr,2}/p_0$ are different. The relative difference for $k_{I(1)}$ and $k_{I(2)}$ is 11% for β=60° and 19% for 90°, and for $p_{cr,1}/p_0$ and $p_{cr,2}/p_0$ is 19% and 25% for β=60° and 90° respectively, d/a=4 and λ=3 ($|f_1 - f_2|/f_1 \times 100\%$ – f_1 and f_2 are the corresponding fracture parameters for cracks 1 and 2).

The results in Fig. 3 demonstrate the shielding effect for crack 2, that is, the $k_{I(2)}$ are smaller than $k_{I(1)}$, and the $p_{cr,2}/p_0$ are greater than $p_{cr,1}/p_0$. As the distance d decreases, the k_I values decrease and the p_{cr}/p_0 values increase. Outer cracks 1 and 3 suppress the propagation of crack 2. This

Materials Research Forum LLC
https://doi.org/10.21741/9781644902950

means that outer cracks 1 and 3 will start to propagate first. The influence of β on k_I is more pronounced than on p_{cr}/p_0.

Fig. 4 shows normalized SIFs k_I, critical loads p_{cr}/p_0 and fracture angles ϕ as functions of grading parameter λ and inclination angle β for distance d/a =2. The results show that the grading parameter λ has a dominant effect on these fracture characteristics. As the grading parameter λ increases, both k_I and p_{cr}/p_0 are increased, Figs. 4a-d. The dependence of ϕ on λ is more complicated, Fig. 4e and f. A higher grading parameter corresponds to a higher metal content and a lower ceramic content in the FGC.

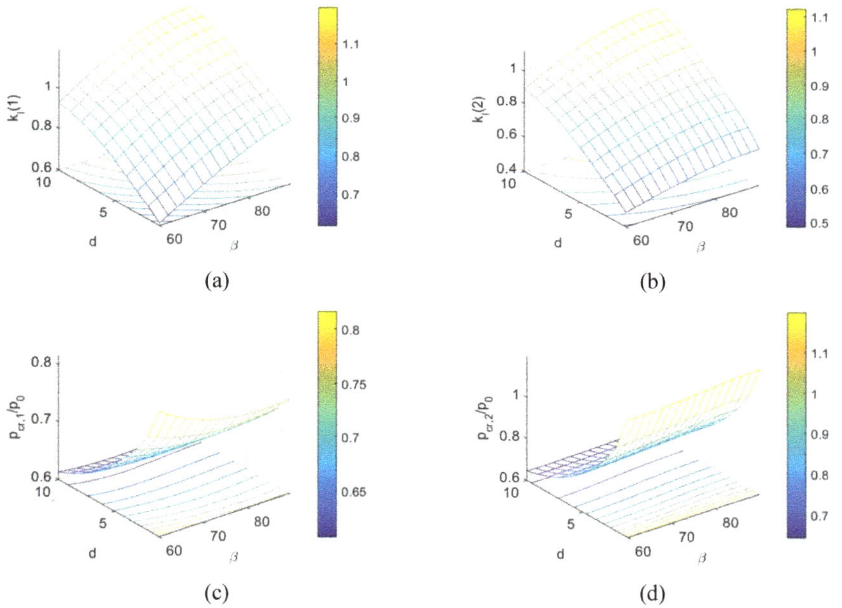

Figure 3. SIFs factors k_I (a, b) and critical loads (c, d) as functions of β and d, λ=3, (a, c) for crack 1, and (b, d) for crack 2, for Fig. 3c.

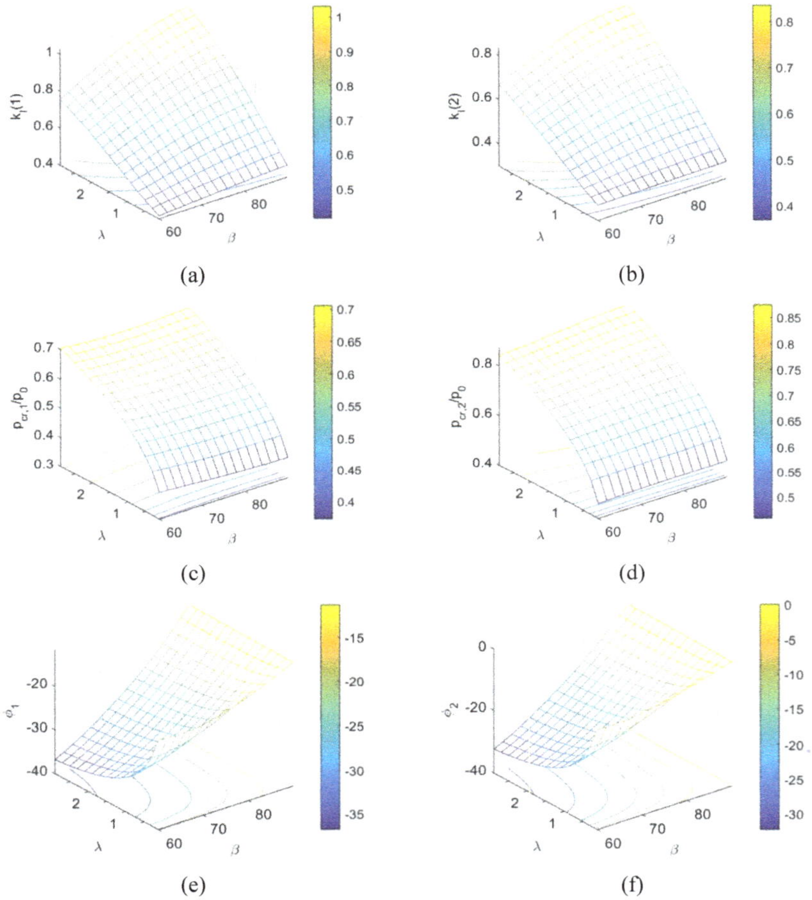

Figure 4. SIFs k_I (a, b), critical loads (c, d) and fracture angles (e, f) as functions of β and λ, d =4; (a, c, e) for crack 1, and (b, d, f) for crack 2, for Fig. 3c.

4.2. Three edge cracks and an internal crack

The presence of an internal crack can enhance or suppress the propagation of edge cracks. Fig. 5 presents critical loads p_{cr}/p_0 and fracture angles ϕ as functions of gradation parameter λ and inclination angle β=$\beta_{1,2,3}$ (d/a=2) for the geometry in Fig. 1d (for three edge cracks and one internal crack with (x_4^0, y_4^0) = (d/a, -3) and β_4=0°). The influence of the internal crack with the center on the line of edge crack 2 is small, which can be seen from a comparison of the corresponding results in Fig. 5 with the results in Fig. 4c-e. However, the closer the crack is to the edge cracks, the stronger the influence. Besides, a crack with $\beta_4 \neq 0$ (a non-zero angle of inclination to the surface) causes a greater effect on the fracture characteristics of edge cracks, see Table 2, where the relative difference (RD) for k_I and p_{cr}/p_{cr0} is calculated by the formula RD $f = (f^{\text{edg}} - f^{\text{edg+int}})/ f^{\text{edg}} \times 100\%$ and is presented for cracks 1 and 2 with $\beta_{1,2}$=90°. The "edge" index means the geometry with only edge cracks (Fig. 1c), and the "edge+int" index – the geometry in Fig. 1d. The propagation of crack 2 is restrained by the internal crack (negative RD k_I and positive RD p_{cr}/p_{cr0}), while the propagation of outer cracks 1 and 3 is enhanced by this internal crack (positive RD k_I and negative RD p_{cr}/p_{cr0}), as seen in Table 2.

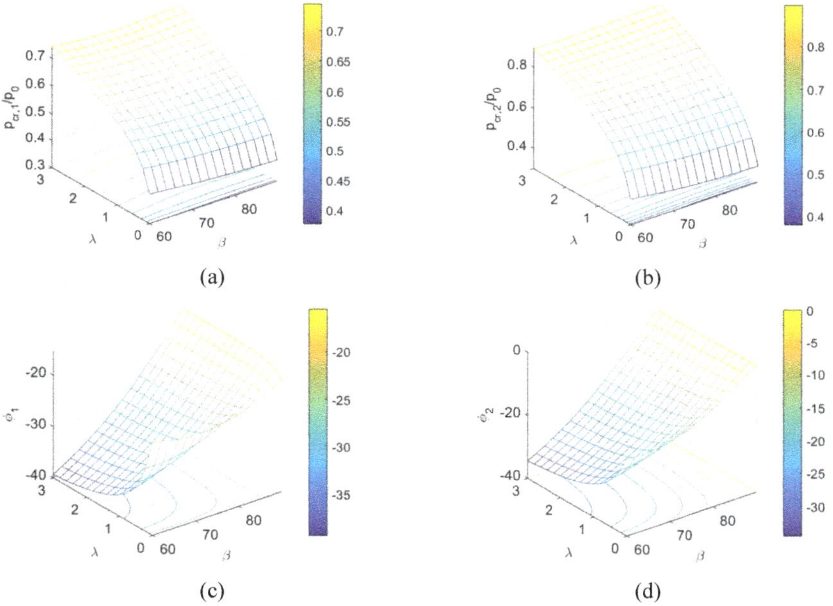

Figure 5. Critical loads (a, b) and fracture angles (c, d) as functions of $\beta = \beta_{1,2,3}$ and λ, d =4, β_4=0; (a, c) for crack 1, (b, d) for crack 2, for Fig. 3d.

Table 2. Influence of an internal crack on edge cracks; relative difference (RD) for k_I and p_{cr}/p_{cr0}.

λ	β_4	RD $k_{I(1)}$ %	RD $p_{cr,1}/p_{cr0}$ %	RD $k_{I(2)}$ %	RD $p_{cr,2}/p_{cr0}$ %
0.2	0°	+7	–6	–22	+18
	45°	+18	–25	–66	+40
3	0°	+2	–3	–7	+7
	45°	+15	–21	–59	+38

4.3 Remark

In TBC systems, ceramics, due to their low thermal conductivity, prevent overheating of the base metal. The appearance of metals degrades the insulation performance; therefore, it is important to have a certain limit on the ceramic content in the FGC in order to still have fully functional and stable coatings. The calculations provide the critical loads for different grading parameters, but what percentage of material content is required from an optimization point of view (balance of the content of ceramics and metal) needs to be determined in future studies, both theoretical and experimental. The proposed model provides a sound basis to optimize FGCs in order to improve the fracture resistance of FGC/H structures.

Conclusions

Based on classical but rather effective analytical and numerical methods, such as the method of analytical functions of complex variables and singular integral equations, the problem is studied for FGC/H structures containing a pre-existing system of multiple cracks. Formulas with application of the RoM model were used to evaluate the properties of FGCs. At first, the stress intensity factors are calculated, then, using the fracture criterion of maximum hoop stresses, the fracture angles and critical stresses are obtained. Illustrative examples are presented to show the influence of material and geometrical parameters of the problem on the fracture characteristics of interacting cracks. The weakest cracks in the considered system of three edge cracks are the outer edge cracks (1 and 3) in the FGC/H structure, (ceramic/metal/) metal. The shielding effect is observed for the middle crack 2. The presence of an internal crack with the center on the line of crack 2 suppresses the propagation of the middle crack 2 and enhances the propagation of the outer edge cracks 1 and 3. That is, an additional internal crack influences on the edge crack interaction, but the shielding effect is generally not disturbed. The structural variation of fracture toughness for the FGM was taken into account in the calculations. For the grading parameter λ in the RoM model, it was obtained: variation of λ leads to a significant change in the fracture parameters, in particular, critical loads.

On the basis of the developed theoretical model, which allows to correlate the material parameters, geometrical parameters and the main fracture characteristics, an optimization problem can be solved, e.g. to find a gradation parameter and crack arrangements that maximize the thermal critical stresses, i.e. increasing the fracture toughness for FGC.

Acknowledgements

The authors would like to acknowledge the financial support of the German Research Foundation under Grant SCHM 746/209-1.

References

[1] D. Clarke, M. Oechsner, N. Padture, Thermal-barrier coatings for more efficient gas-turbine engines, MRS Bulletin 37(2012) 891-941. https://doi.org/10.1557/mrs.2012.232

[2] J.R. Zuiker, Functionally graded materials: Choice of micromechanics model and limitations in property variation, Composites Eng. 5(7) (1995) 807-819. https://doi.org/10.1016/0961-9526(95)00031-H

[3] Noda, N., Ishihara, M., Yamamoto, N., 2004. Two-crack propagation paths in a

functionally graded material plates subjected to thermal loadings, J. Therm. Stresses 27, 457-469. https://doi.org/10.1080/01495730490451422

[4] Z.-H. Jin, R.C. Batra, Some basic fracture mechanics concepts in functionally graded materials, J. Mech. Phys. Solids 44 (1996) 1221-1235. https://doi.org/10.1016/0022-5096(96)00041-5

[5] K. Tohgo, T. Suzuki, H. Araki, Evaluation of R-curve behavior of ceramic–metal functionally graded materials by stable crack growth, Eng. Fract. Mech. 72 (2005) 2359-2372. https://doi.org/10.1016/j.engfracmech.2005.03.006

[6] K. Tohgo, M. Iizuka, H. Araki, Y. Shimamura, Influence of microstructure on fracture toughness distribution in ceramic–metal functionally graded materials, Eng. Fract. Mech. 75 (2008) 4529-4541. https://doi.org/10.1016/j.engfracmech.2008.05.005

[7] Y. Zhang, L. Guo, X. Wang, R. Shen, K. Huang, Thermal shock resistance of functionally graded materials with mixed-mode cracks, Int. J. Solids Struct. 164 (2019) 202-211. https://doi.org/10.1016/j.ijsolstr.2019.01.012

[8] V. Petrova, S. Schmauder, Fracture of functionally graded thermal barrier coating on a homogeneous substrate: Models, methods, analysis, J. Phys.: Conf. Ser. 973(1) (2018) 012017. https://doi.org/10.1088/1742-6596/973/1/012017

[9] V. Petrova, S. Schmauder, Analysis of interacting cracks in functionally graded thermal barrier coatings, Procedia Struct. Integr. 28 (2020) 608-618. https://doi.org/10.1016/j.prostr.2020.10.071

[10] V. Petrova, S. Schmauder, A. Shashkin, Modeling of edge cracks interaction, Frat. ed Integrita Strutt. 36 (2016) 8-26. https://doi.org/10.3221/IGF-ESIS.36.02

[11] V. Petrova, S. Schmauder, A theoretical model for the study of thermal fracture of functionally graded thermal barrier coatings with a system of edge and internal cracks, Theor. Appl. Fract. Mech. 108 (2020a) 102605. https://doi.org/10.1016/j.tafmec.2020.102605

[12] V. Petrova, S. Schmauder, 2021. Thermal fracture of functionally graded thermal barrier coatings with pre-existing edge cracks and multiple internal cracks imitating a curved interface, Contin. Mech. Thermodyn. 33 (2021) 1487-1503. https://doi.org/10.1007/s00161-021-00994-5

[13] A.M. Afsar, H. Sekine, Crack spacing effect on the brittle fracture characteristics of semi-infinite functionally graded materials with periodic edge cracks, Int. J. Fract. 102 (2000) L61- L66.

[14] V.V. Panasyuk, M.P. Savruk, A.P. Datsyshin, Stress Distribution near Cracks in Plates and Shells (in Russian), Naukova Dumka, Kiev, 1976.

[15] F. Erdogan, G. Gupta, On the numerical solution of singular integral equations, Quart. Appl. Math. 29 (1972) 525-534. https://doi.org/10.1090/qam/408277

[16] F. Erdogan, G.C. Sih, On the crack extension in plates under plane loading and transverse shear, J. Basic Eng. 85 (1963) 519-527. https://doi.org/10.1115/1.3656897

[17] S. Rangaraj, K. Kokini, Multiple surface cracking and its effect on interface cracks in

functionally graded thermal barrier coatings under thermal shock, Trans. ASME J. Appl. Mech. 70 (2003) 234-245. https://doi.org/10.1115/1.1533809

[18] Li, B., Fan, X., Okada, H., Wang, T. 2018. Mechanisms governing the failure modes of dense vertically cracked thermal barrier coatings, Eng. Fract. Mech. 189, 451-480. https://doi.org/10.1016/j.engfracmech.2017.11.037

www.ingramcontent.com/pod-product-compliance
Lightning Source LLC
Chambersburg PA
CBHW071336210326
41597CB00015B/1471
* 9 7 8 1 6 4 4 9 0 2 9 4 3 *